ANALYSIS AND DESIGN OF ENERGY SYSTEMS

PRENTICE-HALL SERIES IN ENERGY
W. J. Kennedy and W. C. Turner, *editors*

ANALYSIS AND DESIGN OF ENERGY SYSTEMS

B. K. HODGE
Mechanical and Nuclear Engineering Department
Mississippi State University

PRENTICE-HALL, INC., ENGLEWOOD CLIFFS, NEW JERSEY 07632

Library of Congress Cataloging in Publication Data

Hodge, B. K. (date)
Analysis and design of energy systems.

(Prentice-Hall series in energy)
Includes bibliographical references and index.
1. Power (Mechanics) 2. Fluid mechanics. 3. Heat exchangers. 4. Thermodynamics. I. Title. II. Series.
TJ163.9.H6 1985 621.402 84-11568
ISBN 0-13-032814-6

Editorial/production supervision: *Raeia Maes*
Cover design: *Photo Plus Art*
Manufacturing buyer: *Anthony Caruso*

Printed in the United States of America

10 9 8 7 6 5 4 3 2 1

ISBN 0-13-032814-6 01

Prentice-Hall International, Inc., *London*
Prentice-Hall of Australia Pty. Limited, *Sydney*
Editora Prentice-Hall do Brasil, Ltda., *Rio de Janeiro*
Prentice-Hall of Canada Inc., *Toronto*
Prentice-Hall of India Private Limited, *New Delhi*
Prentice-Hall of Japan, Inc., *Tokyo*
Prentice-Hall of Southeast Asia Pte. Ltd., *Singapore*
Whitehall Books Limited, Wellington, *New Zealand*

TO MY WIFE GAYLE

FOR OUR CHILDREN
BENJAMIN who will do things better, and
LAUREN who will do things with more perseverance

CONTENTS

3

HEAT EXCHANGERS II 108

4

PRIME MOVERS 155

5

SYSTEM SIMULATION 216

PREFACE

This textbook is intended primarily for senior-level engineering students who have completed a first course in both fluid mechanics and heat transfer. The review information included on basic fluid mechanics and basic heat transfer is not of sufficient depth or breadth to constitute an engineer's only exposure to these fundamental subjects. The material presented herein combines fluid mechanics, heat transfer, and thermodynamics concepts and develops application methodologies (analysis and design) useful for a wide variety of energy system components.

Except for elective technical courses, formal instruction in the thermal sciences in engineering often does not go beyond fluid mechanics, heat transfer, and thermodynamics; yet many engineers pursue careers related to the design and application of energy system components. Typical undergraduate offerings in heat transfer and fluid mechanics provide reasonable coverage of fundamentals, but do not pursue in detail either thermal systems design concepts or the intimate relationship between heat transfer and fluid mechanics in design. For example, most heat transfer courses include at least a brief study of heat exchangers, but typically such coverage is limited to simple calculations requiring use of the LMTD or NTU concepts. Heat exchanger pressure losses and pumping power are usually not introduced, and frequently the behavior of the overall heat transfer coefficient U with flow rate or number of tubes is not explored. Isolated study in courses often does not allow the student to "put together" or build up systems using multiple interdependent component devices.

The basic goals of the book are formulated to enhance the student's realization of the design process in energy systems and can be formallly listed as follows:

1. Presentation and study of fluid mechanics and heat transfer principles and techniques beyond those assimilated in undergraduate fluid mechanics and heat transfer courses.

2. Development of an awareness and understanding of the relationship between fluid mechanics, thermodynamics, and heat transfer in design consideration.
3. Introduction to the importance of codes and standards and examination of their effects on the design process.
4. Reinforcement of the thermal systems design viewpoint in the student and development of an appreciation of the relationship between system components and the entire system.

The reasons for the particular choice of topics in this book have been heavily influenced by my belief that most energy systems are composed of piping, pumps, and heat exchangers and that modeling of these components in systems is necessary. This is a textbook and was not written to be an all-inclusive reference. It espouses the solution of rather complex problems by starting with fundamentals.

Since ABET stresses design in accrediting engineering curricula, and will apparently do so for the foreseeable future, this book has been structured with a design emphasis. Only a relatively few problems are given at the end of each chapter, since I feel that most instructors using this text will want to develop their own design problems.

To accomplish the aforementioned goals, this book is divided into six chapters. Chapters 1 and 2 both present brief reviews of fundamentals and use these fundamental concepts to develop more advanced methods and techniques. Chapter 1 is devoted to piping networks and has as its primary goal the analysis of series, parallel, and series–parallel fluid networks. The Hardy–Cross approach is examined and extended for use in complex piping systems and networks.

Chapter 2, Heat Exchangers I, develops both the LMTD and NTU methods of analysis for heat exchangers, presents a summary of heat transfer correlations for diverse arrangements, explores fins and finned surfaces, and introduces an accepted methodology for evaluating heat exchanger pressure drops. This chapter presents all the background and basic information needed to consider design and analysis problems for shell-and-tube and cross-flow heat exchangers.

Chapter 3, Heat Exchangers II, considers in turn shell-and-tube and cross-flow heat exchangers. For shell-and-tube devices the *Tubular Exchanger Manufacturers Association* (TEMA) *Standards* are previewed and general design strategies are considered that allow the satisfaction of both rating (Btu/h) and pressure drop constraints. Cross-flow heat exchanger coverage makes extensive use of the existing friction and heat transfer data base for various heat transfer surfaces (as given by Kays and London, for example). Design strategy for rating and pressure drop constraints are examined for cross-flow devices. Example problems are provided for both shell-and-tube and cross-flow devices. The chapter concludes with a brief overview of heat exchanger mechanical problems.

Chapter 4, Prime Movers, is devoted primarily to pumps and fans. Although the emphasis in this chapter is on centrifugal pumps, rotary and reciprocating devices are also examined. Typical manufacturer's data are presented and discussed, as are pumps in series and parallel. The methodology for the operating point of a given pump in a given system is developed. Fans are briefly studied, as are nozzles.

Chapter 5, System Simulation, is concerned with modeling and simulation of steady-state systems. A brief examination of curve fitting is also made. The generalized

Hardy–Cross method, which was developed in Chapter 1, is reviewed for use as a general simulation technique. The multivariable Newton–Raphson procedure is introduced as the general simulation procedure, and several examples are presented. The chapter closes with a general survey of the influence coefficient method for generalized one-dimensional compressible flow. This compressible flow modeling approach is presented because of its general utility and because so many energy systems use a compressible fluid as the working medium.

Chapter 6, Fluid Transients, contains an introduction and overview of unsteady flow in pipes. Rigid theory is examined and several examples presented. Waterhammer is discussed, first from a qualitative standpoint and then via the method of characteristics. A waterhammer example is presented.

Although only one computer program listing is presented in the text, information and techniques are developed for use in programs. I firmly believe that over the next decade the availability of computing power in the form of microcomputers will virtually revolutionize thermal and fluids engineering practices. I purposely refrained from providing listings of programs used in many of the example problems because it has been my observation over the past few years that the pedagogic impact is lessened if listings that present the logic are available. The student learns by doing; and, while I am not opposed to the use of "canned" programs of great sophistication, student understanding is greatly increased if the logic is not provided for fundamental programs such as appear in this book. All problems, both examples and end-of-chapter assignments, can be worked on a 16K microcomputer with "extended" BASIC. I am a strong believer in Hamming's* statement: "The purpose of computing is insight, not numbers."

Many people have, directly or indirectly, played a role in this undertaking. At Mississippi State University, Professor D. M. Eastland suggested that a course entitled "Energy Systems Design" should be taught; and Dr. C. T. Carley, the Department Head, provided me the opportunity, as well as some release time, to develop the course and some of the material. The Mechanical Engineering Faculty offered both sound advice on content and encouragement. Much of the material was assimilated through NSF Local Course Improvement Grant 80-01005. In no single instance was I denied permission to use previously published information from either books, standards, or company brochures. The Hydraulic Institute, the Crane Co., and Goulds Pumps, Inc., as well as John Wiley and McGraw-Hill were gracious with their permissions. All permissions are duly noted and appreciated. To Matt Fox of Prentice-Hall, Inc., I am appreciative for both his help and his willingness to chance a textbook in a developing area. To Mrs. Teressa Yeatman who typed and retyped and corrected *my* many errors over a three-year period I am indebted. Without her skill, patience, and cheerfulness, this task would have been much more difficult.

To my wife Gayle and to our children Benjamin and Lauren, something more than I can give is due. For nights and days when I wasn't home, and for when I was home but was mumbling about this book, and for understanding why, I can say it was as much your effort as my effort.

B. K. Hodge

*R. W. Hamming, *Numerical Methods for Scientists and Engineers* (New York: McGraw-Hill, 1962).

1

PIPING SYSTEMS

1-1 INTRODUCTION

Except for electrical transmission systems, all energy systems have in common some means of moving, transforming, or transferring energy. Energy systems using a fluid as the working medium must always utilize some form of conduit or pipe. Piping systems used in conjunction with energy systems can range in complexity from simple to complicated. But regardless of the complexity, an amazingly short list of simple principles is required for analysis and design.

This chapter examines in a logical order the concepts needed to quantify various aspects of piping networks. Starting from fundamental principles, this topic is developed to the point where complex fluid-conveying systems can be examined.

The ideas presented herein are not novel or new. Many excellent fluid mechanics texts develop some or most of the major ideas. This chapter attempts to concisely present techniques applicable to many different situations. The technical material presented within this chapter was taken from the sources listed in the References (1–7) for this chapter. Most of the development is flavored by Streeter and Wylie (1) and Bober and Kenyon (2). By far the most complete book dealing with just pipe flow is that of Benedict (8). For supplementary reading beyond what is presented here, Benedict is recommended. The Crane Company Technical Paper 410 (7) is a most convenient compilation of data and information and is highly respected by industry, especially the petrochemical sector. White's (9) readable book treats details of viscous flow; and while on a level much more mathematical than is pursued here, real understanding of the nature of viscous losses can come only through reading and understanding of such presentations.

Modern engineering practice is built around the essentially universal availability of digital computers. The mode of presentation in this chapter, while not providing program listings, does present material in such a fashion as to be utilized for computer programs.

1-2 FUNDAMENTAL EQUATIONS OF FLUID MECHANICS

The mathematical formulation for all our quantitative understanding of fluid mechanics rests on the concept of conservation of physically meaningful quantities as applied to control volumes. Typically, one can anticipate a well-posed problem by considering conservation of mass, linear momentum, and energy in conjunction with other ancillary expressions such as an equation of state. Analyses for rotating fluid machinery or turbomachines require consideration of conservation of angular momentum. Because of the composition of fluids, liquids or gases, in regimes of engineering interest, a control volume analysis for each of these quantities is required.

A generalized control volume with schematic indications of mass, momentum, and energy transfer is illustrated by Fig. 1-1. Formally, we define a *control volume* (CV) as some fixed region in space and a *control surface* (CS) as the boundary of that fixed region. The need to approach fluid mechanics through a control volume methodology arises because of the large number of

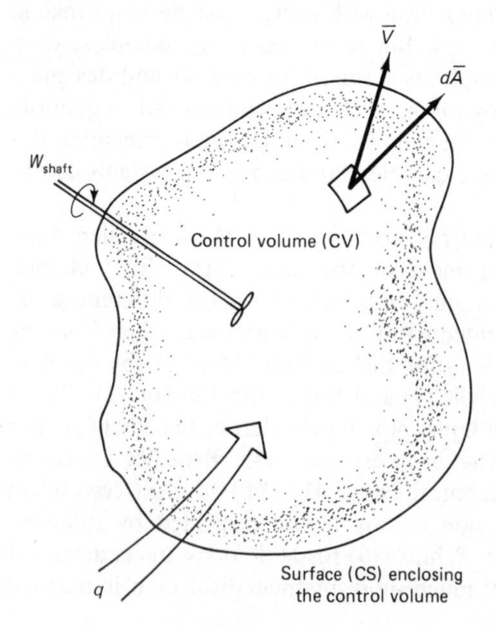

Figure 1-1 Generalized control volume.

discrete fundamental particles involved in any continuum fluid mechanics situation. Two approaches are possible when studying particles in motion: (1) the *Lagrangian* approach in which the dynamics and kinematics of individual fluid particles are observed and (2) the *Eulerian* approach in which fluid fluxes are observed as they cross the control surface of the control volume of interest. The Lagrangian viewpoint is both fundamental and appealing since it rests directly on the ideas of conservation of mass, momentum, and energy as applied to a single particle. Unfortunately, this approach is untenable when applied to the multitude of particles in the typical problem. The kinetic theory of gases or liquids is basically developed using the Lagrangian viewpoint, but it is far too complex for general use.

The Eulerian viewpoint is concerned not with the dynamics and kinematics of particles but with the mass, momentum, and energy crossing into or out of the control volume. The mathematical formulation for the Eulerian viewpoint is derived by applying the aforementioned conservation principles to the particles contained within the control volume at time t. The results appear in the form of integral expressions for the various quantities of interest:

1. Conservation of mass ([≡]* lbm/s).

$$0 = \frac{\partial}{\partial t}\int_{\mathrm{CV}} \rho \, d\mathrm{Vol} + \int_{\mathrm{CS}} \rho \overline{V} \cdot d\overline{A} \tag{1-1}$$

2. Conservation of linear momentum ([≡] lbf).

$$\sum \overline{F} = \frac{\partial}{\partial t}\int_{\mathrm{CV}} \rho \frac{\overline{V}}{g_c} \, d\mathrm{Vol} + \int_{\mathrm{CS}} \rho \frac{\overline{V}\overline{V}}{g_c} \cdot d\overline{A} \tag{1-2}$$

3. Conservation of energy ([≡] ft-lbf/s).

$$\frac{\delta q}{\delta t} - \frac{\delta W_S}{\delta t} = \frac{\partial}{\partial t}\int_{\mathrm{CV}} \rho e \, d\mathrm{Vol} + \int_{\mathrm{CS}} \left(e + \frac{P}{\rho}\right) \rho \overline{V} \cdot d\overline{A} \tag{1-3}$$

4. Conservation of angular momentum ([≡] ft-lbf).

$$\sum \overline{M} = \frac{\partial}{\partial t}\int_{\mathrm{CV}} \rho \left(\frac{\overline{r} \times \overline{V}}{g_c}\right) d\mathrm{Vol} + \int_{\mathrm{CS}} \rho \left(\frac{\overline{r} \times \overline{V}}{g_c}\right) \overline{V} \cdot d\overline{A} \tag{1-4}$$

Each of these expressions is of similar form. The terms on the right-hand side of the equations represent either the time rate of accumulation of a particular quantity in the control volume or the net rate of efflux of that quantity across the control surface. For example, in the case of the linear momentum conservation expression the term $\partial/\partial t \int_{\mathrm{CV}} \rho \overline{V} \, d\mathrm{Vol}$ is the time rate of accumulation of linear momentum within the control volume, and $\int_{\mathrm{CS}} \overline{V} \rho \overline{V} \cdot d\overline{A}$ is the net efflux of linear momentum across the control surface.

*[≡] signifies the units of an equation.

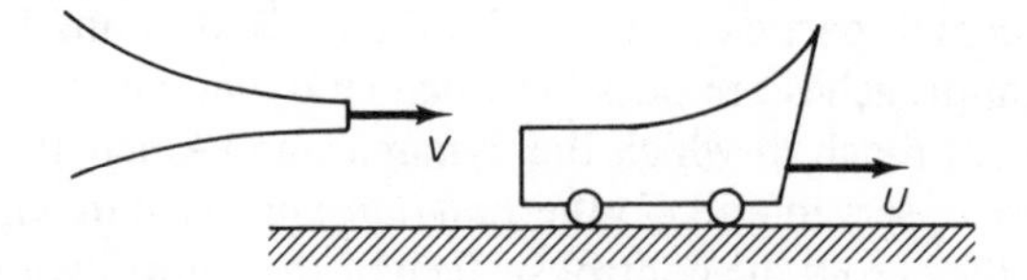

Figure 1-2 Fluid jet impacting a moving vane.

Since linear momentum is a vector quantity, Eq. (1-2) is basically a statement of Newton's law as applicable to a control volume. This should serve to remind us that the reference frame for this expression must be inertial. This might seem like a subtle point, but it is extremely important if correct results are to be achieved.

Consider the case of a fluid jet leaving a nozzle with absolute velocity V and impacting a vane moving with absolute velocity U. A schematic representation is given in Fig. 1-2. The maximum power delivered to the wheeled vane occurs when $U/V=\frac{1}{3}$. Now consider the same situation except that the vane is part of an impulse turbine and that the "wheel" speed U is $r\omega$. Equation (1-2) cannot be applied since the reference frame is now a rotating frame and is not inertial. Equation (1-4) can be applied with the results that the maximum power delivered occurs when $U/V=\frac{1}{2}$. Attempts to apply Eq. (1-2) to the turbine would lead to the incorrect results $U/V=\frac{1}{3}$. This classical example reminds us of the importance of choosing the appropriate conservation equation and reference frame.

A *streamline* is a line that is tangent to the velocity vector of a flowing fluid. A *pathline* is the loci of the spatial positions of a fluid particle as it traverses the flow field. If the flow is steady, the pathline of a particle is also a streamline. Consider, as illustrated in Fig. 1-3, a fluid element moving along a streamline in steady flow. If the additional assumption of frictionless flow is made, then an application of Eq. (1-2) yields

$$\frac{dP}{\rho}+\frac{g}{g_c}dz+\frac{V}{g_c}dV=0 \tag{1-5}$$

Integration of Eq. (1-5) with the additional constraint of incompressible flow

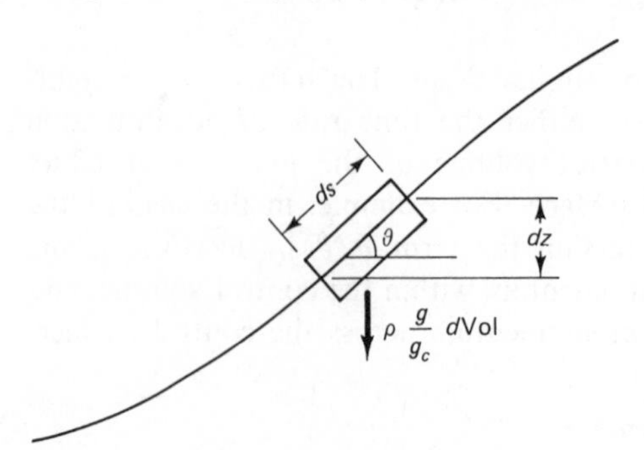

Figure 1-3 Fluid element moving along a streamline.

yields the *Bernoulli equation*

$$P + \rho \frac{g}{g_c} z + \frac{1}{2} \rho \frac{V^2}{g_c} = \text{Const} \tag{1-6}$$

The constant of integration is called the Bernoulli constant and in general varies from one streamline to another, but it remains constant along a given streamline in steady, frictionless, incompressible flow. These restrictions are important, and we should exercise due care in applying Eq. (1-6). The implications of the assumption of frictionless flow are rather far-reaching as the shearing stresses become zero and an inviscid flow results. Thus, Bernoulli's equation is said to apply to an ideal flow or to a potential flow, both denoting a flow that is reversible in the thermodynamic sense.

Each of these four assumptions may, under special circumstances, be waived:

1. When all streamlines originate from a constant energy reservoir, the Bernoulli constant does not change from one streamline to another and Eq. (1-6) may be applied on different streamlines.
2. Bernoulli's equation may be applied to systems involving the flow of compressible gas as long as the gas velocities and hence static pressure changes are small. If local Mach numbers are less than 0.3, then the changes in static pressures are less than 2.25 percent. This is the usually accepted Mach number above which compressibility effects must be considered.
3. For gradually changing conditions (such as the emptying of a reservoir), Bernoulli's equation may be applied to unsteady flows without appreciable error.
4. Some viscous flows can be analyzed by using Bernoulli's equation and experimentally determined coefficients to account for viscous effects. The discharge through an orifice in a holding tank is an example.

Bernoulli's equation cannot handle general losses in piping systems.

An equation that can account for general losses is derivable from Eq. (1-3), conservation of energy. Considering the specific total energy e to be composed of the specific internal energy u, the potential energy $(g/g_c)z$, and the kinetic energy $V^2/2g_c$, the energy equation in the absence of heat transfer becomes

$$dW_s + \frac{dP}{\rho} + \frac{V}{g_c} dV + \frac{g}{g_c} dz + d(\text{losses}) = 0 \tag{1-7}$$

Integration of Eq. (1-7) yields

$$\frac{P_1}{\gamma} + \frac{V_1^2}{2g} + z_1 = \frac{P_2}{\gamma} + \frac{V_2^2}{2g} + z_2 + W_s \frac{g_c}{g} + \text{losses}_{1-2} \tag{1-8}$$

where $\gamma = \rho(g/g_c)$ (specific weight) and W_s is the work per unit mass. Equation (1-8) is generally referred to as the *energy equation* and is the fundamental equation for analyzing pipe flow. Information about losses must come from experimental data. The energy equation is normally written as

$$\frac{P_1}{\gamma} + \frac{V_1^2}{2g} + z_1 = \frac{P_2}{\gamma} + \frac{V_2^2}{2g} + z_2 + W_s\frac{g_c}{g} + h_f\frac{g_c}{g} \tag{1-9}$$

where $h_f(g_c/g)$ is the *total head* loss between positions (1) and (2). The sign convention we have used for W_s in Eq. (1-9) takes work extraction to be positive and work addition to be negative.

1-3 HEAD LOSS REPRESENTATION

Friction Factor

The total head loss, h_f, is usually viewed as being composed of major and minor losses. By definition, *major losses* are losses associated with the pipe wall skin friction over the length of the pipe, and *minor losses* are losses incurred because of fluid flowing through fittings, valves, and process equipment. Minor losses are not necessarily smaller than major losses; they are so named to distinguish them from losses due purely to pipe wall skin friction.

The head loss due to the flow of a fluid at an average velocity V through a length L of pipe with a diameter D is

$$h_f = f_{\text{D-W}}\frac{L}{D}\frac{V^2}{2g_c} \qquad \text{Darcy–Weisbach} \tag{1-10}$$

or

$$h_f = 4f_{\text{F}}\frac{L}{D}\frac{V^2}{2g_c} \qquad \text{Fanning} \tag{1-11}$$

where $f_{\text{D-W}}$ is the Darcy–Weisbach friction factor and f_{F} is the Fanning friction factor. Equating Eqs. (1-10) and (1-11), we have

$$4f_{\text{F}} = f_{\text{D-W}} \tag{1-12}$$

The difference between the two friction factor definitions is a manifestation of the original approaches used. The Darcy–Weisbach friction factor results essentially from dimensional analysis in the hydraulics literature and is the usual friction factor associated with pipe flows. The Fanning friction factor results when Eq. (1-2) is applied to a duct flow and is more commonly used in gas dynamic applications than in simple pipe flows. We shall use the

Darcy–Weisbach formulation in this chapter. When friction factors are numerically stated, care must be exercised to ascertain which definition [Eq. (1-10) or (1-11)] is used since the two differ by a factor of 4.

Basically, each definition infers the following: (1) head loss is directly proportional to pipe length L, (2) head loss is inversely proportional to pipe diameter D, and (3) head loss is proportional to the square of the average fluid velocity V. Dimensional analysis in conjunction with the preceding observations suggests

$$f = f\left(\mathrm{Re}_D, \frac{\varepsilon}{D}\right) \tag{1-13}$$

where Re_D is the Reynolds number based upon the pipe diameter and ε/D is the relative roughness of the pipe. The graphical representation of the function $f(\mathrm{Re}_D, \varepsilon/D)$ is called a *Moody diagram* and is presented in Fig. 1-4. The relative roughnesses for various materials and pipe diameters are presented in Fig. 1-5.

As is evident from Fig. 1-4, the function $f(\mathrm{Re}_D, \varepsilon/D)$ is complex. Further examination of the Moody diagram is germane for our purposes. Generally, we classify a pipe flow as either laminar or turbulent, and basically the Moody diagram is divided into a laminar region and a turbulent region. The laminar friction factor is a single straight line on the log–log plot. The laminar flow friction factor is not influenced by the relative roughness. Increasing pipe roughness may cause an earlier transition to turbulent flow, but as long as the flow is laminar relative roughness is not a parameter. Observation of turbulent flows shows the losses for a given pipe to be proportional to V^2; laminar flow losses can be shown to be proportional to the average velocity. The straight line with the negative slope results from trying to represent the laminar flow friction factor using the Darcy–Weisbach or Fanning expressions, which have built into them a V^2 dependence. Reynolds numbers in the range from 2000 to 4500 lie in a critical region where the flow can be either laminar or turbulent or anywhere in between.

For Reynolds numbers larger than those in the critical region, turbulent flow exists in the pipe. The nature of the turbulent flow depends on the interaction between the wall roughness and the layer of boundary layer fluid adjacent to the wall. The turbulent boundary layer is often viewed as being composed of three regions: (1) a region adjacent to the wall where the turbulent fluctuations are suppressed by the wall and viscous (molecular) shear dominates, (2) an outer region in which turbulent shear dominates, and (3) an overlap region that bridges the gap between the inner and outer regions. The layer adjacent to the wall (the inner layer) is called the viscous sublayer and has some attributes of a laminar flow because of the wall-suppressed turbulent fluctuations. The two regions, transition and complete turbulence, into which the turbulent zone is divided categorize the state of the viscous sublayer as influenced by roughness.

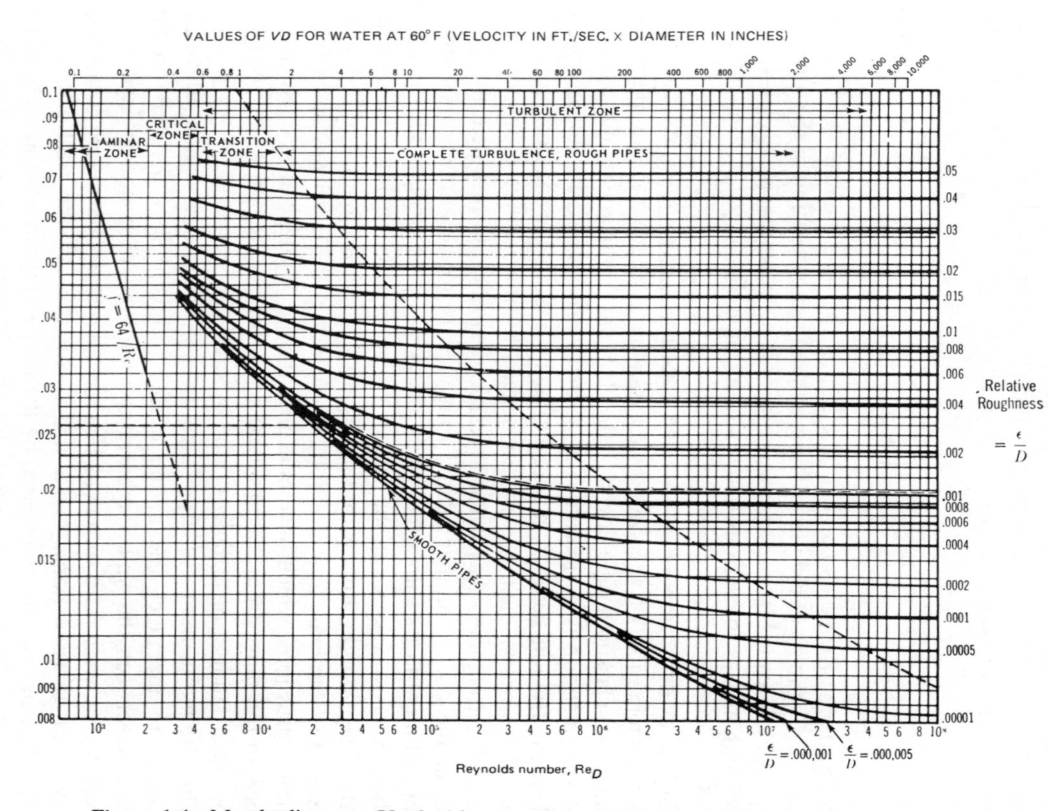

Figure 1-4 Moody diagram. (Used with permission of ASME, from L. F. Moody, "Friction Factors for Pipe Flow," *ASME Transactions*, 1944.)

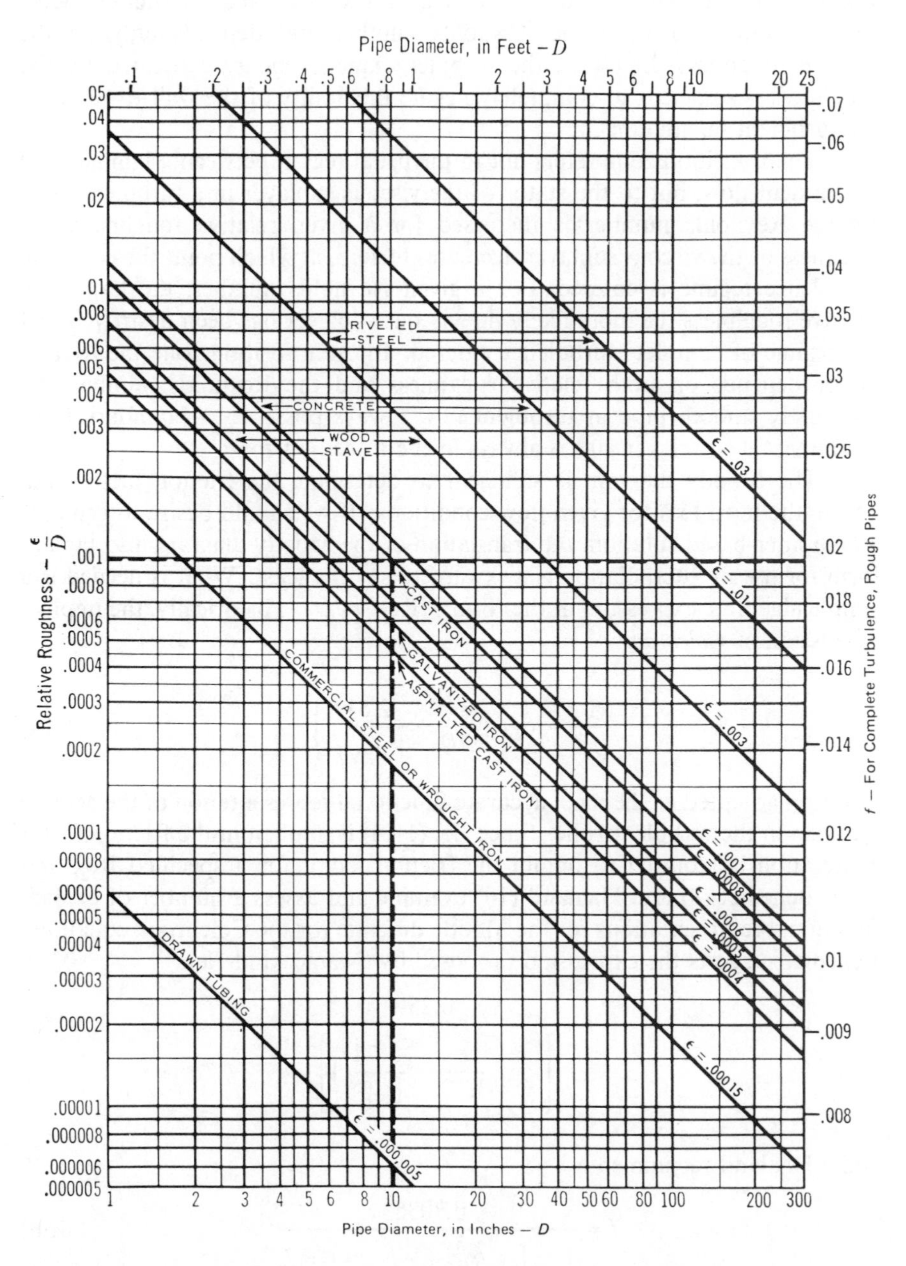

Figure 1-5 Relative roughness of pipe material. (Used with permission of ASME, from L. F. Moody, "Friction Factors for Pipe Flow," *ASME Transactions*, 1944.)

The complete turbulence region refers to a rough-wall turbulent flow in which the roughness is of sufficient magnitude to completely destroy the viscous sublayer. All attributes of laminar flow are lost, and the friction factor becomes independent of the Reynolds number and depends only on the relative roughness. In effect, the turbulent kinetic energy generated by the roughness is sufficient to completely negate the ability of the wall to suppress the turbulent fluctuations.

The transition zone refers not to the usual region between a laminar and a turbulent flow, but to the status of the viscous sublayer in a turbulent flow. As the Reynolds number is increased for a given relative roughness, the thickness of the viscous sublayer decreases to zero, at which point the complete turbulence regime is entered. For a given Reynolds number, the larger the relative roughness the more likely the viscous sublayer has been destroyed and the regime of complete turbulence entered. In the transition zone the friction factor depends on both relative roughness and the Reynolds number. An absolutely smooth pipe in turbulent flow always possesses a boundary layer with a viscous sublayer and is always in the transition zone.

The Moody diagram is sufficient to determine the friction factor and, hence, the head loss for given flow conditions. If we should desire to generate a computer-based solution, the translation of the Moody diagram into tabular form for use in interpolation is awkward to say the least. What is needed is a simple algebraic expression in the form $f(\mathrm{Re}_D, \varepsilon/D)$. Historically, the implicit expression of *Colebrook*

$$\frac{1}{\sqrt{f}} = \log\left(\frac{\varepsilon}{3.7D} + \frac{2.51}{\mathrm{Re}_D\sqrt{f}}\right)^{-2} \tag{1-14}$$

has been accepted as the most accurate functional representation of the Moody diagram in the turbulent zone. Since Eq. (1-14) is implicit in the friction factor f, iteration is required to obtain the friction factor for a specified Re_D and ε/D. Benedict (8) and Haaland (10) examine and assess a number of explicit functional representations for the Moody diagram (or the Colebrook equation). Benedict suggests the expression proposed by Swamee and Jain,

$$f = \frac{0.25}{\left[\log\left(\frac{\varepsilon}{3.7D} + \frac{5.74}{\mathrm{Re}_D^{0.9}}\right)\right]^2} \tag{1-15a}$$

while Haaland recommends

$$f = \frac{0.3086}{\left\{\log\left[\frac{6.9}{\mathrm{Re}_D} + \left(\frac{\varepsilon}{3.7D}\right)^{1.11}\right]\right\}^2} \tag{1-15b}$$

for $\varepsilon/D > 10^{-4}$. For situations where ε/D is very small, as in natural gas pipelines, Haaland proposes

$$f = \frac{0.3086n^2}{\left\{\log\left[\left(\frac{7.7}{\text{Re}_D}\right)^n + \left(\frac{\varepsilon}{3.7D}\right)^{1.11n}\right]\right\}^2} \tag{1-15c}$$

where $n \sim 3$. Overall, Eqs. (1-15b) and (1-15c) are preferred, although the Swamee–Jain expression, Eq. (1-15a), is the best known of the modern explicit representations. Equation (1-15a) has less than ± 1.5 percent error when compared with the Colebrook equation for $4 \times 10^3 \leq \text{Re}_D \leq 10^8$ and $0 \leq \varepsilon/D \leq 5 \times 10^{-2}$. The use of the explicit Eq. (1-15b) yields results comparable to the implicit Colebrook equation. As a matter of fact, the friction factor can be computed from Eqs. (1-15) more accurately than it can be read from the Moody diagram. The use of the Swamee–Jain or Haaland equations is thus the preferred technique.

In our discussion so far we have been concerned only with circular pipes, but for a variety of reasons conduit cross sections often deviate from circular. The appropriate characteristic length to use in evaluating the Reynolds number for noncircular cross-sectional areas is the hydraulic diameter. The hydraulic diameter is defined as

$$D_H \equiv 4\,\frac{\text{cross-sectional area}}{\text{wetted perimeter}}$$

White (9) gives an excellent discussion of the hydraulic diameter. To use the hydraulic diameter concept, the Reynolds number is defined as

$$\text{Re}_{D_H} = \frac{\rho V D_H}{\mu} \tag{1-16}$$

and is then used as Re_D in either the Moody diagram or the Haaland expressions to obtain the friction factor. For all flows the hydraulic diameter as just given is only approximate; the approximation is much more accurate for turbulent flow than for laminar flow. Friction factors for turbulent flows evaluated using this concept generally agree to within ± 2.0 percent of the experimentally determined values; laminar flow friction factors evaluated via this concept exhibit greater errors. The use of the hydraulic diameter in evaluating Reynolds numbers for conduit friction factor determination from circular pipe data fails for eccentric annuli because the eccentricity markedly affects conduit losses and the hydraulic diameter is invariant with eccentricity. The hydraulic diameter concept also has problems for conduit cross sections that have high aspect ratios or are configured such that "corner" viscous effects are dominant. The preceding has given us a remarkably simple and accurate approach for quantitative evaluation of pipe and conduit head losses; it has said nothing about the so-called minor losses.

Minor Losses

Minor losses are expressed in a form similar to the head loss expression concept except that, instead of fL/D, a loss coefficient C is defined for various fittings. The head loss due to a fitting is given by

$$h_f = C\frac{V^2}{2g_c} \tag{1-17}$$

Equation (1-17) is the defining expression for the *loss coefficient* C; but once C is known for a fitting, the head loss across the fitting can be calculated. Two approaches are used for obtaining loss coefficients for different fittings.

The simpler of the two methods prescribes a single value of C for a given fitting (such as valves, couplings, elbows, etc.) that is invariant with size and Reynolds number. Table 1-1 lists loss coefficients for the more commonly used fittings. We should realize that in this approach the data scatter can be large, and that in systems where minor losses represent a significant portion of the total losses, some inaccuracy is to be expected.

The other approach is to specify the C for a given fitting in terms of the value of the complete turbulence friction factor f_T for the nominal pipe size. This method implicitly accounts for pipe size and, where information is available, is preferable over the invarient C of the simple method. For example, for standard 90° elbows the recommended loss coefficient is $C = 30f_T$. Figure 1-6 gives the accepted values of f_T for clean commercial steel pipe for various nominal sizes.

For a standard 90° elbow, $C = 0.81$ for a nominal 1/2-in. pipe, and $C = 0.45$ for a nominal 6-in. pipe, as compared to $C = 0.75$ from Table 1-1. The Crane Company's Technical Paper 410 (7) is the best single source of information on accepted industrial piping information. Figure 1-7 is taken directly from the Crane Company's report and is reproduced here for completeness. Where possible this information should be used. In the reproduced sections (Fig. 1-7), K is used instead of C, and $\overline{V}$ is the specific volume of the fluid in ft^3/lbm. In addition to the usual fittings, the loss coefficients for sudden and gradual contractions and enlargements are given, as are pipe entrance and exit loss coefficients. Benedict (8) extends this method and provides yet another extensive source for loss coefficients. We will not use Benedict's results, although they are probably the most accurate available.

The preceding provides sufficient information to quantify and estimate the head losses for an amazing variety and complexity of piping and conduit systems. The basis for any analysis or design is the energy equation written between any two points and incorporating multiple pumps, turbines, and

TABLE 1-1 Loss Coefficients for Fittings and Valves

Type of Fitting or Valve	Loss Coefficient (C)
45° elbow, standard	0.35
45° elbow, long radius	0.2
90° elbow, standard	0.75
Long radius	0.45
Square or miter	1.3
180° bend, close return	1.5
Tee, standard, along run, branch blanked off	0.4
Used as elbow, entering run	1.3
Used as elbow, entering branch	1.5
Branching flow	1
Coupling	0.04
Union	0.04
Gate valve, open	0.17
$\frac{3}{4}$ open	0.9
$\frac{1}{2}$ open	4.5
$\frac{1}{4}$ open	24.0
Diaphragm valve, open	2.3
$\frac{3}{4}$ open	2.6
$\frac{1}{2}$ open	4.3
$\frac{1}{4}$ open	21.0
Globe valve, bevel seat, open	6.4
$\frac{1}{2}$ open	9.5
Composition seat, open	6.0
$\frac{1}{2}$ open	8.5
Plug disk, open	9.0
$\frac{3}{4}$ open	13.0
$\frac{1}{2}$ open	36.0
$\frac{1}{4}$ open	112.0
Angle valve, open	3.0
Y or blowoff valve, open	3.0
Plug cock, $\theta =$ 5°	0.05
10°	0.29
20°	1.56
40°	17.3
60°	206.0
Butterfly valve, $\theta =$ 5°	0.24
10°	0.52
20°	1.54
40°	10.8
60°	118.0
Check valve, swing	2.0
Disk	10.0
Ball	70.0
Foot valve	15.0
Water meter, disk	7.0
Piston	15.0
Rotary (star-shaped disk)	10.0
Turbine wheel	6.0

Used with permission, from J. H. Perry and C. H. Chilton, *Chemical Engineers' Handbook*, McGraw-Hill Book Company, 1963.

Nominal size	1/2″	3/4″	1″	1 1/4″	1 1/2″	2″	2 1/2, 3″	4″	5″	6″	8–10″	12–16″	18–24″
Friction factor (f_T)	0.027	0.025	0.023	0.022	0.021	0.019	0.018	0.017	0.016	0.015	0.014	0.013	0.012

Figure 1-6 Complete turbulence friction factors for clean commercial steel pipe. (Used with permission, from Technical Paper 410, Crane Co.)

"K" FACTOR TABLE—SHEET 1 of 4

Representative Resistance Coefficients (K) for Valves and Fittings

FORMULAS FOR CALCULATING "K" FACTORS FOR VALVES AND FITTINGS WITH REDUCED PORT

• **Formula 1**

$$K_2 = \frac{0.8 \sin\frac{\theta}{2}(1 - \beta^2)}{\beta^4}$$

• **Formula 2**

$$K_2 = \frac{0.5\,(1 - \beta^2)\sqrt{\sin\frac{\theta}{2}}}{\beta^4}$$

• **Formula 3**

$$K_2 = \frac{2.6 \sin\frac{\theta}{2}(1 - \beta^2)^2}{\beta^4}$$

• **Formula 4**

$$K_2 = \frac{(1 - \beta^2)^2}{\beta^4}$$

• **Formula 5**

$$K_2 = \frac{K_1}{\beta^4} + \text{Formula 1} + \text{Formula 3}$$

$$K_2 = \frac{K_1 + \sin\frac{\theta}{2}[0.8\,(1 - \beta^2) + 2.6\,(1 - \beta^2)^2]}{\beta^4}$$

• **Formula 6**

$$K_2 = \frac{K_1}{\beta^4} + \text{Formula 2} + \text{Formula 4}$$

$$K_2 = \frac{K_1 + 0.5\sqrt{\sin\frac{\theta}{2}}(1 - \beta^2) + (1 - \beta^2)^2}{\beta^4}$$

• **Formula 7**

$$K_2 = \frac{K_1}{\beta^4} + \beta\,(\text{Formula 2} + \text{Formula 4}) \text{ when } \theta = 180^\circ$$

$$K_2 = \frac{K_1 + \beta\left[0.5\,(1 - \beta^2) + (1 - \beta^2)^2\right]}{\beta^4}$$

$$\beta = \frac{d_1}{d_2}$$

$$\beta^2 = \left(\frac{d_1}{d_2}\right)^2 = \frac{a_1}{a_2}$$

Subscript 1 defines dimensions and coefficients with reference to the smaller diameter.

Subscript 2 refers to the larger diameter.

SUDDEN AND GRADUAL CONTRACTION

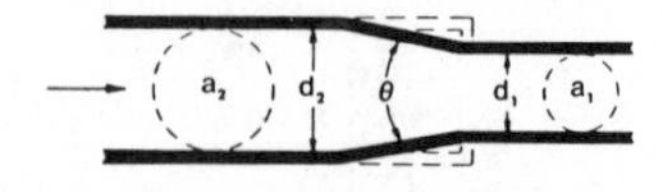

If: $\theta \leq 45^\circ$ . . . . K_2 = Formula 1

$\theta > 45^\circ \leq 180^\circ$ . . . K_2 = Formula 2

SUDDEN AND GRADUAL ENLARGEMENT

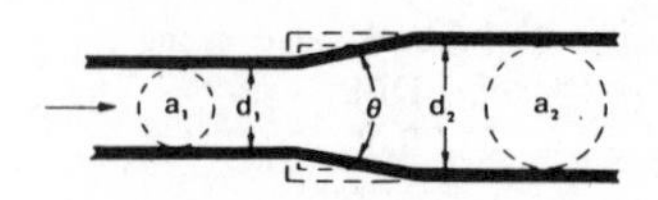

If: $\theta \leq 45^\circ$ . . . . K_2 = Formula 3

$\theta > 45^\circ \leq 180^\circ$ . . . K_2 = Formula 4

Figure 1-7 Crane Co. loss coefficients. (Used with permission, from Technical Paper 410, Crane Co.)

"K" FACTOR TABLE—SHEET 2 of 4

Representative Resistance Coefficients (K) for Valves and Fittings

GATE VALVES
Wedge Disc, Double Disc, or Plug Type

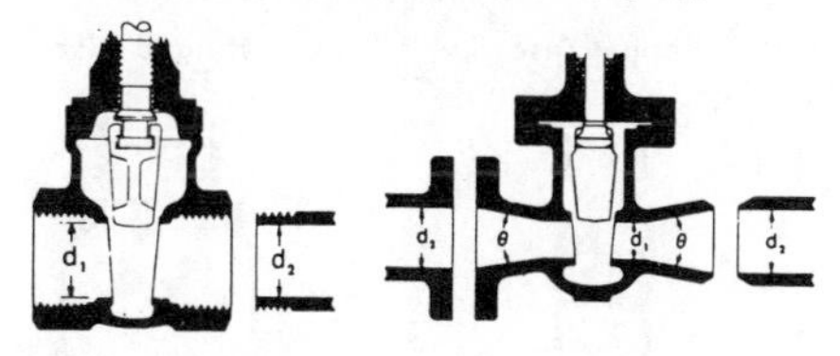

If: $\beta = 1$, $\theta = 0$. . . . .$K_1 = 8\,f_T$

$\beta < 1$ and $\theta \leq 45°$. . . . . .K_2 = Formula 5

$\beta < 1$ and $\theta > 45° \leq 180°$. . .K_2 = Formula 6

SWING CHECK VALVES

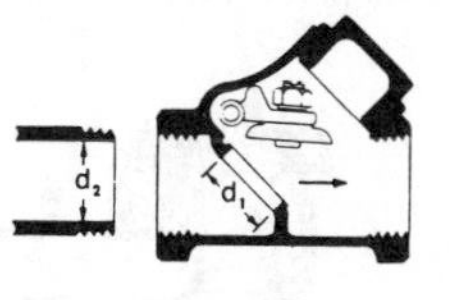

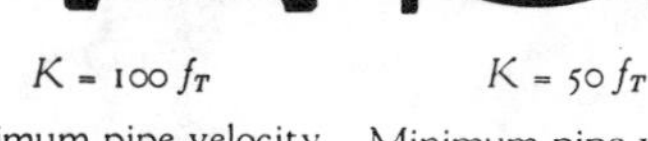

$K = 100\,f_T$

Minimum pipe velocity (fps) for full disc lift $= 35\sqrt{\bar{V}}$

$K = 50\,f_T$

Minimum pipe velocity (fps) for full disc lift $= 48\sqrt{\bar{V}}$

GLOBE AND ANGLE VALVES

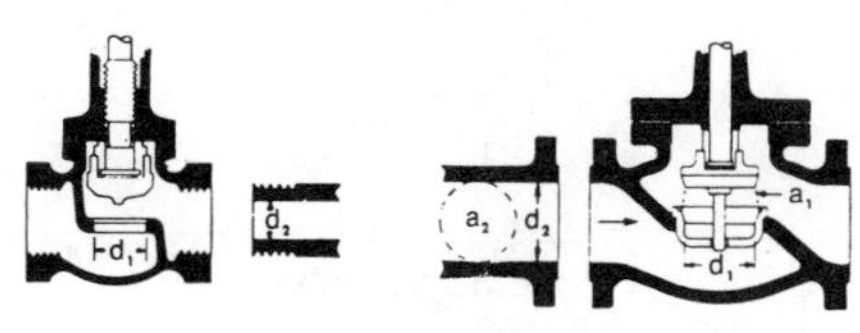

If: $\beta = 1 \ldots K_1 = 340\,f_T$

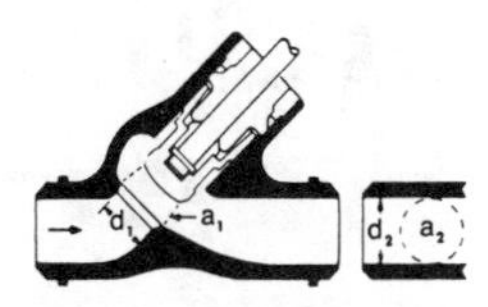

If: $\beta = 1 \ldots K_1 = 55\,f_T$

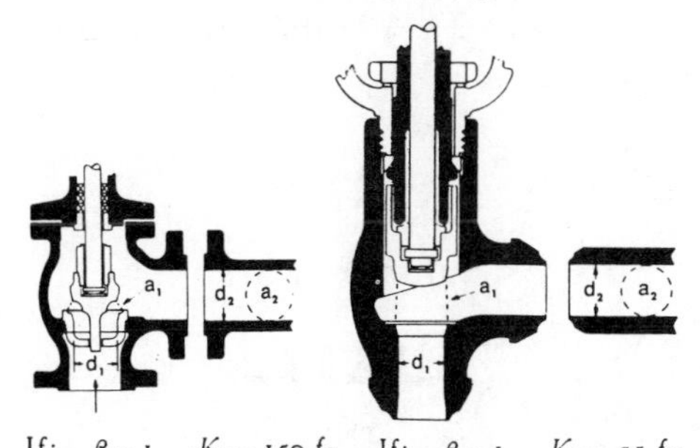

If: $\beta = 1 \ldots K_1 = 150\,f_T$ If: $\beta = 1 \ldots K_1 = 55\,f_T$

All globe and angle valves, whether reduced seat or throttled,

If: $\beta < 1 \ldots K_2$ = Formula 7

LIFT CHECK VALVES

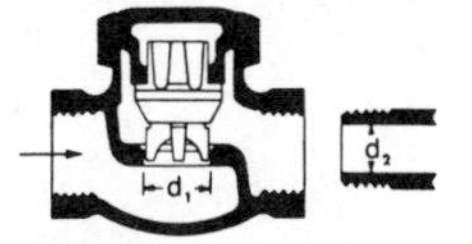

If: $\beta = 1 \ldots K_1 = 600\,f_T$

$\beta < 1 \ldots K_2$ = Formula 7

Minimum pipe velocity (fps) for full disc lift $= 40\,\beta^2\sqrt{\bar{V}}$

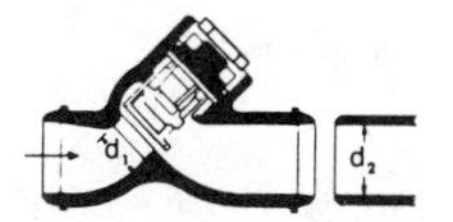

If: $\beta = 1 \ldots K_1 = 55\,f_T$

$\beta < 1 \ldots K_2$ = Formula 7

Minimum pipe velocity (fps) for full disc lift $= 140\,\beta^2\sqrt{\bar{V}}$

TILTING DISC CHECK VALVES

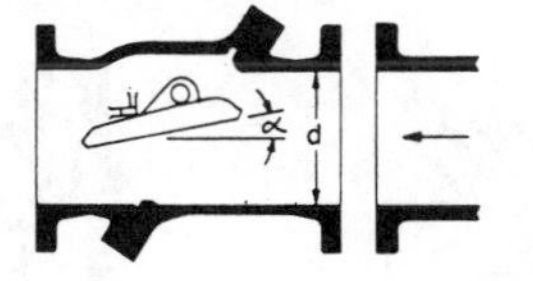

	$\alpha = 5°$	$\alpha = 15°$
Sizes 2 to 8"... K =	$40\,f_T$	$120\,f_T$
Sizes 10 to 14"... K =	$30\,f_T$	$90\,f_T$
Sizes 16 to 48"... K =	$20\,f_T$	$60\,f_T$
Minimum pipe velocity (fps) for full disc lift =	$80\sqrt{\bar{V}}$	$30\sqrt{\bar{V}}$

Figure 1-7 (continued)

"K" FACTOR TABLE—SHEET 3 of 4

Representative Resistance Coefficients (K) for Valves and Fittings

STOP-CHECK VALVES (Globe and Angle Types)

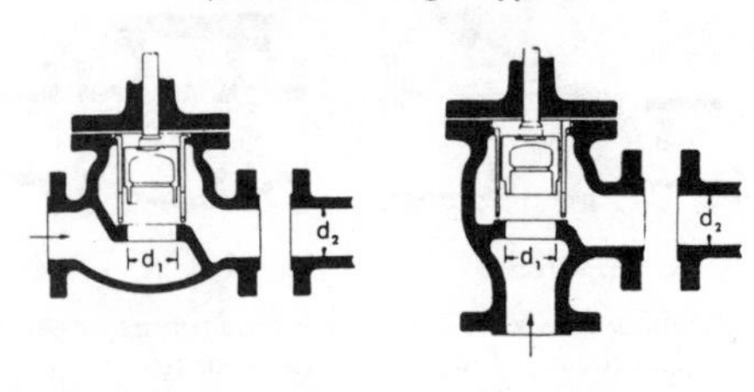

If:
$\beta = 1 \ldots K_1 = 400\, f_T$
$\beta < 1 \ldots K_2$ = Formula 7

Minimum pipe velocity for full disc lift
$= 55\, \beta^2 \sqrt{\bar{V}}$

If:
$\beta = 1 \ldots K_1 = 200\, f_T$
$\beta < 1 \ldots K_2$ = Formula 7

Minimum pipe velocity for full disc lift
$= 75\, \beta^2 \sqrt{\bar{V}}$

If:
$\beta = 1 \ldots K_1 = 350\, f_T$
$\beta < 1 \ldots K_2$ = Formula 7

If:
$\beta = 1 \ldots K_1 = 300\, f_T$
$\beta < 1 \ldots K_2$ = Formula 7

Minimum pipe velocity (fps) for full disc lift
$= 60\, \beta^2 \sqrt{\bar{V}}$

If:
$\beta = 1 \ldots K_1 = 55\, f_T$
$\beta < 1 \ldots K_2$ = Formula 7

If:
$\beta = 1 \ldots K_1 = 55\, f_T$
$\beta < 1 \ldots K_2$ = Formula 7

Minimum pipe velocity (fps) for full disc lift
$= 140\, \beta^2 \sqrt{\bar{V}}$

FOOT VALVES WITH STRAINER

Poppet Disc

$K = 420 f_T$

Minimum pipe velocity (fps) for full disc lift
$= 15 \sqrt{\bar{V}}$

Hinged Disc

$K = 75 f_T$

Minimum pipe velocity (fps) for full disc lift
$= 35 \sqrt{\bar{V}}$

BALL VALVES

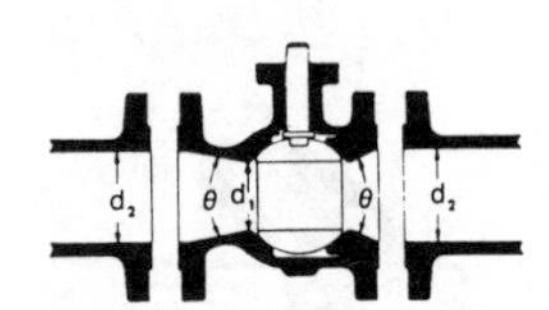

If: $\beta = 1,\ \theta = 0 \ldots\ldots\ldots\ldots K_1 = 3\, f_T$
$\beta < 1$ and $\theta \leq 45° \ldots\ldots\ldots K_2$ = Formula 5
$\beta < 1$ and $\theta > 45° \leq 180° \ldots K_2$ = Formula 6

BUTTERFLY VALVES

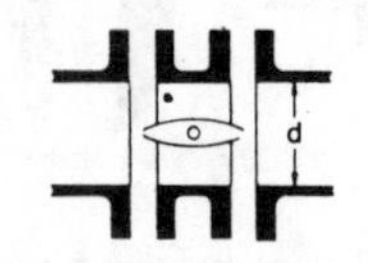

Sizes 2 to 8"... $K = 45\, f_T$
Sizes 10 to 14"... $K = 35\, f_T$
Sizes 16 to 24"... $K = 25\, f_T$

Figure 1-7 (continued)

"K" FACTOR TABLE—SHEET 4 of 4

Representative Resistance Coefficients (K) for Valves and Fittings

PLUG VALVES AND COCKS

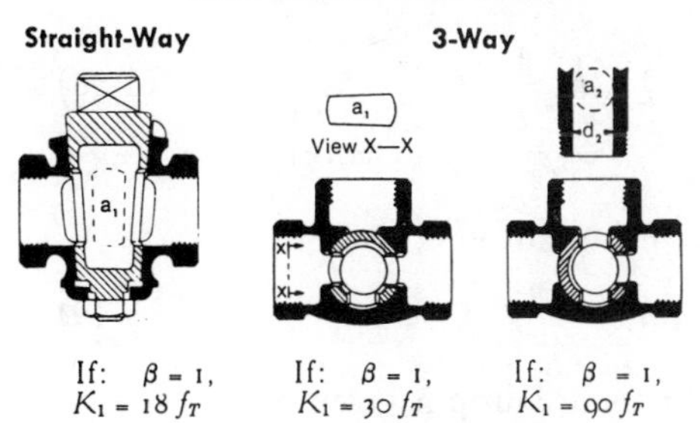

If: $\beta = 1$, $K_1 = 18\, f_T$ — If: $\beta = 1$, $K_1 = 30\, f_T$ — If: $\beta = 1$, $K_1 = 90\, f_T$

If: $\beta < 1 \ldots K_2$ = Formula 6

MITRE BENDS

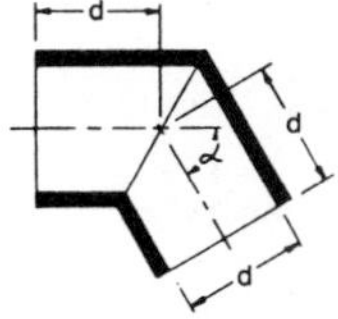

α	K
0°	$2\, f_T$
15°	$4\, f_T$
30°	$8\, f_T$
45°	$15\, f_T$
60°	$25\, f_T$
75°	$40\, f_T$
90°	$60\, f_T$

90° PIPE BENDS AND FLANGED OR BUTT-WELDING 90° ELBOWS

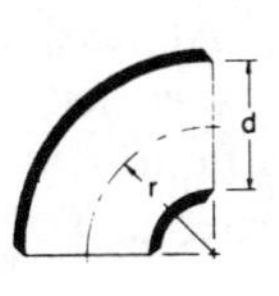

r/d	K	r/d	K
1	$20\, f_T$	10	$30\, f_T$
2	$12\, f_T$	12	$34\, f_T$
3	$12\, f_T$	14	$38\, f_T$
4	$14\, f_T$	16	$42\, f_T$
6	$17\, f_T$	18	$46\, f_T$
8	$24\, f_T$	20	$50\, f_T$

The resistance coefficient, K_B, for pipe bends other than 90° may be determined as follows:

$$K_B = (n - 1)\left(0.25\, \pi\, f_T \frac{r}{d} + 0.5\, K\right) + K$$

n = number of 90° bends
K = resistance coefficient for one 90° bend (per table)

CLOSE PATTERN RETURN BENDS

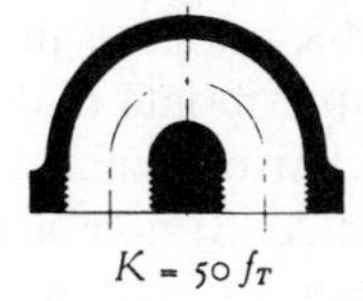

$K = 50\, f_T$

STANDARD ELBOWS

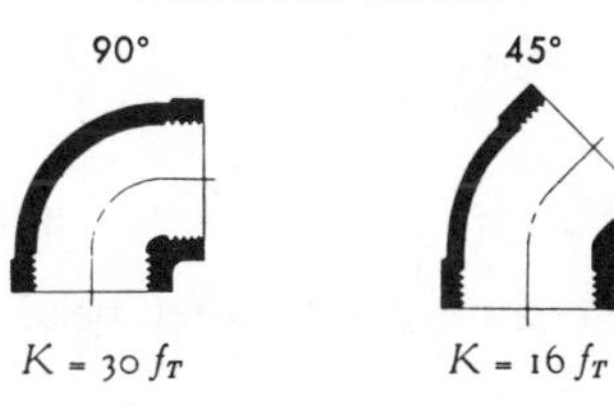

90°: $K = 30\, f_T$ — 45°: $K = 16\, f_T$

STANDARD TEES

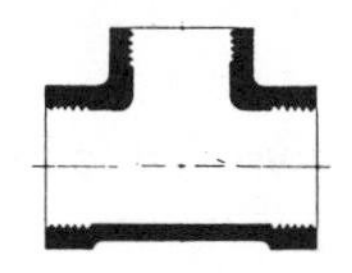

Flow thru run. . . . . . . $K = 20\, f_T$
Flow thru branch. . . . $K = 60\, f_T$

PIPE ENTRANCE

Inward Projecting

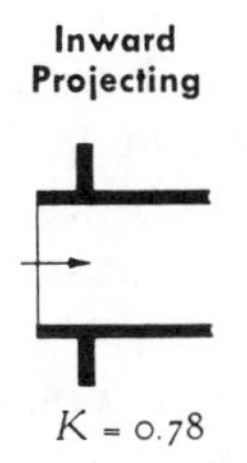

$K = 0.78$

r/d	K
0.00*	0.5
0.02	0.28
0.04	0.24
0.06	0.15
0.10	0.09
0.15 & up	0.04

*Sharp-edged

Flush

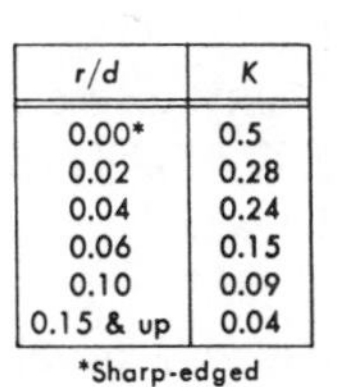

For K, see table

PIPE EXIT

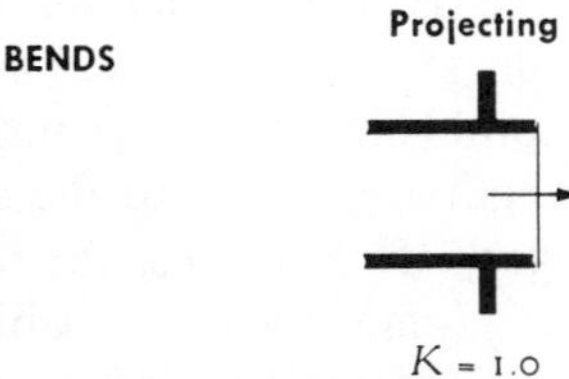

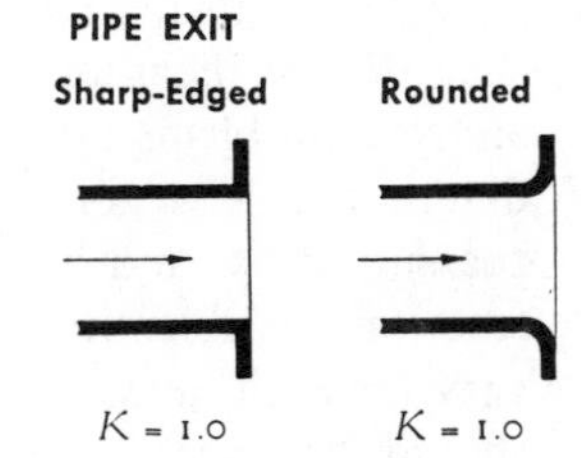

Projecting: $K = 1.0$ — Sharp-Edged: $K = 1.0$ — Rounded: $K = 1.0$

Figure 1-7 (continued)

major and minor losses. The general representation that we shall use is

$$\frac{P_1}{\gamma}+\frac{V_1^2}{2g}+z_1=\frac{P_2}{\gamma}+\frac{V_2^2}{2g}+z_2+\underset{\text{(turbines)}}{\sum_{i=1}^{I}} W_{s_i}\frac{g_c}{g}$$

$$-\underset{\text{(pumps)}}{\sum_{j=1}^{J}} W_{s_j}\frac{g_c}{g}+\underset{\text{(major)}}{\sum_{k=1}^{K}} h_{f_k}\frac{g_c}{g}+\underset{\text{(minor)}}{\sum_{l=1}^{L}} h_{f_l}\frac{g_c}{g} \qquad (1\text{-}18)$$

The relationship between head loss and pressure drop is given by

$$\Delta P=\rho h_f \qquad (1\text{-}19)$$

and the relationship between power and pressure drop is given by

$$\text{power}=Q\,\Delta P=\rho Q h_f=\dot{m}h_f \qquad (1\text{-}20)$$

In the sections that follow we shall be concerned with applying Eq. (1-18) to a variety of situations.

1-4 PIPING NETWORKS

General Considerations

Piping networks vary in complexity from the simple series system illustrated in Fig. 1-8(a), to the more complex parallel system shown in Fig. 1-8(b), to the very complex series–parallel network presented in Fig. 1-8(c). At least in principle, any discrete line or line segment can be analyzed or specified by using Eq. (1-18). Three categories of problems are possible when Eq. (1-18) is applied between any two points on a single line. These are delineated in Table 1-2.

Category I problems are obtainable by direct solution and have as given conditions both the geometry (L, D, ε) and the flow rate (Q). The head loss solved for can be used to compute the power required to pump the specified flow rate Q through the specified system.

Problems in *categories II* and *III* are iterative and require use of either the Moody diagram or the Haaland equation. Category II problems are analysis problems in that the system is specified and the flow rate Q is to be determined. Category III problems are design problems, since the flow rate, pressure drop, and line length are given and the appropriate diameter is the solution variable. In category III problems, the relative roughness ε/D can vary for each iterative pass. Commercially available ANSI nominal pipe sizes should be used for all three categories. ANSI B36.10 wall nominal pipe information is given in Appendix C.

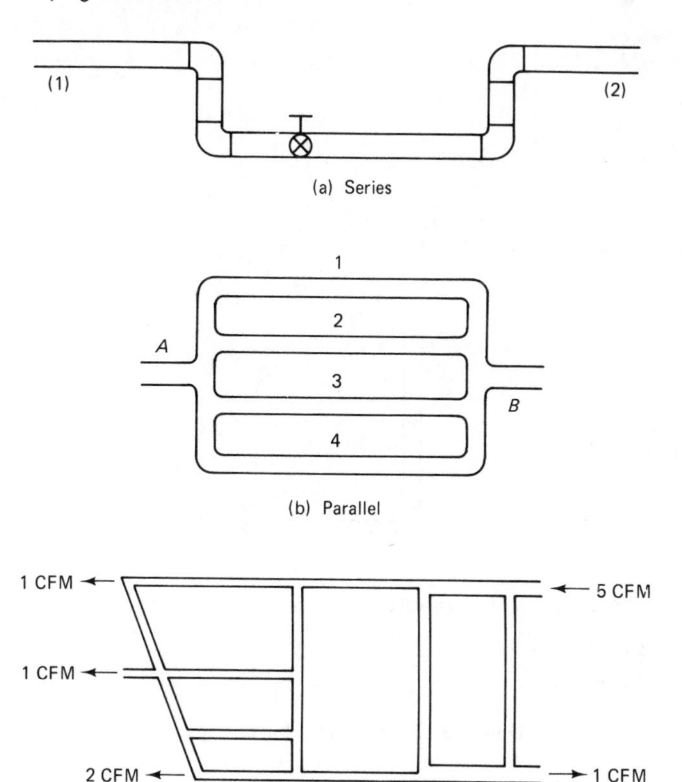

Figure 1-8 Piping network classification.

TABLE 1-2 Classification of Energy Equation Problems

Category	Given Conditions	Solution Variable	Comment
I	$Q, L, D, \nu, \varepsilon$	$h_f(\Delta P)$	Direct
II	$h_f(\Delta P), L, D, \nu, \varepsilon$	Q	Iterative
III	$h_f(\Delta P), Q, L, \nu, \varepsilon$	D	Iterative

Series Piping Systems

As indicated by its name, series piping systems are composed of components in series with each other. Steady flow series piping systems must have the same volume or mass flow rate at the exit as at the inlet. The series systems can be analyzed by simply applying the energy equation, Eq. (1-18), in a systematic, sequential manner from the initial station 1 to the exit station 2.

Example 1-1

Apply the energy equation to the situation sketched in Fig. 1-9.

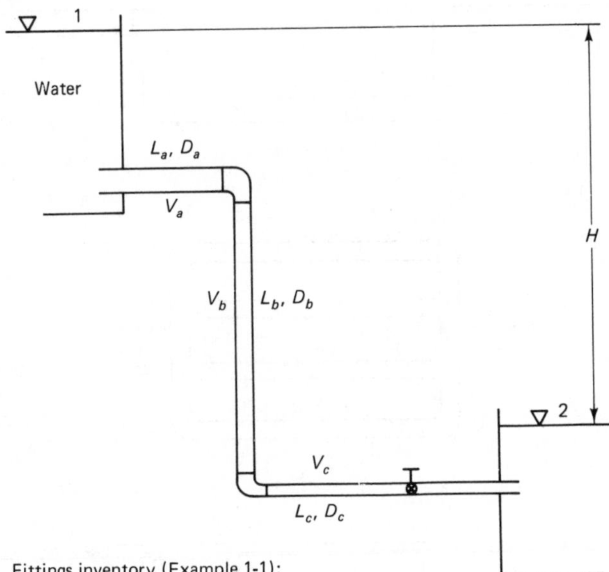

Figure 1-9 System sketch for Example 1-1.

Solution. The energy equation will be applied from the free surface at position 1 of the upper reservoir to the free surface at position 2 of the lower reservoir. The results, with each loss term identified, are

$$\frac{P_1}{\gamma} + \frac{V_1^2}{2g} + z_1 = \frac{P_2}{\gamma} + \frac{V_2^2}{2g} + z_2 + \underset{\text{(entrance)}}{C_a \frac{V_a^2}{2g}} + \underset{\text{(pipe friction)}}{f_a \frac{L_a}{D_a}\frac{V_a^2}{2g}} + \underset{\text{(elbow)}}{C_b \frac{V_b^2}{2g}}$$

$$+ \underset{\text{(pipe friction)}}{f_b \frac{L_b}{D_b}\frac{V_b^2}{2g}} + \underset{\text{(elbow)}}{C_c \frac{V_c^2}{2g}} + \underset{\text{(pipe friction)}}{f_c \frac{L_c}{D_c}\frac{V_c^2}{2g}} + \underset{\text{(valve)}}{C_V \frac{V_c^2}{2g}} + \underset{\text{(exit)}}{C_e \frac{V_c^2}{2g}}$$

From Fig. 1-9, we obtain the following:

$$P_1 = P_2$$

$$V_1 = V_2 \sim 0$$

$$z_1 - z_2 = H$$

and from the continuity equation for incompressible steady flow, we find

$$V_a A_a = V_b A_b = V_c A_c$$

which for circular pipes becomes

$$V_a D_a^2 = V_b D_b^2 = V_c D_c^2$$

Substitution of the preceding into the energy equation, after some algebra, yields

$$H=\frac{V_a^2}{2g}\left[C_a+\frac{f_aL_a}{D_a}+\frac{D_a^4}{D_b^4}\left(C_b+\frac{f_bL_b}{D_b}\right)+\frac{D_a^4}{D_c^4}\left(C_c+f_c\frac{L_c}{D_c}+C_V+C_e\right)\right] \tag{1-21}$$

Until additional information is specified, we really cannot proceed any further than this.

Example 1-2

For the system illustrated in Example 1-1, specify the nominal size of clean commercial steel pipes required for a flow rate of 0.2 ft^3/s if the following are given:

$$D_a=D_b=D_c$$
$$L_a=100\text{ ft}$$
$$L_b=20\text{ ft}$$
$$L_c=80\text{ ft}$$
$$H=75\text{ ft}$$

Solution. This is a category III problem since the system is specified except for the pipe size required for a given flow rate. Using the values specified, Eq. (1-21) becomes, for this system,

$$75\text{ ft}=\frac{V^2}{2g}\left[C_a+\frac{f}{D}(L_a+L_b+L_c)+2C_b+C_V+C_c\right]$$

For the loss coefficients, Fig. 1-7 yields

entrance	$C_a=0.78$
elbows	$C_b=30f_T$
valve	$C_v=55f_T$
exit	$C_e=1.0$

The equation then finally reduces to

$$75\text{ ft}=\frac{V^2}{2g}\left(200\frac{f}{D}+115f_T+1.78\right)$$

The solution of this equation must be by iteration since V and f are functions of D. For this particular example we shall use the diameter D as the iteration variable and compute the head (H) necessary to deliver 0.2 ft^3/s through the system. The iterative sequence is given in Table 1-3. The results show us that a pipe 0.146 ft (1.758 in.) would satisfy the requirements. If we are using schedule 40 pipe, the nominal $1\frac{1}{2}$ in. has an ID of 1.61 in. and the nominal 2 in. has an ID of 2.067 in,; so we must choose the 2-in. schedule 40. The next smaller size, the $1\frac{1}{2}$ in., will not satisfy the flow rate unless a pump is provided. For the 2-in. nominal not to exceed the flow rate requirement, additional resistance must be placed in the series system (i.e., the angle valve must be closed slightly).

TABLE 1-3 Iteration for Category III Problem

D (ft)	ε/D	V (ft/s)	Re_D	f	f_T	$\left(200\frac{f}{D} + 115\, f_T + 1.78\right)$	H (ft)
0.1	0.00150	25.46	1.819×10^5	0.0228	0.023	50.036	504.20
0.2	0.00075	6.37	9.095×10^4	0.0213	0.018	25.126	15.82
0.15	0.00100	11.32	1.213×10^5	0.0216	0.020	32.894	65.48
0.14	0.00107	12.99	1.299×10^5	0.0218	0.020	35.152	92.91
0.145	0.00103	12.10	1.254×10^5	0.0217	0.020	33.959	77.42
0.147	0.00102	11.78	1.237×10^5	0.0216	0.020	33.534	72.37
0.146	0.00103	11.95	1.246×10^5	0.0217	0.020	33.770	74.90

Parallel Systems

Parallel piping systems, such as indicated by Fig. 1-8(b), can be analyzed by applying two principles: (1) the pressure drops across lines in loops with common inlet and exit manifolds must be equal and (2) the total flow rate is the sum of the individual flow rates in each line of the commonly manifolded lines. For example, in Fig. 1-8(b) the pressures at points A and B are common to lines 1 through 4, and the pressure drops across each individual line must be equal. The total flow rate in the system illustrated in Fig. 1-8(b) is the sum of the individual flow rates in lines 1 through 4. Parallel systems are analyzed by considering equal pressure drop with additive flow rates, whereas series systems are analyzed by considering cumulative pressure drops with a constant flow rate. The example parallel system is governed by

$$\Delta P_1 = \Delta P_2 = \Delta P_3 = \Delta P_4 \tag{1-22}$$

$$Q_T = Q_1 + Q_2 + Q_3 + Q_4 \tag{1-23}$$

Parallel system analysis problems can be divided into two types: (1) given the pressure drop across the parallel lines, find the total flow rate or (2) given the total flow rate, find the pressure drop across the parallel lines and the flow rate in each parallel line. *Type 1* problems are straightforward as they can be solved by the category II problem methodology. For each parallel line the flow rate can be obtained, since the line pressure drop is known, and the flow rates of the individual lines added to obtain the total flow rate of the parallel system.

Type 2 problems are more complex since the total flow rate must be apportioned to each of the parallel lines in such a manner that the pressure drops across each line are equal. We can list a simple sequence to systematically accomplish this apportionment:

1. Assume a discharge Q_i' through pipe i of the parallel system. Solve for the head loss h_{f_i}' (or pressure drop $\Delta p_i'$) through pipe i; this is a category I pipe flow solution.
2. Using $h_{f_i}' = h_{f_j}'$ $(i \neq j)$, solve for the Q_j' $(i \neq j)$. This is a category II pipe flow solution.
3. Redistribute the total flow rate Q_T by the simple ratio process of

   $$Q_k = \frac{Q_k'}{\sum_l Q_l'} Q_T \qquad (k = 1, K,\ l = 1, K)$$

   when K is the total number of pipes.
4. Check the h_{f_k} $(k = 1, K)$ for equality using the Q_k $(k = 1, K)$ obtained from Step 3. Repeat steps 1 through 4 until convergence is obtained.

Example 1-3

Given the parallel flow network of Figure 1-10 with the specifications indicated:

$$L_1 = 3000 \text{ ft} \qquad P_A = 80 \text{ psia} \qquad L_2 = 3000 \text{ ft}$$

$$D_1 = 1 \text{ ft} \qquad Z_A = 100 \text{ ft} \qquad D_2 = 8 \text{ in.}$$

$$\varepsilon_1 = 0.001 \text{ ft} \qquad Z_B = 80 \text{ ft} \qquad \varepsilon_2 = 0.0001$$

$$\frac{\varepsilon_1}{D_1} = 0.001 \qquad \frac{\varepsilon_2}{D_2} = 0.00015$$

We are to find Q_1, Q_2, and P_B.

Solution. This is a type 2 problem.

Step 1:

Assume $Q_1' = 3 \text{ ft}^3/\text{s}$. Apply the energy equation along line 1 from A to B to obtain

$$P_B = \rho \frac{g}{g_c}(z_A - z_B) - \rho h_{f_1} + P_A$$

assuming $V_A = V_B$. Finding h_{f_1}' with Q_1' specified is a category I problem, and we have

$$V_1' = \frac{Q_1'}{A_1} = 3.82 \text{ ft/s}$$

$$\text{Re}_{D_1} = \frac{V_1' D_1}{\nu_1} = 1.273 \times 10^5$$

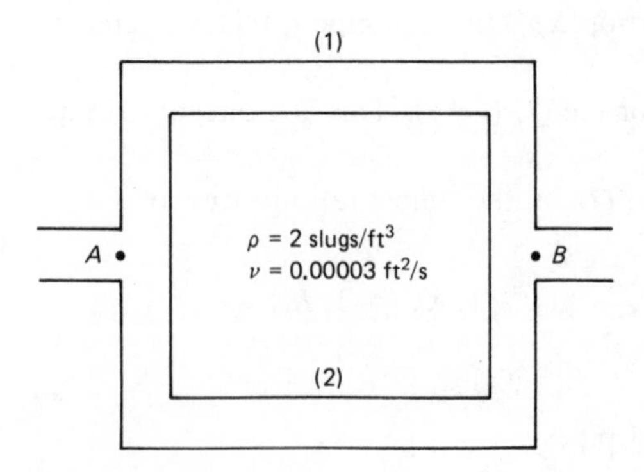

Figure 1-10 Parallel flow network for Example 1-3.

So $f_1' = 0.022$ and

$$h_f' = 14.97 \frac{\text{ft-lbf}}{\text{lbm}}$$

$$= f_1' \frac{L_1}{D_1} \frac{V_1'^2}{2g_c}$$

Step 2:

The loss for pipe 2 must then become

$$h_{f_2}' = h_{f_1}' = 14.97 \frac{\text{ft-lbf}}{\text{lbm}}$$

Finding Q_2' for $h_{f_2}' = 14.97$ ft-lbf/lbm is a category II problem. We shall use V_2' as the iteration variable. The results are given in Table 1-4.

Step 3:

Then $Q_2' = V_2' A_2 = 1.141$ ft^3/s and $\sum_{l=1}^{2} Q_l' = 1.141 + 3 = 4.141$ ft^3/s. We find the corrected values by using

$$Q_1 = \frac{3}{4.141} 5.3 = 3.84 \text{ ft}^3/\text{s}$$

$$Q_2 = \frac{1.141}{4.141} 5.3 = 1.46 \text{ ft}^3/\text{s}$$

Step 4:

Using $Q_1 = 3.84$ ft^3/s and $Q_2 = 1.46$ ft^3/s, we compute V_1, V_2, Re_{D_1}, Re_{D_2}, f_1, f_2, h_{f_1}, and h_{f_2} to find

$$h_{f_1} = 23.85 \frac{\text{ft-lbf}}{\text{lbm}}$$

$$h_{f_2} = 23.98 \frac{\text{ft-lbf}}{\text{lbm}}$$

The head losses in the pipes then agree to 0.54 percent, which is sufficient.

Basically, what we are doing in the type II problem is to take advantage of the rather weak dependence of friction factor on Reynolds number in many

TABLE 1-4 Category II Problem for Pipe 2

V_2' (ft/s)	Re_D	f_2' (−)	h_{f_2}' (ft-lbf/lbm)
1	22233	0.0265	1.859
4	88932	0.0192	21.47
3.27	72702	0.0200	14.94

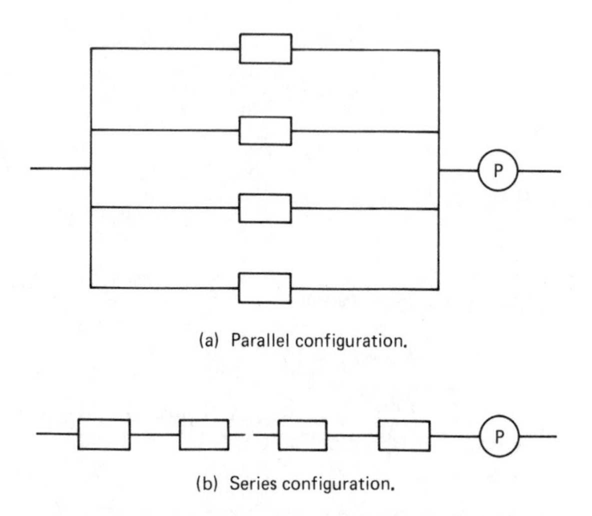

Figure 1-11 Schematic representation of process system.

piping networks to form ratios of pipe flows to the total flow rate, which are also relatively invariant with the pipe Reynolds numbers. Convergence for this methodology is faster if all flows are in the complete turbulence regime. Variations of this technique can also be used for the design of parallel systems.

Parallel systems are often used to reduce the pumping power required for process control systems. Let us consider a system composed of four devices each with a pressure drop ΔP. Two alternate systems are considered, configuration 1 [shown in Fig. 1-11(a)] is parallel and configuration 2 [shown in Fig. 1-11(b)] is series. The parallel configuration has a pressure drop ΔP across each pump and a total flow rate of Q. The pumping power required is $Q\Delta P$ for this configuration. For the series configuration, the pressure drop is at least $4\Delta P$* and the pumping power required is $4Q\Delta P$. In the series system, the total flow rate must be pumped through each device, while for the parallel system only $Q/4$ goes through a device. This simple example demonstrates one very important reason for designing parallel systems.

Series–Parallel Networks

By extending the concepts previously developed for series and parallel piping systems to the more complex series–parallel networks, we can develop analysis techniques for very complex series–parallel networks. Most analysis methods for series–parallel networks are devised around loops and nodes. A *loop* is defined as a series of pipes forming a closed path; a *node* is defined as a point where two or more lines are joined. The sign conventions typically used are

*If the loss through each device is quadratic in flow rate, the pressure drop is $64\Delta P$ for the series situation.

that the head loss h_f in a line is taken as positive for flow in the counterclockwise direction around a loop and that flow toward a node is positive. The loops and nodes are important for the two concepts we adopt from series and parallel systems:

1. Conservation of mass at a node, which for node α is

$$\sum_{\beta=1} Q_{\alpha\beta} = 0 \tag{1-24a}$$

   That is, the node cannot accumulate mass.

2. The pressure at a node must be single valued, which for the ith loop becomes

$$\sum_{j=1} h_{f_{ij}} = 0 \tag{1-24b}$$

   That is, the sum of the pressure drops around a loop must be zero.

These concepts, taken with the method developed for category II problems, are in principle all that is needed to analyze any series–parallel system. Such an approach to complex series–parallel networks is very awkward and can easily lead to ill-posed systems of equations. For these reasons, we shall develop a systematic approach to complex fluid flow networks. This approach is called the Hardy–Cross method and has been extensively used for the analysis of fluid conveying networks. Because of the importance of the Hardy–Cross method, it will be covered as a primary topic in this chapter.

1-5 HARDY–CROSS METHOD

The *Hardy–Cross* formulation is an iterative method for obtaining the steady-state solution for any generalized series–parallel flow network. The great advantage the method offers over the approach described in the previous section is systematism. The Hardy–Cross method can be systematically applied to any fluid flow network; and, if the guidelines are followed, a converged solution will always be obtained. The approach is readily adapted to the computer; in fact, a number of software firms have generalized Hardy–Cross programs available for sale, and Messal (11) suggests that the approach is ideal for computer education.

The basis for any Hardy–Cross analysis technique is the same as for any series–parallel flow network analysis: (1) conservation of mass at a node and (2) uniqueness of pressure at a given point in the loop. Category I problems always arise in this application; and to reduce the numerical evaluations necessary per step, a slightly modified version of the Darcy–Weisbach expression is developed. This modified version of the head loss representation is called the *Hazen–Williams* expression. Since flow rate rather than velocity is usually of primary interest, the conventional head loss expression is written in terms of the flow rate as

$$h_f = \frac{fL}{D}\frac{1}{2g_c}\frac{Q^2}{A^2} \tag{1-25}$$

For circular pipes, this becomes

$$h_f = \frac{16}{2\pi^2 g_c}\frac{fL}{D^5}Q^2 \tag{1-26}$$

or with $K_1 = 8/(g_c\pi^2)$

$$h_f = K_1 \frac{fL}{D^5}Q^2 \tag{1-27}$$

The friction factor f is a function of Reynolds number, hence of flow rate, diameter, and relative roughness. Hazen and Williams suggested that for a general fluid the head loss could then be written as

$$h_f = KQ^n \tag{1-28}$$

where K and n are determined either by experiment or by curve fits using the Moody diagram. For water flow through pipes, the head loss is usually written as

$$h_f = \frac{k_1 L}{C^{1.852}D^{4.8704}}Q^{1.852} \tag{1-29}$$

where C is a dimensionless number indicative of the roughness of the pipe and is called the *Hazen–Williams coefficient*. Values of the Hazen–Williams coefficient for various pipe surfaces are given in Table 1-5. The smoother the pipe is, the larger the value of the Hazen–Williams coefficient. The constant k_1 is

TABLE 1-5 Hazen–Williams Coefficients

Types of Pipe	C
Extremely smooth and straight pipes	140
New, smooth cast iron pipes	130
Average cast iron, new riveted steel pipes ...	110
Vitrified sewer pipes..........................	110
Cast iron pipes, some years in service	100
Cast iron pipes, in bad condition	80

Used with permission, from R. V. Giles, *Fluid Mechanics and Hydraulics*, Schaum's Outline Series, McGraw-Hill, 1962.

dependent on the dimensions of Q; various values of k_1 are given in Table 1-6. Then, in accordance with the preceding,

$$K = \frac{k_1 L}{C^{1.852} D^{4.8704}} \tag{1-30}$$

For a given line, K is a constant and need only be evaluated once. Situations arise for which the Hazen–Williams coefficient is not known or in which water at typical temperatures is not the flowing fluid. In either of these cases, the constant K and, if necessary, the exponent n can be estimated by parametrically solving category I problems for a given (or a given set) pipe and curve fitting the results for the "best" values of K and n in the least-squares sense. The prior determination of the K's and n's is necessary for the Hardy–Cross procedure.

The basic idea of the Hardy–Cross methodology is that conservation of mass at each node can be established initially without consideration of uniqueness of pressure. Uniqueness of pressure can then be used to calculate correction factors for each loop. This procedure maintains conservation of mass at each node and iterates using uniqueness of pressure to drive the iteration. The fundamental development needed for the Hardy–Cross approach is a systematic scheme for calculating the correction factor for a given loop for a system in which nodal conservation is always enforced.

Let us consider the triangular network of Fig. 1-12. The system is divided into two loops, and each pipe in each loop is assigned a number. The flow rate in each pipe is then identified as Q_{ij}, where i is the loop and j the pipe number; thus, pipe 3 in loop 1 has the flow rate identified as Q_{13}. Notice that under this scheme, and with the sign convention previously declared, pipes that are common between two loops have flow rates that are identified with and relatable to each loop; thus, Q_{11} and Q_{23} refer to the same pipe and, in fact, $Q_{11} = -Q_{23}$ according to the sign convention. The initial guesses for all Q_{ij} are made such that conservation of mass is enforced at each node; the superscript 0 denotes the current iterate or the first guess. For Q_{ij}^0 we have

TABLE 1-6 Values of k_1 for Different Units

Units of Q	k_1
CFS (ft^3/s)	4.727
MGD (million gals/day)	10.63
CMS (m^3/s)	10.466

Used with permission, from W. Bober and R. A. Kenyon, *Fluid Mechanics*, John Wiley & Sons, Inc. 1980.

forced

$$\sum_{\beta} Q^0_{\alpha\beta} = 0 \qquad \text{(node } \alpha\text{)} \tag{1-31}$$

but we in general have

$$\sum_{j} h^0_{f_{ij}} \neq 0 \qquad \text{(loop } i\text{)} \tag{1-32}$$

We need to develop a method for calculating the correction factor ΔQ_i for each loop such that uniqueness of pressure at a node is obtained.

Since the sign convention yields either a positive or a negative Q_{ij}, some preliminary relations taking these signs into account need to be examined. Consistency suggests that a negative Q_{ij} yield a negative $h_{f_{ij}}$ so that

$$h_f = \begin{cases} KQ^n & Q \geq 0 \\ -K(-Q)^n & Q < 0 \end{cases} \tag{1-33}$$

and then

$$\frac{dh_f}{dQ} = \begin{cases} nKQ^{n-1} & Q \geq 0 \\ nK(-Q)^{n-1} & Q < 0 \end{cases} \tag{1-34}$$

A Taylor series can be used to expand the head loss about Q such that

$$h_f(Q+\Delta Q) = h_f(Q) + \frac{dh_f}{dQ}\Delta Q + \frac{d^2h_f}{dQ^2}\frac{\Delta Q^2}{2!} + \cdots \tag{1-35}$$

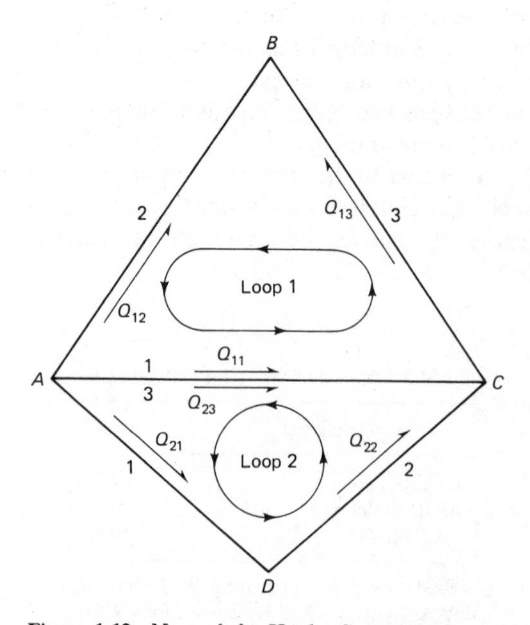

Figure 1-12 Network for Hardy–Cross development.

As the iteration approaches convergence, ΔQ for a given loop decreases and higher-order terms of the expansion can be neglected, with the results

$$h_f(Q+\Delta Q)=\begin{cases} K\left[Q^n+nQ^{n-1}\Delta Q\right] & Q\geq 0\\ -K\left[(-Q)^n-n(-Q)^{n-1}\Delta Q\right] & Q<0\end{cases} \tag{1-36}$$

Equations (1-36) can now be used to evaluate the correction factor ΔQ_1 needed to achieve uniqueness of pressure for any given node in loop 1 of Fig. 1-12. The figure indicates $Q_{11}^0>0$, $Q_{13}^0>0$, and $Q_{12}^0<0$. The initial guesses Q_{ij}^0 yield

$$h_{f_{11}}^0+h_{f_{13}}^0-h_{f_{12}}^0\neq 0 \tag{1-37}$$

while

$$\pm h_{f_{11}}^1\pm h_{f_{13}}^1\pm h_{f_{12}}^1=0 \tag{1-38}$$

is desired; the superscript 1 indicates the next iterate. Substituting the results for $h_f(Q+\Delta Q)$ into Eq. (1-38), we have

$$\begin{aligned} K_{11}\left[\left(Q_{11}^0\right)^n+n\left(Q_{11}^0\right)^{n-1}\Delta Q_1\right]+K_{13}\left[\left(Q_{13}^0\right)^n+n\left(Q_{13}^0\right)^{n-1}\Delta Q_1\right]\\ -K_{12}\left[\left(-Q_{12}^0\right)^n-n\left(-Q_{12}^0\right)^{n-1}\Delta Q_1\right]=0\end{aligned} \tag{1-39}$$

Solving for ΔQ_1,

$$\Delta Q_1=-\frac{K_{11}\left(Q_{11}^0\right)^n+K_{13}\left(Q_{13}^0\right)^n-K_{12}\left(-Q_{12}^0\right)^n}{K_{11}n\left(Q_{11}^0\right)^{n-1}+K_{13}n\left(Q_{13}^0\right)^{n-1}+K_{12}n\left(-Q_{12}^0\right)^{n-1}}, \tag{1-40}$$

or, using absolute values,

$$\Delta Q_1=-\frac{K_{11}Q_{11}^0|Q_{11}^0|^{n-1}+K_{12}Q_{12}^0|Q_{12}^0|^{n-1}+K_{13}Q_{13}^0|Q_{13}^0|^{n-1}}{n\left(K_{11}|Q_{11}^0|^{n-1}+K_{12}|Q_{12}^0|^{n-1}+K_{13}|Q_{13}^0|^{n-1}\right)}, \tag{1-41}$$

Use of the absolute value signs in Eq. (1-41) allows the correct sense of the sign on $h_{f_{ij}}$ to be maintained. The summation notation for loop 1 permits

$$\Delta Q_1=-\frac{\sum_{j=1}^{3}K_{1j}Q_{1j}^0|Q_{1j}^0|^{n-1}}{n\sum_{j=1}^{3}k_{1j}|Q_{1j}^0|^{n-1}} \tag{1-42}$$

and for any loop i with J lines

$$\Delta Q_i = -\frac{\sum_{j=1}^{J} K_{ij} Q_{ij}^0 |Q_{ij}^0|^{n-1}}{n \sum_{j=1}^{J} K_{ij} |Q_{ij}^0|^{n-1}} \tag{1-43}$$

This equation allows the evaluation of the correction factor for any loop i. The methodology of Bober and Kenyon (2) can then be utilized to analyze any complex parallel–series networks.

The Bober and Kenyon procedure* for implementing the Hardy–Cross method is as follows:

1. Subdivide the network into a number of loops. Be sure that all pipes are included in at least one loop.
2. Determine the zeroth estimate for the flow rate $Q_{\alpha\beta}^0$ for each line according to the following procedure. Let s equal the total number of nodes in the system and r equal the total number of lines. Invariably, r will be greater than s. Write a node equation for each node, using the sign convention for the node rule. Since $r > s$, there will be more unknowns than equations. If we assume $(r - s)Q_{\alpha\beta}^0$ values, then the system should reduce to s linear algebraic equations in s unknowns. If the resulting set is linearly independent, then a solution for all other $Q_{\alpha\beta}^0$ values can be determined. This would give the set of values $\{Q_{\alpha\beta}^0\}$. Linear independence can be tested by taking the determinant of the coefficient matrix. In most problems, the system of equations turns out to be linearly dependent, and one additional $Q_{\alpha\beta}^0$ value must be assumed before a set of values $\{Q_{\alpha\beta}^0\}$ can be established. Note that if the established $Q_{\alpha\beta}^0$ values are reasonably accurate the convergence will be quick.
3. Relabel the set of $Q_{\alpha\beta}^0$ values obtained in step 2 according to the loop rule, giving a set of values $\{Q_{ij}^0\}$. The index (ij) designates the jth line in the ith loop, where the positive direction is counterclockwise around the loop.
4. Now determine a correction factor, ΔQ_i, for each loop according to the equation

$$\Delta Q_i = -\frac{\sum_{j=1}^{J} K_{ij} Q_{ij}^0 |Q_{ij}^0|^{n-1}}{n \sum_{j=1}^{J} K_{ij} |Q_{ij}^0|^{n-1}} \tag{1-43}$$

*Adapted from and used with permission, W. Bober and R. A. Kenyon, *Fluid Mechanics* (New York: John Wiley & Sons Inc., 1980).

where J is the total number of lines in the loop and K_{ij} equals the constant appearing in Eq. (1-30) for the jth line in the ith loop.

5. After obtaining a ΔQ_i for each loop, obtain algebraically a new value for the flow rate in each line; that is,

$$Q_{ij}^1 = Q_{ij}^0 + \Delta Q_i \tag{1-44}$$

This method is best explained by considering a specific example.

Example 1-4

Use the Hardy–Cross method to obtain the flow rates in each of the lines of the network shown in Fig. 1-13(a) ($C = 130$).

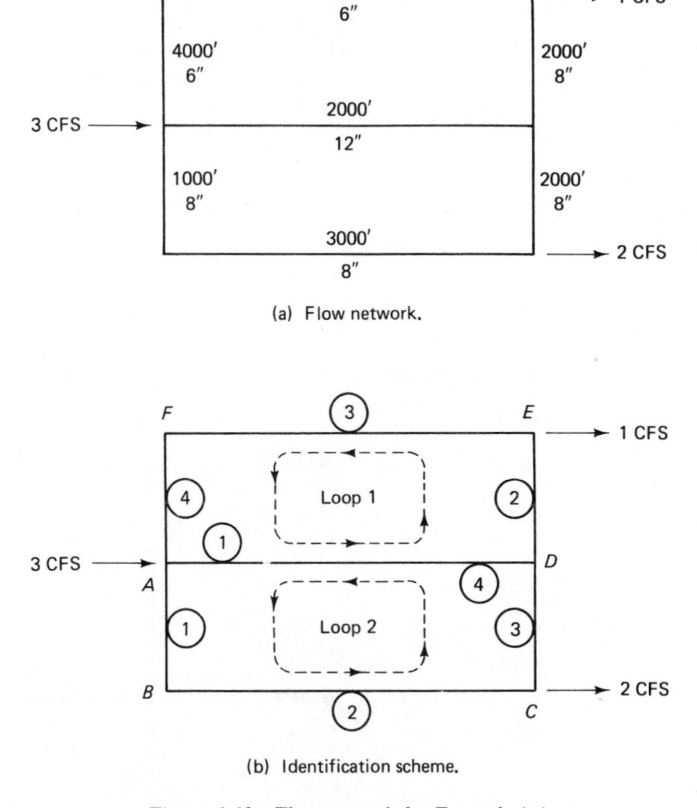

Figure 1-13 Flow network for Example 1-4.

Solution

Step 1:

We begin by dividing the network into loops and numbering all pipes and nodes in each loop. See Fig. 1-13(b) for the results of step 1. The choices are arbitrary as long as every pipe is included in at least one loop.

Steps 2 and 3:

Determine the zeroth estimate. This system has 6 nodes ($s = 6$) and 7 lines ($r = 7$). If we write a conservation equation for each node, we will have 6 equations with 7 unknowns; one ($r - s = 1$) $Q^0_{\alpha\beta}$ will have to be assumed. The resulting system of equations will be linearly dependent, and a second $Q^0_{\alpha\beta}$ must be assumed. Bober and Kenyon (2) state that ($r - s + 1$) values of $Q^0_{\alpha\beta}$ must be chosen. We shall follow their conclusions and accept 2 (since $7 - 6 + 1 = 2$). An equation will be written at each node. To start the process, we shall assume $Q_{21} = 1.0$; then visiting each node in term:

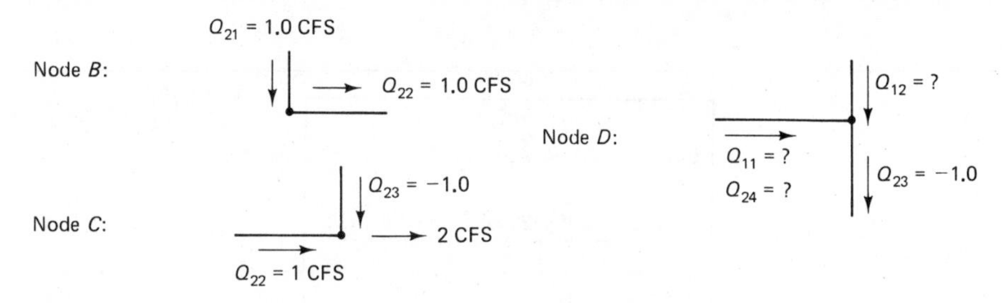

Here we make the second assumption,

$$Q_{11} = 0.8 \text{ CFS} \quad \text{or} \quad Q_{24} = -0.8 \text{ CFS}$$

from which

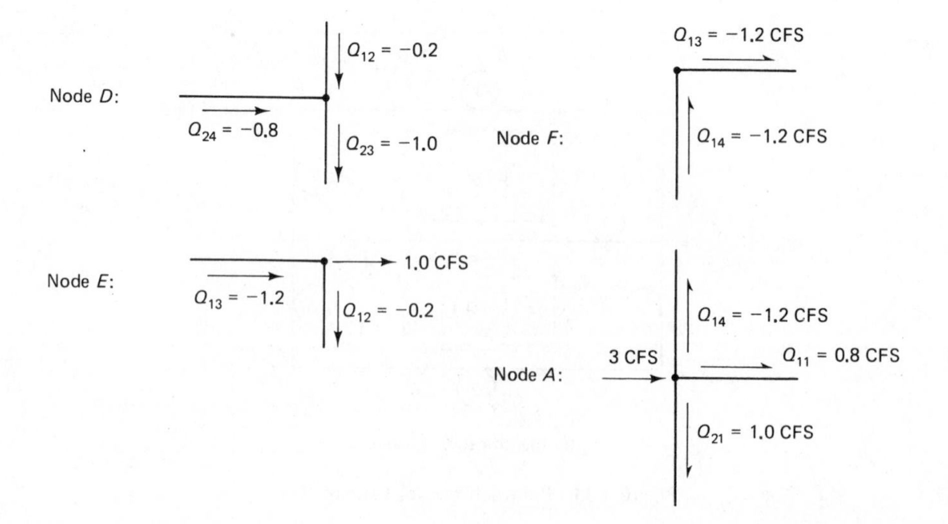

This procedure has produced $Q^0_{\alpha\beta}$ for the network in such a manner that conservation of mass is satisfied at each node.

Steps 4 and 5:

Determine the correct factor ΔQ_i for each loop.

The equation derived previously for the correction factors is very convenient to use either in a computer program or by hand calculation. Table 1-7 gives some details of the iteration process and provides a useful format for hand calculations. The advantage of the form of the equation is clearly evident. After iteration number 1, the correction factor for loop 1 is $\Delta Q_1 = 0.6326$ CFS, and for loop 2 it is $\Delta Q_2 = -0.1574$ CFS. These factors are added to the assumed flow rates for each loop according to the loop sign convention.

Flow rates for lines that are contained in a single loop are corrected simply by

$$Q^1_{ij} = Q^0_{ij} + \Delta Q_i \tag{1-44}$$

For example, for line 3 in loop 1

$$\begin{aligned} Q^1_{13} &= Q^0_{13} + \Delta Q_1 \\ &= -1.20 + 0.6362 = -0.5674 \text{ CFS} \end{aligned}$$

Flow rates for lines that are contained in more than a single loop are corrected by considering the correction factors for common loops. For example, for line 1 in loop 1 and line 4 in loop 2

$$\begin{aligned} Q^1_{11} &= Q^0_{11} + \Delta Q_1 - \Delta Q_2 \\ &= 0.80 + 0.6326 - (-0.1574) \\ &= 1.590 \text{ CFS} \end{aligned}$$

The negative sign is required since positive flow rates in one loop are negative flow rates in other common loops. The converged solution is presented in Fig. 1-14. With the flow rates and K_{ij}'s known, the pressure drop between any two nodes can be calculated from $K_{ij}Q^n_{ij}$.

The procedures are repeated for each iteration. Table 1-7 gives a summary of each iteration until convergence is reached. Convergence is rapid if reasonable estimates are made for the initial flow rates. Examination of the ΔQ_i's from various iterations illustrates the typical approach to convergence.

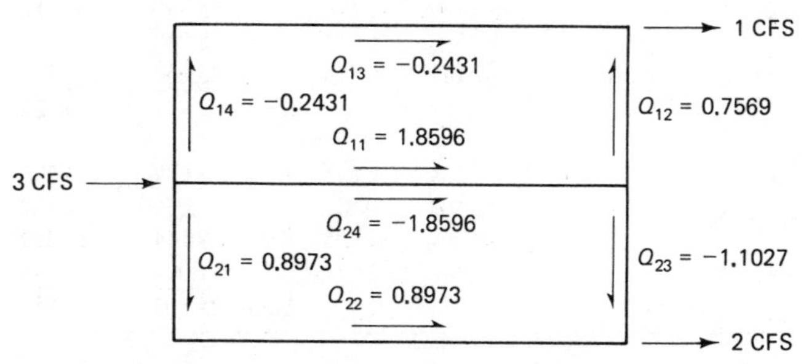

Figure 1-14 Converged solution for Hardy–Cross example problem.

TABLE 1-7 Hardy–Cross Iteration Sequence

Iteration 1

Loop No.	Pipe No.	Pipe Diameter (in.)	Pipe Length (ft)	K	Q		$K\|Q\|^{0.852}$	$KQ\|Q\|^{0.852}$	ΔQ
1	1	12	2000	1.148	0.8		0.949	0.759	
	2	8	2000	8.269	−0.2		2.099	−0.420	
	3	6	3000	50.360	−1.2		58.823	−70.587	
	4	6	4000	67.146	−1.2		78.430	−94.116	
						Sum	140.301	−164.364	0.6326
2	1	8	1000	4.134	1.0		4.134	4.134	
	2	8	3000	12.403	1.0		12.403	12.403	
	3	8	2000	8.269	−1.0		8.269	−8.269	
	4	12	2000	1.148	−0.8		.949	−0.759	
						Sum	25.755	7.509	−0.1574

Iteration 2

Loop No.	Pipe No.	Pipe Diameter (in.)	Pipe Length (ft)	K	Q		$K\|Q\|^{0.852}$	$KQ\|Q\|^{0.852}$	ΔQ
1	1			1.148	1.590		1.704	2.710	
	2			8.269	0.433		4.049	1.751	
	3			50.360	−0.567		31.076	−17.633	
	4			67.146	−0.567		41.434	−23.511	
						Sum	78.263	−36.683	0.2531
2	1			4.134	0.843		3.573	3.010	
	2			12.403	0.843		10.719	9.031	
	3			8.269	−1.157		9.365	−10.840	
	4			1.148	−1.590		1.704	−2.710	
						Sum	25.361	−1.509	0.0321

Iteration 3

Loop No.	Pipe No.	Pipe Diameter (in.)	Pipe Length (ft)	K	Q		$K\|Q\|^{0.852}$	$KQ\|Q\|^{0.852}$	ΔQ
1	1			1.148	1.811		1.904	3.448	
	2			8.269	0.686		5.995	4.111	
	3			50.360	−0.314		18.788	−5.906	
	4			67.146	−0.314		25.050	−7.875	
						Sum	51.737	−6.222	0.0649
2	1			4.134	0.875		3.689	3.226	
	2			12.403	0.875		11.066	9.679	
	3			8.269	−1.125		9.144	−10.290	
	4			1.148	−1.811		1.904	−3.448	
						Sum	25.803	−0.833	0.0174

TABLE 1-7 (Continued)

Iteration 4

Loop No.	Pipe No.	Pipe Diameter (in.)	Pipe Length (ft)	K	Q		$K\|Q\|^{0.852}$	$KQ\|Q\|^{0.852}$	ΔQ
1	1			1.148	1.858		1.947	3.618	
	2			8.269	0.751		6.475	4.860	
	3			50.360	−0.249		15.426	−3.847	
	4			67.146	−0.249		20.569	−5.130	
						Sum	44.417	−0.499	0.0061
2	1			4.134	0.892		3.751	3.346	
	2			12.403	0.892		11.253	10.039	
	3			8.269	−1.108		9.023	−10.000	
	4			1.148	−1.858		1.947	−3.618	
						Sum	25.974	−0.233	0.0048

Iteration 5

Loop No.	Pipe No.	Pipe Diameter (in.)	Pipe Length (ft)	K	Q		$K\|Q\|^{0.852}$	$KQ\|Q\|^{0.852}$	ΔQ
1	1			1.148	1.86		1.948	3.623	
	2			8.269	0.757		6.520	4.933	
	3			50.360	−0.243		15.106	−3.676	
	4			67.146	−0.243		20.141	−4.901	
						Sum	43.715	−0.021	0.0003
2	1			4.134	0.897		3.768	3.379	
	2			12.403	0.897		11.304	10.139	
	3			8.269	−1.103		8.990	−9.917	
	4			1.148	−1.860		1.948	−3.622	
						Sum	26.010	−0.021	0.0005

Iteration 6

Loop No.	Pipe No.	Pipe Diameter (in.)	Pipe Length (ft)	K	Q		$K\|Q\|^{0.852}$	$KQ\|Q\|^{0.852}$	ΔQ
1	1			1.148	1.86		1.948	3.622	
	2			8.269	0.757		6.522	4.937	
	3			50.360	−0.243		15.092	−3.669	
	4			67.146	−0.243		20.123	−4.892	
						Sum	43.685	−0.002	0.0000
2	1			4.134	0.897		3.770	3.383	
	2			12.403	0.897		11.309	10.148	
	3			8.269	−1.103		8.987	−9.910	
	4			1.148	−1.860		1.948	−3.622	
						Sum	26.014	−0.009	0.0000

The process has converged to ΔQ_i within ± 0.00005 CFS.

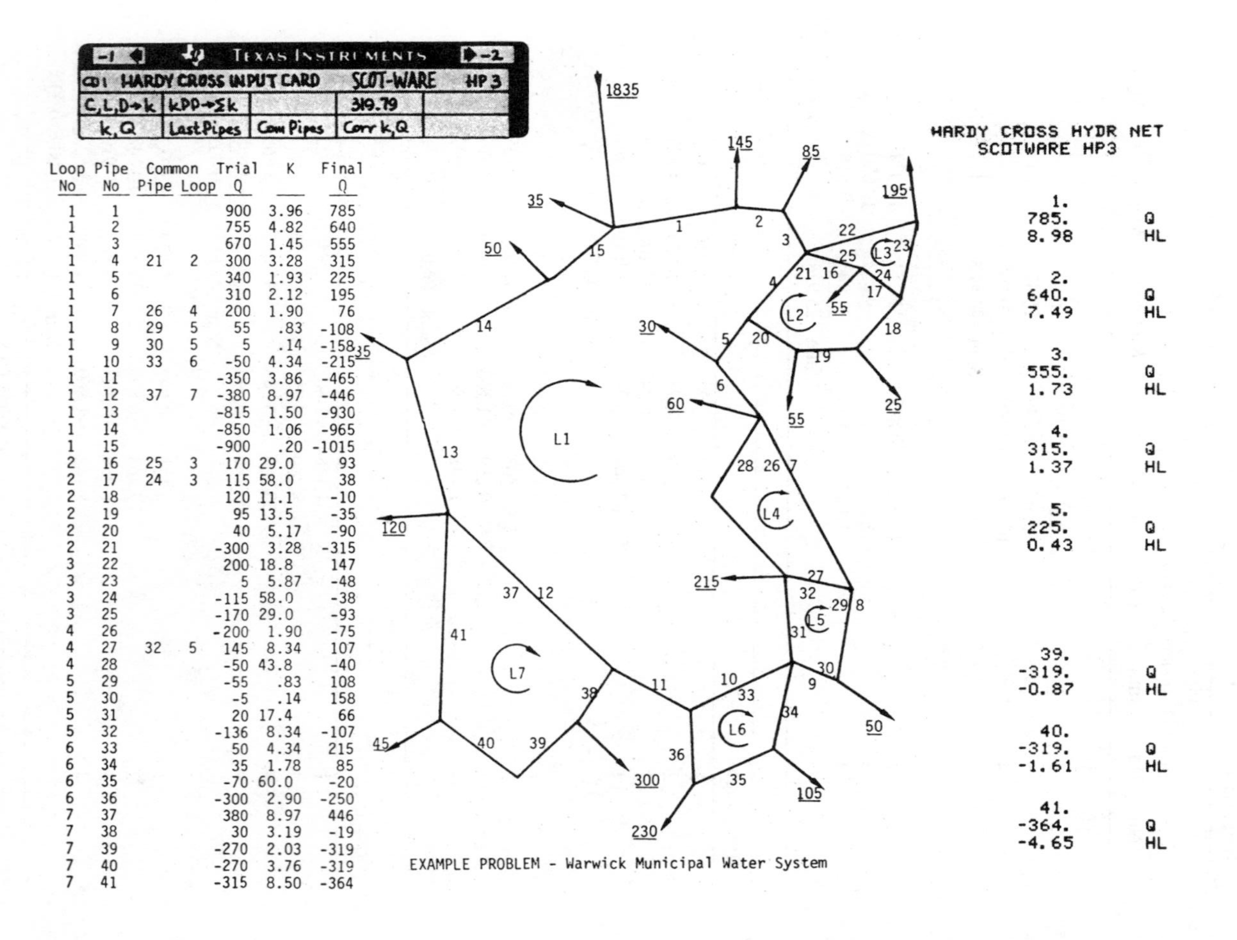

Loop No	Pipe No	Common Pipe	Common Loop	Trial Q	K	Final Q
1	1			900	3.96	785
1	2			755	4.82	640
1	3			670	1.45	555
1	4	21	2	300	3.28	315
1	5			340	1.93	225
1	6			310	2.12	195
1	7	26	4	200	1.90	76
1	8	29	5	55	.83	-108
1	9	30	5	5	.14	-158
1	10	33	6	-50	4.34	-215
1	11			-350	3.86	-465
1	12	37	7	-380	8.97	-446
1	13			-815	1.50	-930
1	14			-850	1.06	-965
1	15			-900	.20	-1015
2	16	25	3	170	29.0	93
2	17	24	3	115	58.0	38
2	18			120	11.1	-10
2	19			95	13.5	-35
2	20			40	5.17	-90
2	21			-300	3.28	-315
3	22			200	18.8	147
3	23			5	5.87	-48
3	24			-115	58.0	-38
3	25			-170	29.0	-93
4	26			-200	1.90	-75
4	27	32	5	145	8.34	107
4	28			-50	43.8	-40
5	29			-55	.83	108
5	30			-5	.14	158
5	31			20	17.4	66
5	32			-136	8.34	-107
6	33			50	4.34	215
6	34			35	1.78	85
6	35			-70	60.0	-20
6	36			-300	2.90	-250
7	37			380	8.97	446
7	38			30	3.19	-19
7	39			-270	2.03	-319
7	40			-270	3.76	-319
7	41			-315	8.50	-364

```
HARDY CROSS HYDR NET
    SCOTWARE HP3

     1.
   785.        Q
   8.98        HL

     2.
   640.        Q
   7.49        HL

     3.
   555.        Q
   1.73        HL

     4.
   315.        Q
   1.37        HL

     5.
   225.        Q
   0.43        HL

    39.
  -319.        Q
  -0.87        HL

    40.
  -319.        Q
  -1.61        HL

    41.
  -364.        Q
  -4.65        HL
```

EXAMPLE PROBLEM - Warwick Municipal Water System

Program HP 3

Hardy Cross Pipe Network Analysis
(for TI-59 Calculators)

Abstract: The program performs a Hardy Cross solution for water flows and pressure drops in pipe networks of up to 51 pipes and 9 loops, with up to 10 pipes common to more than one loop. The program is useful in the calculation of Fire Sprinkler Systems, Fire Water Distribution and Municipal Water Distribution Systems. Inputs are the "k" value for each pipe and a trial flow quantity. Outputs are the equilibrium flow quantity and the head loss for each pipe.

Routines are provided for calculation of the "k" value from length, diameter and the Hazen-Williams "C" factor, which can be varied for each pipe. A routine is provided to determine the equivalent "k" for arrays of similar or dissimilar parallel pipes, with no limit on the number of pipes. Instructions are given for the reduction of other sub-systems to equivalent pipes if necessary to bring the overall system within the size limits of the program.

A simple, straightforward pipe and loop numbering and data entering system is used. Inputs are prompted and outputs are labeled in alphanumeric words. Once inputs have been made, the calculation if completely automatic. The program iterates through the calculation and correction routine for all the loops and common pipes until the correction factors have been reduced to a size pre-set by the user. The user may interact with the program while it is running to moniter convergence, and stop the run early if he wishes. Inputs, intermediate outputs and final outputs of flow and pressure drop may be recorded on magnetic cards for reference and later use.

The program runs on a Texas Instruments TI-59 calculator with PC-100C printer. It can also be run, although less conveniently, without the printer.

Figure 1-15 Example of TI-59 Hardy–Cross advertisement. (Used with permission of the publisher, McClintock Corp.)

The Hardy–Cross method of pipe network analysis is a potent method for complex networks. It is easily programmed for use on the digital computer. The Hardy–Cross analysis is even adaptable to programmable hand-held calculators. A number of software specialty houses market versions executable on a wide variety of hand-held calculators and microcomputers. As an example, Fig. 1-15 is a reproduction of an ad for a Hardy–Cross analysis suitable for the TI-59 calculator. This version requires as input the initial guesses on all line flow rates, as well as the system geometry (diameters, lengths, K values). Thus it effectively accomplishes steps 4 and 5 of the Bober–Kenyon process.

1-6 GENERALIZED HARDY–CROSS ANALYSIS

The Hardy–Cross analysis with which we have dealt has been restricted to flow networks in which the pipe wall friction represented the only loss (i.e., minor loss has been neglected). If line lengths are short enough so that minor losses are important, the equivalent-length approach can be used to include the losses due to fittings. The *equivalent length* is the additional length of pipe needed to give the same head loss (or pressure drop) as a fitting at a given flow rate. Hence the equivalent length L_{eq} is obtained by equating the loss coefficient expression to the conventional head loss expression:

$$C\frac{V^2}{2g_c} = f\frac{L_{\text{eq}}}{D}\frac{V^2}{2g_c} \tag{1-45}$$

or

$$L_{\text{eq}} = C\frac{D}{f} \tag{1-46}$$

However, the flexibility of the Hardy–Cross method makes a generalized formulation possible. The *generalized Hardy–Cross analysis* can be used for piping networks in which the lines can contain devices that result in either additional pressure drop (a heat exchanger or turbine, for example) or in a pressure increase (a pump, for example).

Consider a typical line, line j of loop i, with a length L_{ij} and some device in the line that causes either a head decrease or increase. Figure 1-16 schematically illustrates such a line. In general, the change in head h_{f_D} across the device will depend on the flow rate

$$h_{f_{D_{ij}}} = g_{ij}(Q) \tag{1-47}$$

This functional dependence can be represented by a polynomial expression:

$$h_{f_{D_{ij}}} = A_{ij} + \sum_{m=1}^{M} B_{ijm}Q^m \tag{1-48}$$

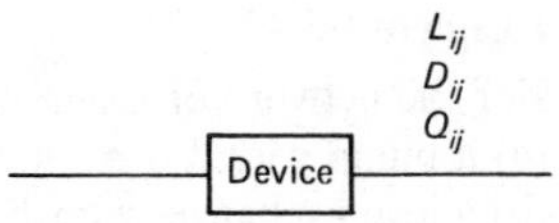

Figure 1-16 Active device in line j of loop i.

where M represents the degree of the polynomial. The A_{ij}'s and B_{ijm}'s can have either positive, negative, or zero values depending on the particular device being described.

The head loss through a fitting is typically described by

$$h_{f_{D_{ij}}} = C_{ij}\frac{V_{ij}^2}{2g_c} = C_{ij}\frac{1}{2g_c}\frac{Q_{ij}^2}{A_{ij}^2} \tag{1-49}$$

The coefficients in the polynomial expression then take the values

$$\begin{aligned} A_{ij} &= 0 \\ B_{ij1} &= 0 \qquad\qquad M = 2 \\ B_{ij2} &= C_{ij}\frac{1}{2g_c A_{ij}^2} \end{aligned} \tag{1-50}$$

Since this represents a *positive* head loss for loop flow in the positive direction, $B_{ij2} > 0$. If an ideal backward curved-blade centrifugal pump is used, the representation is

$$h_{f_{D_{ij}}} = -(A_{ij} - B_{ij1}Q) \tag{1-51}$$

where

$$A_{ij} > 0 \qquad B_{ij1} > 0$$

These must be less than zero because a pump represents a *negative* head loss (i.e., an increase in head).

Incorporation of this polynomial representation in the Hardy–Cross method as previously derived is straightforward and results in

$$\Delta Q_i = -\frac{\sum_{j=1}^{J}\left[K_{ij}Q_{ij}|Q_{ij}|^{n-1} + \text{SGN}(Q_{ij})A_{ij} + \sum_{m=1}^{M} B_{ijm}Q_{ij}|Q_{ij}|^{m-1}\right]}{\sum_{j=1}^{J}\left(K_{ij}n|Q_{ij}|^{n-1} + \sum_{m=1}^{M} B_{ijm}m|Q_{ij}|^{m-1}\right)} \tag{1-52}$$

where $\text{SGN}(Q_{ij}) = 1$ when $Q_{ij} > 0$ and $\text{SGN}(Q_{ij}) = -1$ when $Q_{ij} < 0$. This is easily implemented on the digital computer once the values of A_{ij} and B_{ijm} are known. A_{ij} and B_{ijm} must come from curve fits developed from experimental data or from analytical results.

Example 1-5

For the network of Example 1-4, investigate the effects of adding
(a) a pump with $h_{f_D} = -(50 - 0.4Q)$ ft-lbf/lbm to line 2 in loop 2,
(b) a heat exchanger with $h_{f_D} = 50Q^2$ ft-lbf/lbm to loop 2 in line 2, and (c) a very large pump with $h_{f_D} = -1000$ ft-lbf/lbm to line 2 of loop 2.

Solution

(a) Pump with $h_{f_D} = -(50 - 0.4Q)$ in line 2 of loop 2:

A centrifugal pump with backward curved blades has an ideal characteristic curve of the form

$$H = A - BQ$$

where A is called the shut-off head and represents the increase in pressure through the device if the flow rate Q is zero. The centrifugal pump of Figure 1-17(a) has

$$H = 50 - 0.4Q$$

Since this represents a negative head loss, an increase in the head

$$h_{f_D} = -(50 - 0.4Q)$$

In terms of the generalized Hardy–Cross expression, all A_{ij}'s and B_{ijm}'s are zero except

$$A_{22} = -50.0 \qquad B_{221} = +0.4 \qquad M = 1$$

The same initial estimates can be used for Example 1.5(a) as were used for Example 1.4. The converged results of the iteration are given in Figure 1-17(a). As expected, the addition of the pump results in an increase in the flow rate through lines 1 and 2 of loop 2. Loop 1 is affected very little.

(b) Heat exchanger with $h_{f_D} = 50Q^2$ in line 2 of loop 2:

Many devices, including heat exchangers and fittings, have losses proportional to Q^2. In terms of the generalized Hardy–Cross formulation, all A_{ij}'s and B_{ijm}'s are zero except

$$A_{22} = 0$$
$$B_{221} = 0 \qquad M = 2$$
$$B_{222} = 50$$

The converged results of the iteration are given in Fig. 1-17(b). The addition of the heat exchanger reduces the flow rate in lines 1 and 2 of loop 2. As in part (a), loop 1 is not affected much.

(c) Pump with $h_{f_D} = -1000$ ft-lbf/lbm in line 2 of loop 2:

This purely hypothetical example has a very large pump placed in line 2 of loop 2. All A_{ij}'s and B_{ijm}'s are zero except

$$A_{22} = -1000.0 \qquad M = 0$$

The converged results of the Hardy–Cross iterations are given in Fig. 1-17(c). Such a large pump completely changes the flow rate distributions. Pipe 11 (or pipe 24) now flows in a direction opposite from the previous cases (Example 1-3 included). The size of the head increase dominates loop 2 and forces a useless "circulation" on the loop.

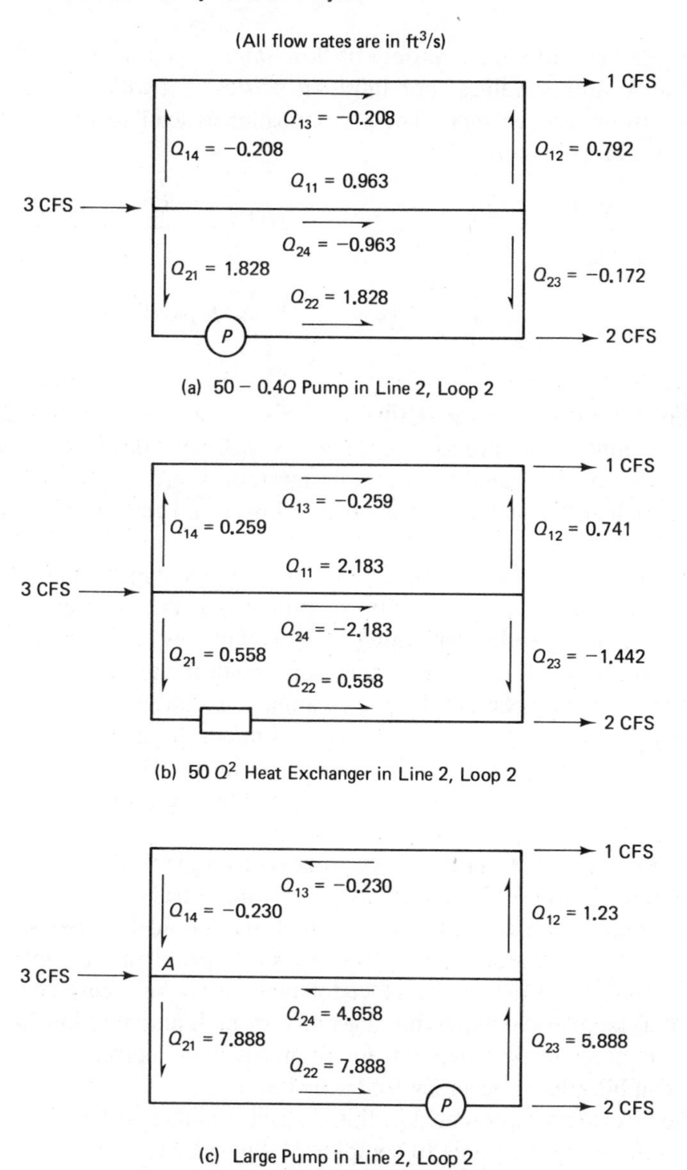

(a) $50 - 0.4Q$ Pump in Line 2, Loop 2

(b) $50\,Q^2$ Heat Exchanger in Line 2, Loop 2

(c) Large Pump in Line 2, Loop 2

Figure 1-17 Generalized Hardy–Cross examples.

This is a hypothetical example that we would never consider using; nonetheless, the Hardy–Cross convergence was swift (eight iterations) and stable.

Several avenues are open for use of this generalized technique when specific Hazen–Williams coefficients are not known or when a different fluid is used. The simplest is to use the Moody diagram or the Haaland equations and compute the head loss for a variety of flow rates. Then a curve fit is used to obtain the K and n values. For fluids in process networks, the exponent n may vary from pipe to pipe. These circumstances lead to the most general Hardy–Cross expression

$$\Delta Q_i = - \frac{\sum_{j=1}^{J} \left[K_{ij} Q_{ij} |Q_{ij}|^{n_{ij}-1} + \mathrm{SGN}(Q_{ij}) A_{ij} + \sum_{m=1}^{M} B_{ijm} Q_{ij} |Q_{ij}|^{m-1} \right]}{\sum_{j=1}^{J} \left(K_{ij} n_{ij} |Q_{ij}|^{n_{ij}-1} + \sum_{m=1}^{M} B_{ijm} m |Q_{ij}|^{m-1} \right)} \tag{1-53}$$

The generalized Hardy–Cross analysis allows us to see what effects process equipment (or additions or deletions) will have on the flow of a given network. It is both an analysis and a design tool of great flexibility. We have hardly scratched the surface of the potential uses and results in this presentation.

The Hardy–Cross method is one of five techniques that have been developed and used for the solution of piping network problems [see Wood and Rayes (12)]. Of the remaining four techniques, two are based upon different procedures for satisfying the loop equations given continuity at the nodes; the remaining two are developed about the concept of computing head adjustment factors for assumed heads at each node. All five methodologies use the node–loop nomenclature and conservation ideas. The Hardy–Cross techniques explored herein are the most used of the five piping network procedures.

Wood and Rayes (12) report some convergence difficulties with the Hardy–Cross method. These seem to occur when lines of very high and very low head losses appear in the same loop. Care should be exercised and the solution examined carefully for such category of problems. Double precision arithmetic and very small convergence tolerances may be necessary for systems with several loops each containing lines of very high and very low head losses. Alternative procedures based on simultaneous loop correction calculations seem to exhibit greater accuracy for such situations.

The information presented in this chapter is sufficient for us to use in the analysis and design of simple to complex piping systems.

REFERENCES

1. Streeter, V. L., and Wylie, E. B., *Fluid Mechanics*, 7th ed. New York: McGraw-Hill, 1979.
2. Bober, W., and Kenyon, R. A., *Fluid Mechanics*. New York: John Wiley, 1980.
3. Welty, J. R., Wicks, C. E., and Wilson, R. E., *Fundamentals of Momentum Heat and Mass Transfer*, 2nd ed. New York: John Wiley, 1976.

4. Hughes, W. F., and Brighton, J. A., *Schaums Outline Series: Fluid Dynamics*. New York: McGraw-Hill, 1967.
5. Giles, R. V., *Schaum's Outline Series: Fluid Mechanics and Hydraulics*, 2nd ed. New York: McGraw-Hill, 1962.
6. Lyderson, A. L., *Fluid Flow and Heat Transfer*. New York: John Wiley, 1979.
7. "Flow of Fluids," Technical Paper 410, 4th printing. Chicago: Crane Co., 1957.
8. Benedict, R. P., *Fundamentals of Pipe Flow*. New York: John Wiley, 1980.
9. White, F. M., *Viscous Fluid Flow*. New York: McGraw-Hill, 1974.
10. Haaland, S. E., "Simple and Explicit Formulas for the Friction Factor in Turbulent Pipe Flow," *Trans. ASME J. Fluids Engineering*, Vol. 105, No. 3, March 1983, pp. 89–90.
11. Messal, E. E., "An Ideal Problem for Computer Education." Presented at the ASME Century Two Advances in Computer Technology—1980 Conference, Vol. II, San Francisco, August 12–15, 1980.
12. Wood, D. J., and Rayes, A. G., "Reliability of Algorithms for Pipe Network Analysis," *J. Hydraulics Division, Proc. ASCE*, Vol. 107, No. HY10, October 1981, pp. 1145–1161.

REVIEW QUESTIONS

1. What are the differences between the Lagrangian and Eulerian approaches to fluid mechanics?
2. What are the assumptions for Bernoulli's equation:

$$P + \rho\frac{g}{g_c}Z + \frac{1}{2}\rho\frac{V^2}{2g_c} = \text{constant}$$

3. How does the energy equation differ from the Bernoulli equation?
4. What are minor losses and major losses?
5. In the fully rough (complete turbulence) regime, of what is f a function?
6. What two principles are used in solving series–parallel network problems?
7. Why is the pumping power for a parallel network less than the pumping power for a series network for the same flow rate?
8. Why is the Hazen–Williams representation for head loss used rather than the Darcy–Weisbach expression in the Hardy–Cross method?
9. Loop 3, line 2 contains a pump with the characteristic curve $H = 100 + 5Q - 10Q^2$. What are A_{ij}, the B_{ijm}'s, and M for use in the generalized Hardy–Cross ?
10. Why is initial mass conservation at nodes so important for the Hardy–Cross method?

PROBLEMS

1. A cast-iron pipeline carries 4 million gallons of water (at 55°F) per day. Over a 3-mile segment of the pipeline the pressure drop is 400 ft-lbf/lbm. The installed

costs for three pipe sizes are as follows:

12-in. (nominal)	\$44/ft
14-in. (nominal)	\$55/ft
16-in. (nominal)	\$66/ft

Power costs are \$0.075 per kWh over the expected 15-year lifetime. If the pump efficiency is 56 percent, what is the most economical pipe diameter?

2. The cold-water faucet in a house feeds from a water main where the pressure is 50 psig. The system consists of 200 ft of $\frac{1}{2}$-in. ID galvanized pipe, three 90° standard elbows, a fully opened gate valve, and a faucet. The faucet can be modeled by considering it to be a conventional globe valve followed by a nozzle with an ID of 0.25-in. If the water is at 60°F, find the flow rate through the faucet.
3. An old pipe 36 in. in diameter has a roughness of $\varepsilon = 0.1$ in. A $\frac{1}{2}$-in.-thick lining would reduce the roughness to $\varepsilon = 0.0004$ in. How much in pumping cost would be saved per year per 1000 ft of pipe for water at 70°F with a velocity of 8 ft/s? The pump and motor are 54 percent efficient and energy costs 8¢/kWh. Would this be economically feasible if it cost \$50/ft to reline the pipe?
4. A hydraulic turbine requires 2000 gpm flow through a 10,000-ft-long pipe. The available head is 100 ft-lbf/lbm. The pipe is to be schedule 40 commercial steel, and a gate valve ($C = 0.17$ fully open) is available for flow control.
   (a) Find the minimum standard pipe diameter required.
   (b) For gate valve settings of 3/4, 1/2, and 1/4 open, find the resulting flow rate.
   (c) Plot the results of part (b) as flow rate versus valve position.
   (d) What does part (c) say about control of low flow rates through this pipe?
   (e) What is the C value for a closed valve?
5. **(a)** A pump moves 0.4 ft^3/s of light oil through a 2-in ID, cast-iron line 500 ft long, as shown in Fig. P1-5. Determine the pumping power required.
   (b) What power would be required if the oil were heated such that $\mu = 1.6 \times 10^{-3}$ lbm/ft-s?
6. In Problem 5, what would the flow rate be if the pump delivered 5 hp to the fluid?
7. A chilled-water air-conditioning system is to be designed for a four-story office complex, as outlined in Fig. P1-7. Each story requires a chilled-water flow rate of 18,000 lbm/h. All water flow velocities are to be maintained at less than 5 ft/s. A single-size pipe is used for the vertical supply and return pipes, and a single-size pipe is used on each floor. Each floor has 100 ft of pipe and two valves, one check and one gate. The pressure drop across the air handling unit on each floor is 50 psi,

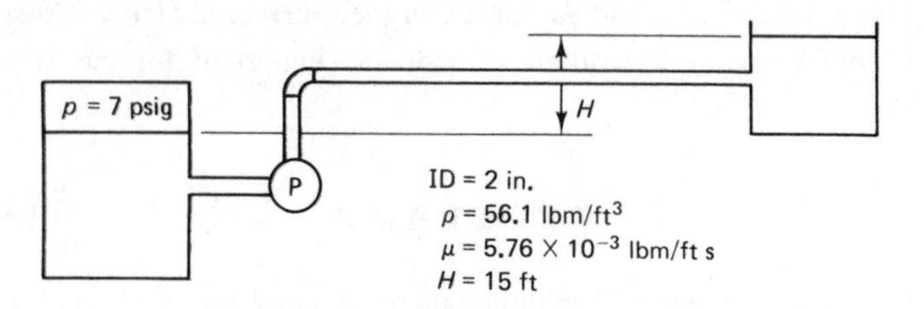

Figure P1-5

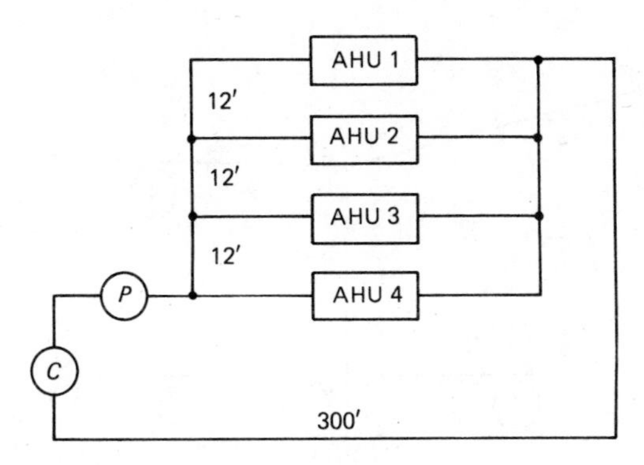

Figure P1-7

and the pressure drop across the chiller is 100 psi. Neglecting the losses of any elbows or tees, specify the following:

(a) Supply and return pipe size (schedule 80)
(b) Floor pipe size (schedule 80)
(c) ΔP required for the pump
(d) Pump power requirements for $\eta = 0.55$
(e) Verification that the flow rate on each floor is 18,000 lbm/h.

Evaluate all properties at 60°F. Be sure to state all assumptions made in the calculations.

8. A light oil at 100°F is to be pumped in a 1000-ft-long nominal 3-in. schedule 40 pipe. Devise a computer program to evaluate the K and n in the Hazen–Williams head loss expression for this pipe.

9. Determine the flow rate in each line and the pressure at each node for the system shown in Fig. P1-9. The Hazen–Williams coefficient is 100, and the pressure at node A is 88 psig. Pipe lengths and diameters are given in the table.

Loop No.	Line No.	Length (ft)	Diameter (in.)
1	1	6000	10
	2	3000	8
	3	3500	8
	4	5000	12
2	1	3500	8
	2	5000	10
	3	3500	16
	4	5000	12
3	1	6000	8
	2	5000	10
	3	3000	8
	4	3000	8
4	1	3000	8
	2	6000	8
	3	5000	10

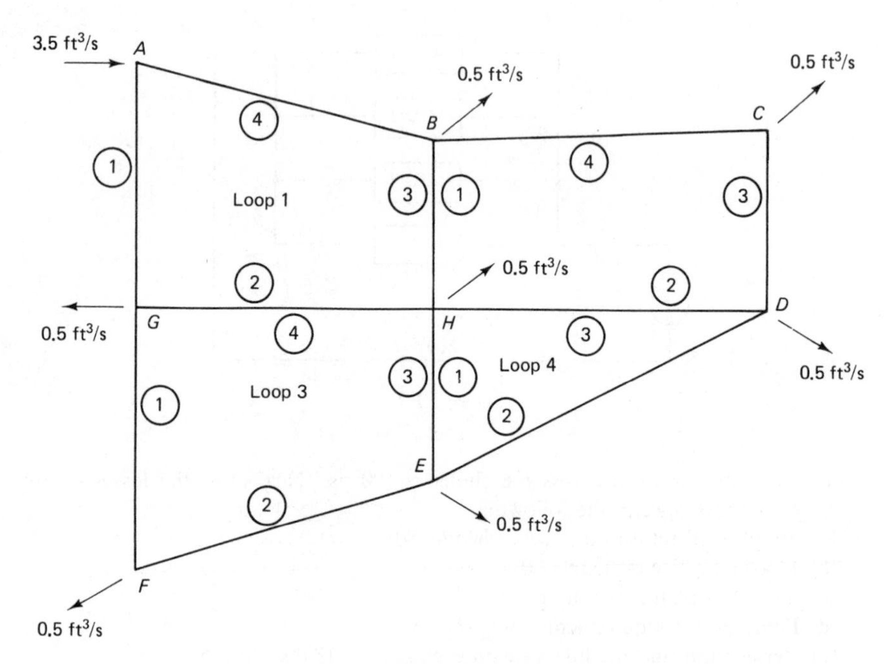

Figure P1-9

10. In Problem 9, what would be the effect on the line flow rates and node pressures if a pump with the pressure flow rate relationship

$$\Delta H(\text{ft-lbf/lbm}) = 75.0 - 16Q^2$$

were placed in line number 2 of loop number 2 (or line 3 of loop 4)?

11. In Problem 9, what would be the effect on the line flow rates and node pressures if a device with a pressure drop relationship of $\Delta H = 23Q^2$ were placed in line 4 of loop 1?

12. A military jet transport uses JP-4, a kerosenelike fuel. The fuel distribution network is shown in Fig. P1-12. What pressure and power must the booster pump at A develop if the minimum inlet pressure at the engines (H, G, F, E) is 40 psig? The head prior to the pump is 4 ft-lbf/lbm.

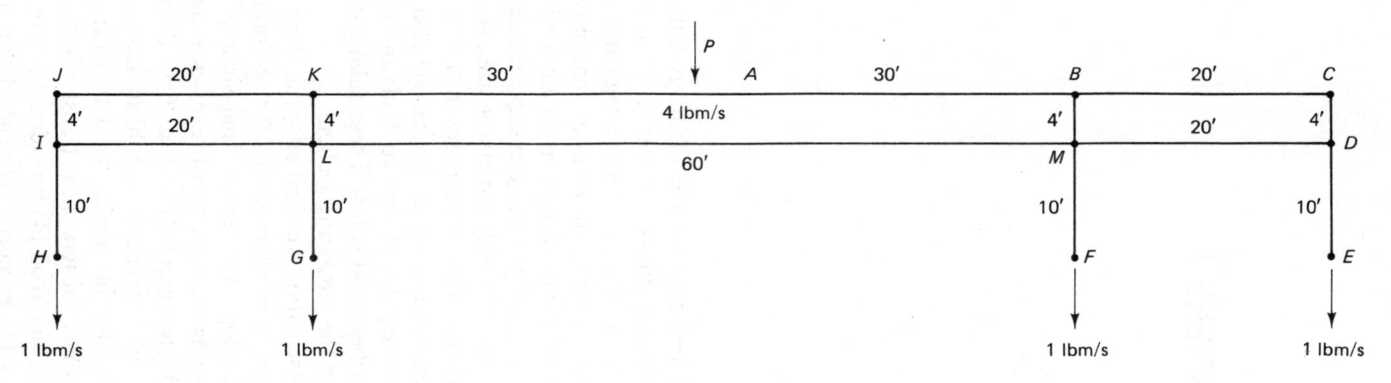
P
J
20′
K
30′
A
30′
B
20′
C
4′
20′
4′
4 lbm/s
4′
20′
4′
I
L
60′
M
D
10′
10′
10′
10′
H
G
F
E
1 lbm/s
1 lbm/s
1 lbm/s
1 lbm/s

Figure P1-12

2

HEAT EXCHANGERS I

2-1 INTRODUCTION

The most frequently used device for transfer of energy is the *heat exchanger*. A heat exchanger is a device that effects the transfer of heat (energy) from one fluid to another. Virtually any energy system imaginable will contain one or more heat exchangers. The radiator in a motor vehicle is one; so is the oil cooler on an aircraft engine, the condenser in an air-conditioning system, the feedwater heater on a boiler, and the waste-energy recovery unit in a packing plant. Heat exchangers come in a wide variety of sizes, shapes, and types and utilize a wide range of fluids. For our purposes in the next two chapters, we shall divide heat exchangers into two categories: (1) shell and tube and (2) cross flow. Depending on the paths of the two fluids and their relative motion, each category can be further subdivided. These subdivisions are useful and will be explored further in the remaining sections.

The sequence of presentation we shall adopt in these chapters will be to examine some temperature–area relationships for several common shell-and-tube heat exchangers, develop two general analysis and design procedures, explore approaches for estimating fluid pressure losses, examine shell-and-tube design situations, provide techniques for cross-flow design problems, and, finally, present an overview of mechanical problem areas. An examination of applicable standards will also be made. As with Chapter 1, the material presented here is not original. General heat transfer texts such as Kreith (1), Incropera and DeWitt (2), and Wolf (4) provided much of the basic information. Kays and London's (3) excellent reference book, *Compact Heat Ex-*

changers, is the source of much of the material presented in the coverage of cross-flow heat exchangers. The increasing cost of energy has resulted in much continued study of various aspects of heat exchanger design and performance. The journal *Heat Transfer Engineering* is devoted entirely to development and research in heat exchangers. Recent developments pertaining to heat transfer aspects are also reported in *Transactions of the ASME: Journal of Heat Transfer* and the *International Journal of Heat and Mass Transfer*.

2-2 METHODS OF ANALYSIS

The simplest heat exchanger considered is usually the *double-pipe* or *annulus* exchanger shown in Figure 2-1. In this arrangement one fluid flows through the center pipe while another flows in the annulus. The exchanger is said to be *parallel* if the two fluids enter at A and C (and exit at B and D) or D and B (and exit at C and A). Conversely, the exchanger is said to be *counterflow* if the two fluids enter at C and B (and exit at D and A) or D and A (and exit at C and B). The notion of parallel or counterflow is quite important in the design or analysis of shell-and-tube heat exchangers.

If we envision the heat transfer surface as extending from C to D in increments of dA, temperature-versus-area diagrams can be sketched for parallel and counterflow arrangements, as well as for condensers and evaporators. Figure 2-2 presents these temperature–area diagrams. The distinction of parallel or counterflow is not needed for the evaporator or condenser since in each case the fluid-changing phase remains at a constant temperature. A counterflow condenser with subcooling, as illustrated in Fig. 2-3(a), can be conveniently viewed as two exchangers, one a condenser and the other simple counterflow, as shown in Fig. 2-3(b). The same representation is obviously possible for an evaporator with superheat. The double-pipe heat exchanger and the temperature–area diagram can be extended to more complex configurations.

If several pipes are placed together within a shell, the classical single-pass shell-and-tube arrangement results. A single-pass shell and tube schematic is

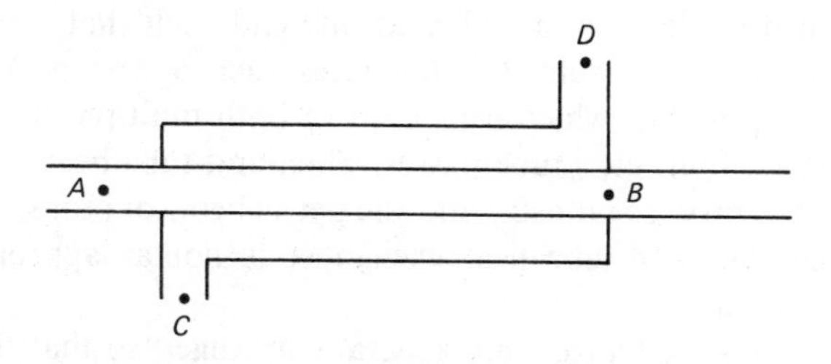

Figure 2-1 Double-pipe heat exchanger.

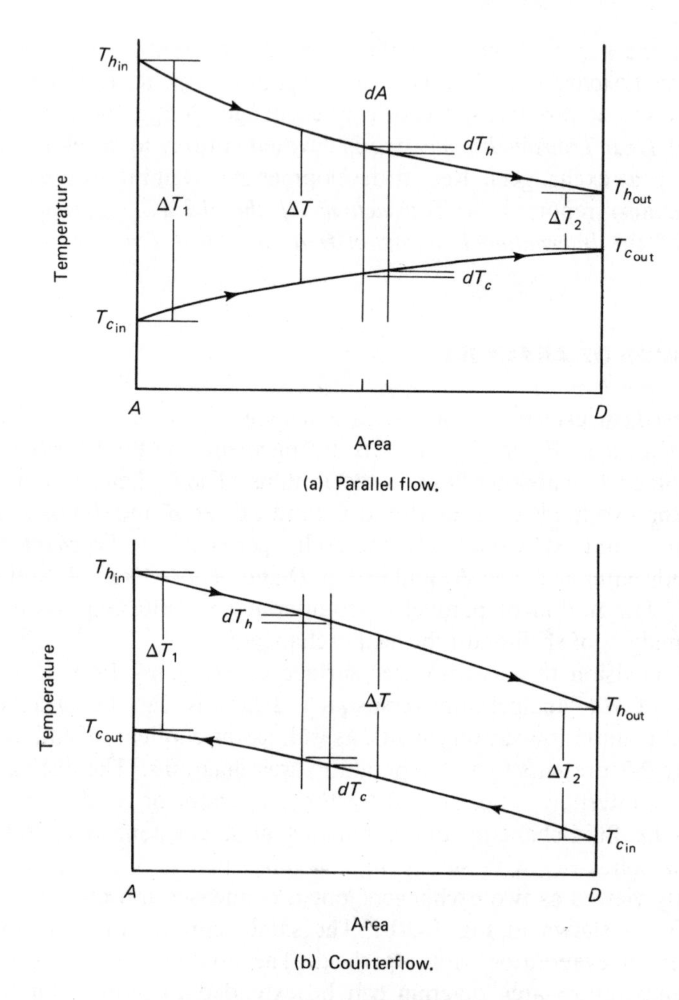

Figure 2-2 Temperature–area diagram.

given in Figure 2-4. Baffles are often placed within the shell to promote higher heat transfer rates and to increase the overall effectiveness of the tubes. A plenum is often provided for the tubes at one end such that a two-tube pass arrangement is secured. Figure 2-5 illustrates such a design. Various other arrangements are possible, which can result in both multiple shell passes and even multiple (two, four, etc.) tube passes. Shell-and-tube heat exchangers are widely used by the process industry and the petrochemical industry. For many applications the shell-and-tube heat exchanger is not as appropriate as the cross-flow heat exchanger.

Cross-flow heat exchangers are generally arranged so that the two fluids flow perpendicular to each other. This arrangement offers some advantages in

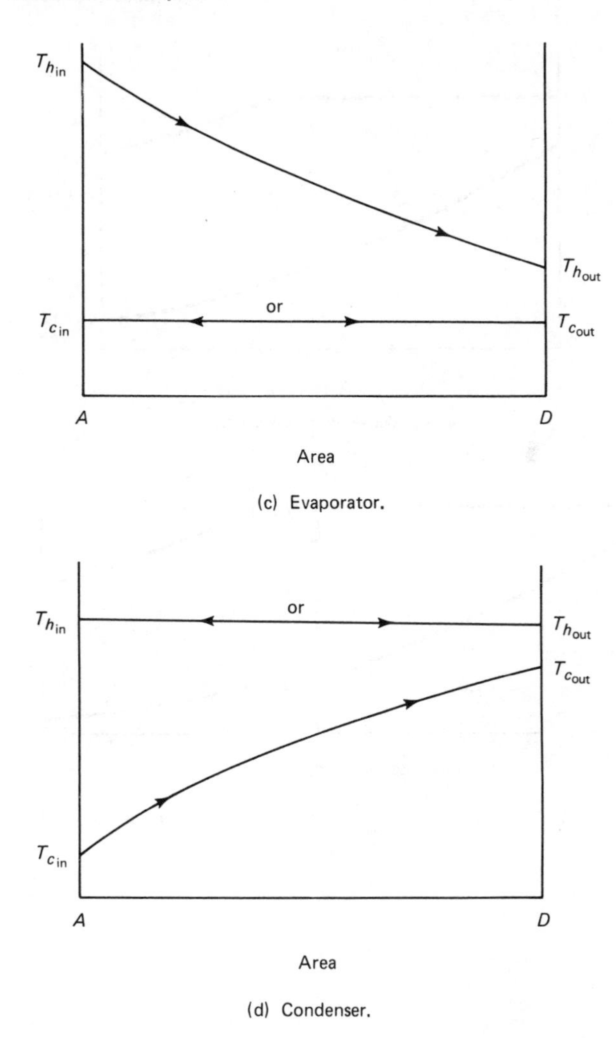

(c) Evaporator.

(d) Condenser.

Figure 2-2 (continued)

terms of compactness and effectiveness and some disadvantages in terms of fabrication and maintenance. A typical cross-flow heat exchanger schematic is presented in Figure 2-6. Cross-flow exchangers are often classified in terms of the flow arrangement for each fluid: (1) both fluids unmixed, (2) both fluids mixed, and (3) mixed and unmixed. A fluid is said to be unmixed if the passageway contains the same discrete portion of the fluid throughout its transverse of the exchanger; if the fluid passageways are such that fluid from one passageway can mix with the fluid from others, the pass is mixed. In Figure 2-6, mixed and unmixed are illustrated.

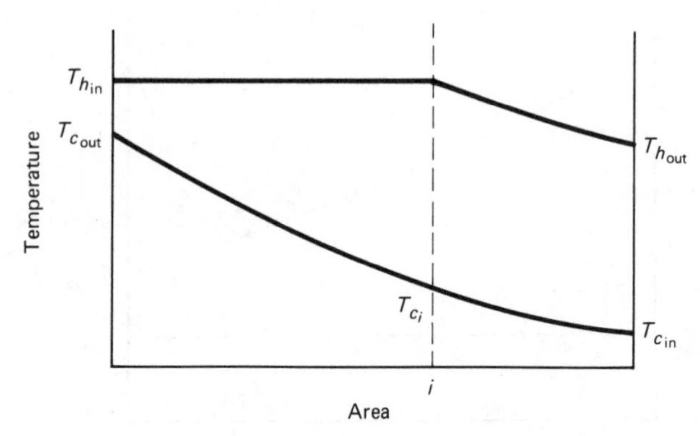

(a) Condenser with subcooling.

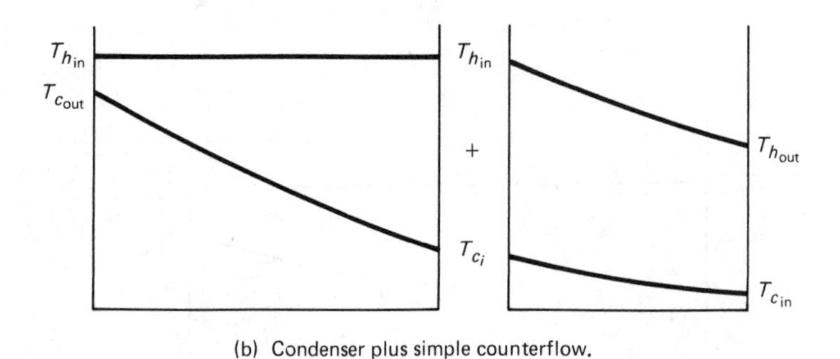

(b) Condenser plus simple counterflow.

Figure 2-3 Temperature–area diagram for a condenser with subcooling.

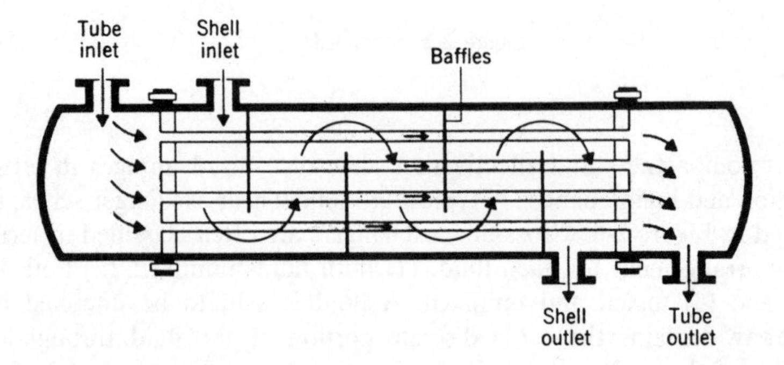

Figure 2-4 Single-pass shell-and-tube heat exchanger. (Used with permission, from F. P. Incropera and D. P. DeWitt, *Fundamentals of Heat Transfer*, John Wiley & Sons, Inc., 1981.)

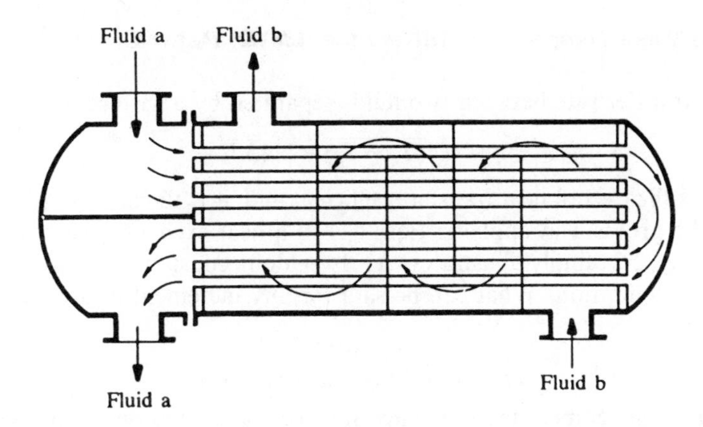

Figure 2-5 Two-tube pass shell-and-tube heat exchanger. (Used with permission, from D. R. Pitts and L. E. Sissom, *Schaum's Outline of Theory and Problems of Heat Transfer*, McGraw-Hill Book Company, 1977.)

The distinctions, that is, parallel or counterflow for shell and tube and mixed, unmixed, or mixed and unmixed for cross flow, are fundamental to both methods of analysis used for heat exchangers. The *log mean temperature difference* (LMTD) method was the first method developed for heat exchanger design and is still widely used. A relatively recent method is the *number of transfer units* (NTU) approach. For general-purpose use, analysis, and design, the NTU method is generally preferred. Both methods will yield the same answer for a given problem, so uniqueness is satisfied.

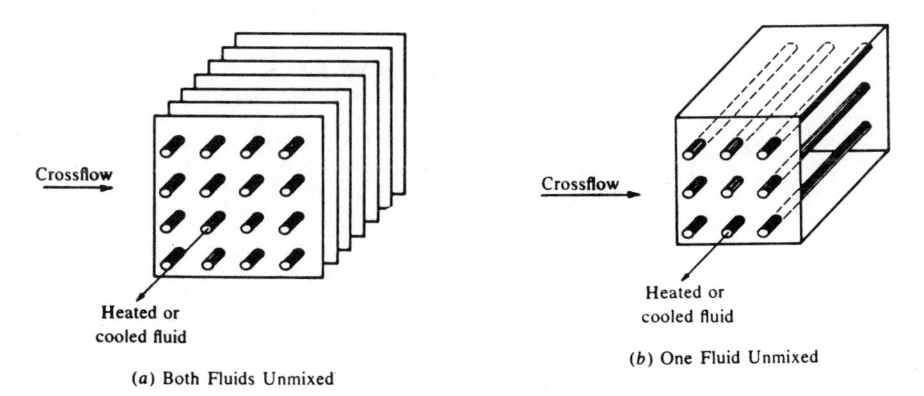

Figure 2-6 Cross-flow heat exchanger core. (Used with permission, from D. R. Pitts and L. E. Sissom, *Schaum's Outline of Theory and Problems of Heat Transfer*, McGraw-Hill Book Company, 1977.)

Log Mean Temperature Difference (LMTD) Method

The heat transfer rate between two fluids separated by a surface is

$$q = UA\overline{(T_h - T_c)} = UA\overline{\Delta T} \tag{2-1}$$

where U is the overall heat transfer coefficient and A is the area on which U is based. The problem in applying Eq. (2-1) to a heat exchanger is what $\overline{\Delta T}$ to use. As Fig. 2-2 amply illustrates, $\overline{\Delta T}$ is seldom constant for any heat exchanger configuration. What can be said for any increment of area dA of an exchanger is

$$dq = U\Delta T\,dA = U\,dA(T_h - T_c) \tag{2-2}$$

where ΔT is the temperature difference at a given area. Energy balances on the hot and cold fluids for a counterflow device [Fig. 2-2(b)] yield

$$\begin{aligned} dq &= \dot{m}_c C_{p_c}\,dT_c = C_c\,dT_c \\ dq &= \dot{m}_h C_{p_h}\,dT_h = C_h\,dT_h \end{aligned} \tag{2-3}$$

and

$$d(\Delta T) = dT_h - dT_c = \frac{dq}{C_h} - \frac{dq}{C_c} \tag{2-4}$$

Equating the integrated energy balance expressions [integrated Eqs. (2-3)] gives

$$C_h\left(T_{h_{\text{in}}} - T_{h_{\text{out}}}\right) = C_c\left(T_{c_{\text{out}}} - T_{c_{\text{in}}}\right) \tag{2-5}$$

from which

$$d(\Delta T) = \frac{dq}{C_h}\left(1 - \frac{T_{c_{\text{out}}} - T_{c_{\text{in}}}}{T_{h_{\text{in}}} - T_{h_{\text{out}}}}\right) \tag{2-6}$$

Substituting Eq. (2-2) for dq and integrating results in

$$\int_{\Delta T_1}^{\Delta T_2} \frac{d(\Delta T)}{\Delta T} = \frac{U}{q}(\Delta T_2 - \Delta T_1)A \tag{2-7}$$

where $\Delta T_2 = T_{h_{\text{out}}} - T_{c_{\text{in}}}$ and $\Delta T_1 = T_{h_{\text{in}}} - T_{c_{\text{out}}}$. So

$$\ln\frac{\Delta T_2}{\Delta T_1} = \frac{UA}{q}(\Delta T_2 - \Delta T_1) \tag{2-8}$$

Hence

$$q = UA\frac{\Delta T_2 - \Delta T_1}{\ln \Delta T_2/\Delta T_1} \tag{2-9}$$

Then

$$\overline{\Delta T} = \frac{\Delta T_2 - \Delta T_1}{\ln \Delta T_2/\Delta T_1} \tag{2-10}$$

Equations (2-9) and (2-10) are valid for parallel flow or counterflow. Equation (2-10) is usually taken as defining a very special type of temperature difference, the log mean (LMTD). If ΔT_1 and ΔT_2 differ by less than 50 percent, the arithmetic temperature difference will be within 1 percent of the LMTD. Equation (2-9) is only an approximation since U is generally not constant across the entire exchanger. U is usually evaluated at a mean section, generally halfway between the ends. If U exhibits a large variation, integration of Eq. (2-2) may be necessary.

For more complex heat exchangers, such as multipass shell-and-tube arrangements or cross-flow types, the effective mean temperature $\overline{\Delta T}$ is usually written as

$$\overline{\Delta T} = \text{LMTD} \times F \tag{2-11}$$

where the F is a correction factor based on the LMTD for a *counterflow arrangement*. The mathematical derivation of the true $\overline{\Delta T}$ for any arrangement other than simple parallel or counterflow is tedious and complex; hence the use of a factor F that can be evaluated via charts or tables. For shell-and-tube exchangers, TEMA (5), the Tubular Exchanger Manufacturers Association, provides a number of charts for evaluating the F for various arrangements and various numbers of shell and tube passes. These correction factors for shell-and-tube heat exchangers are given in Fig. 2-7; part (a) is for one shell pass and a multiple of two tube passes, and part (b) for two shell passes and a multiple of two tube passes. Correction factors for other arrangements are given in the TEMA standards. The parameter Z is defined as

$$Z = \frac{\dot{m}_t C_{p_t}}{\dot{m}_s C_{p_s}} = \frac{T_{s_{\text{in}}} - T_{s_{\text{out}}}}{T_{t_{\text{out}}} - T_{t_{\text{in}}}} \tag{2-12}$$

and the parameter P as

$$P = \frac{T_{t_{\text{out}}} - T_{t_{\text{in}}}}{T_{s_{\text{in}}} - T_{t_{\text{in}}}} \tag{2-13}$$

where the subscript t is for tube and s is for shell.

Just as correction factors are available for various shell-and-tube arrangements, so correction factors are available for various cross-flow exchanger arrangements. Like the shell-and-tube correction factors, the cross-flow correction factors are based on the assumption of a constant U. Figure 2-7 presents the correction factors for the cross-flow arrangements: part (c) is for both fluids mixed, one pass, and part (d) for one fluid mixed and the other unmixed, one pass.

The LMTD approach is well suited for design purposes since the inlet and exit temperatures are either known from the specifications or obtained from an overall energy balance of the hot and cold fluids

$$|\dot{m}_h C_{p_h} \Delta T_h| = |\dot{m}_c C_{p_c} \Delta T_c| \tag{2-14}$$

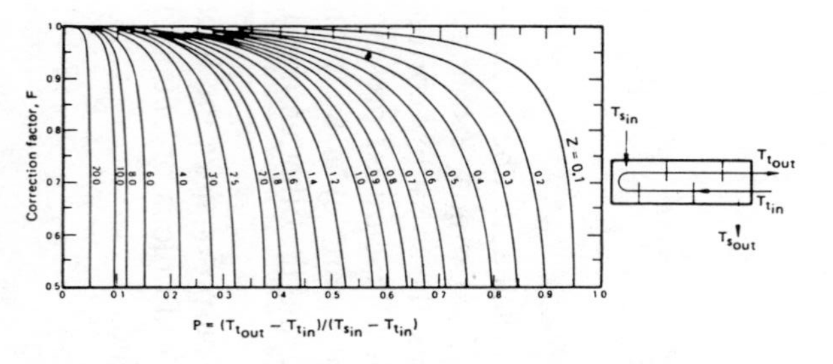

(a) Correction factor to counterflow LMTD for heat exchanger with one shell pass and two, or a multiple of two, tube passes. (Used with permission, from J. H. Lienhard, *A Heat Transfer Textbook*, Prentice-Hall, 1981.)

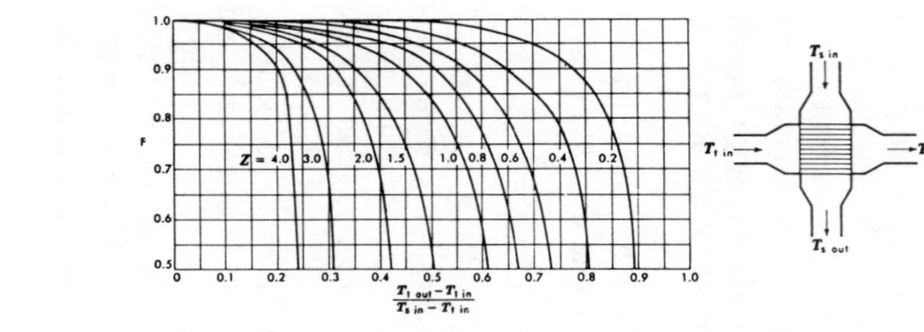

(c) Correction factor to counterflow LMTD for cross-flow heat exchanger, both fluids unmixed, one tube pass. (Extracted from "Mean Temperature Difference in Design," by R. A. Bowman, A. C. Mueller, and W. M. Nagel, published in *Trans. ASME*, Vol. 62, 1940.)

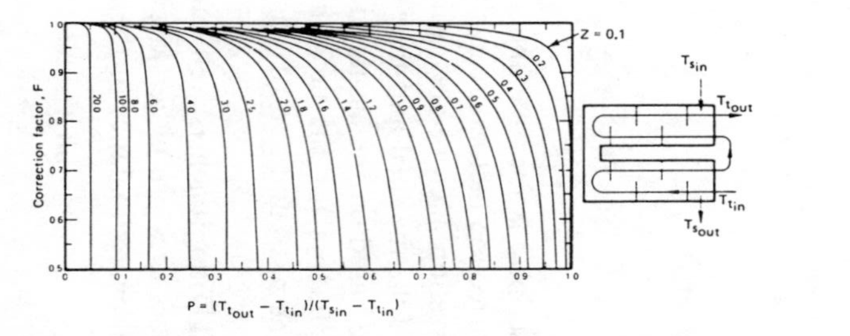

(b) Correction factor for a shell-and-tube heat exchanger with two shell passes and any multiple of four tube passes (four, eight, etc., tube passes). (Used with permission, from J. H. Lienhard, *A Heat Transfer Textbook*, Prentice-Hall, 1981.)

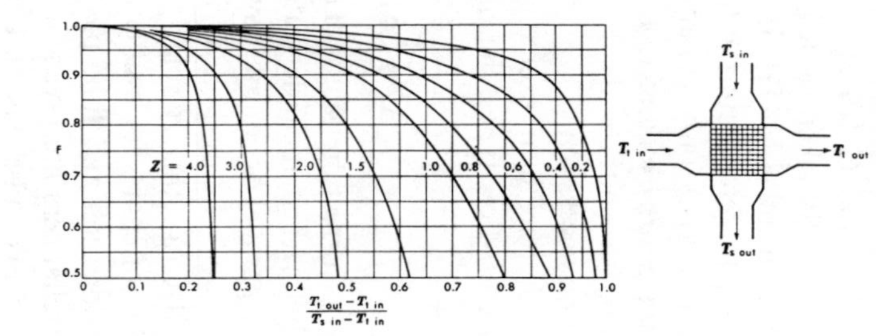

(d) Correction factor to counterflow LMTD for cross-flow heat exchangers, fluid on shell side mixed, other fluid unmixed, one tube pass. (Extracted from "Mean Temperature Difference in Design," by R. A. Bowman, A.C. Mueller, and W. M. Nagel, published in *Trans. ASME*, Vol. 62, 1940.)

Figure 2-7 Correction factors for heat exchangers.

Analysis problems, problems in which the heat exchanger characteristics are given and the exit temperatures required, are generally iterative since P and Z are not known a priori. The NTU method, which will be explored next, does not suffer from this restriction.

Number of Transfer Units (NTU) Method

The NTU method is derived about the concept of a formal definition of *effectiveness*. The heat exchanger effectiveness is defined as the ratio of the actual rate of heat transfer to the maximum possible rate of heat exchange. The maximum possible rate would be obtained in a *counterflow* heat exchanger of infinite heat transfer area. In such a unit, assuming no external heat losses, the outlet temperature of the colder fluid equals the inlet temperature of the hotter fluid when $\dot{m}_c C_{p_c} < \dot{m}_h C_{p_h}$, and the outlet temperature of the warmer fluid equals the inlet temperature of the colder fluid when $\dot{m}_h C_{p_h} < \dot{m}_c C_{p_c}$. Thus the basis for this ratio is a maximum rate as limited by the second law of thermodynamics. The effectiveness definition then depends on which capacity

$$C_h = \dot{m}_h C_{p_h}$$
$$C_c = \dot{m}_c C_{p_c} \tag{2-15}$$

is the smaller, and the effectiveness definitions are

$$\xi = \frac{C_h\left(T_{h_{\text{in}}} - T_{h_{\text{out}}}\right)}{C_{\min}\left(T_{h_{\text{in}}} - T_{c_{\text{in}}}\right)} \tag{2-16}$$

or

$$\xi = \frac{C_c\left(T_{c_{\text{out}}} - T_{c_{\text{in}}}\right)}{C_{\min}\left(T_{h_{\text{in}}} - T_{c_{\text{in}}}\right)} \tag{2-17}$$

with $C_{\min}$ the smaller of C_h and C_c. With the effectiveness known, the rate of heat transfer is given by

$$q = \xi C_{\min}\left(T_{h_{\text{in}}} - T_{c_{\text{in}}}\right) \tag{2-18}$$

which involves only *inlet* conditions. The only question remaining is how to determine ξ.

Consider the case of a parallel-flow heat exchanger where $C_{\min} = C_h$ and $C_{\max} = C_c$. Then Eq. (2-16) reduces to

$$\xi = \frac{C_{\max}\left(T_{c_{\text{out}}} - T_{c_{\text{in}}}\right)}{C_{\min}\left(T_{h_{\text{in}}} - T_{c_{\text{in}}}\right)} \tag{2-19}$$

and

$$\frac{C_{\min}}{C_{\max}} = \frac{C_h}{C_c} = \frac{T_{c_{\text{out}}} - T_{c_{\text{in}}}}{T_{h_{\text{in}}} - T_{h_{\text{out}}}} \tag{2-20}$$

Equation (2-8) can be written as

$$\ln \frac{\Delta T_2}{\Delta T_1} = -UA\left(\frac{1}{C_h} + \frac{1}{C_c}\right) \tag{2-21}$$

and then appears as

$$\ln \frac{T_{h_{out}} - T_{c_{out}}}{T_{h_{in}} - T_{c_{in}}} = -\frac{UA}{C_{min}}\left(1 + \frac{C_{min}}{C_{max}}\right) \tag{2-22}$$

Defining

$$\text{NTU} = \frac{UA}{C_{min}} \tag{2-23}$$

Eq. (2-22) can be written as

$$\frac{T_{h_{out}} - T_{c_{out}}}{T_{h_{in}} - T_{c_{in}}} = \exp\left[-\text{NTU}\left(1 + \frac{C_{min}}{C_{max}}\right)\right] \tag{2-24}$$

The left-hand side of Eq. (2-24) can be expressed as

$$\frac{T_{h_{out}} - T_{c_{out}}}{T_{h_{in}} - T_{c_{in}}} = 1 - \xi\left(1 + \frac{C_{min}}{C_{max}}\right) \tag{2-25}$$

by adding $-T_{h_{in}} + T_{h_{in}}$ and using Eq. (2-20). Solving Eqs. (2-24) and (2-25) for the effectiveness ξ results in

$$\xi = \frac{1 - \exp[-\text{NTU}(1 + C_{min}/C_{max})]}{1 + C_{min}/C_{max}} \tag{2-26}$$

which possesses no dependence on $T_{c_{out}}$ or $T_{h_{out}}$. Had we selected $C_{min} = C_c$, the same equation would have resulted. Thus, in general, we expect for any configuration

$$\xi = \xi\left(\text{NTU}, \frac{C_{min}}{C_{max}}\right) \tag{2-27}$$

Kays and London (3) pioneered in this form of heat exchanger analysis and present results in the form of Eq. (2-27) for many different types.

Pitts and Sissom (6) give a compact summary both graphical and in closed form for many configurations. The ξ versus NTU relations are available for essentially the same heat exchanger arrangements as the correction factors F are available in the LMTD method. Table 2-1 presents a summary of analytical results of these relations and indicates the appropriate figure. These expressions allow the ξ–NTU method to be readily used in computer-based analyses. For rapid hand use, the graphs presented in Fig. 2-8 should be useful. In Fig. 2-8(f), the solid lines represent $C_{mixed}/C_{unmixed}$ with $C_{mixed} < C_{unmixed}$. The dashed lines represent $C_{mixed}/C_{unmixed}$ when $C_{unmixed} < C_{mixed}$. In the latter case, the NTU value is based on the smaller capacity, that is,

TABLE 2-1 Summary of Effectiveness Relations (5) for Heat Exchangers

Exchanger Type	Effectiveness	See Graph in:	Eq.
Parallel flow: single pass	$\xi = \dfrac{1 - \exp[-\text{NTU}(1 + C)]}{1 + C}$	Fig. 2-8(a)	(2.28)
Counterflow: single pass	$\xi = \dfrac{1 - \exp[-\text{NTU}(1 - C)]}{1 - C\exp[-\text{NTU}(1 - C)]}$	Fig. 2-8(b)	(2.29)
Shell and tube (one shell pass; 2, 4, 6, etc., tube passes)	$\xi_1 = 2\left[1 + C + \dfrac{1 + \exp\left[-\text{NTU}(1 + C^2)^{1/2}\right]}{1 - \exp\left[-\text{NTU}(1 + C^2)^{1/2}\right]}(1 + C^2)^{1/2}\right]^{-1}$	Fig. 2-8(c)	(2-30)
Shell and tube (n shell passes; $2n$, $4n$, $6n$, etc., tube passes)	$\xi_n = \left[\left(\dfrac{1 - \xi_1 C}{1 - \xi_1}\right)^n - 1\right]\left[\left(\dfrac{1 - \xi_1 C}{1 - \xi_1}\right)^n - C\right]^{-1}$	Fig. 2-8(d) for $n = 2$	(2-31)
Cross flow (both streams unmixed)	$\xi \approx 1 - \exp\{\overline{C}(\text{NTU})^{0.22}[\exp[-C(\text{NTU})^{0.78}] - 1]\}, \quad \overline{C} = \dfrac{1}{C}$	Fig. 2-8(e)	(2-32)
Cross flow (both streams mixed)	$\xi = \text{NTU}\left[\dfrac{\text{NTU}}{1 - \exp(-\text{NTU})} + \dfrac{(\text{NTU})(C)}{1 - \exp[-(\text{NTU})(C)]} - 1\right]^{-1}$		(2-33)
Cross flow (stream $C_{\min}$ unmixed)	$\xi = \overline{C}\{1 - \exp[-C[1 - \exp(-\text{NTU})]]\}, \quad \overline{C} = \dfrac{1}{C}$	Fig. 2-8(f) dashed curves	(2-34)
Cross flow (stream $C_{\max}$ unmixed)	$\xi = 1 - \exp\{-\overline{C}[1 - \exp[-(\text{NTU})(C)]]\}, \quad \overline{C} = \dfrac{1}{C}$	Fig. 2-8(f) solid curves	(2-35)

Used with permission, from D. R. Pitts and L. E. Sissom, *Schaum's Outline of Theory and Problems of Heat Transfer*, McGraw-Hill Book Company, 1977.

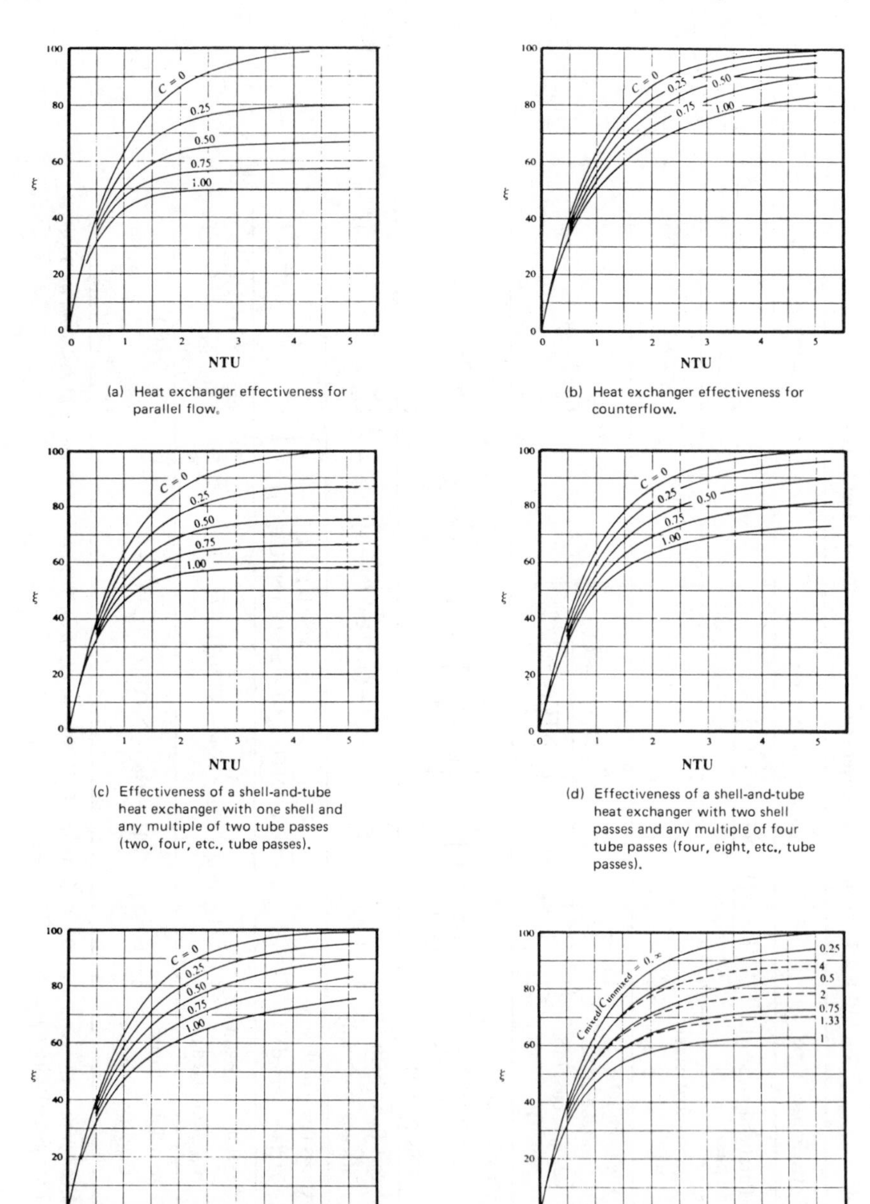

(a) Heat exchanger effectiveness for parallel flow.

(b) Heat exchanger effectiveness for counterflow.

(c) Effectiveness of a shell-and-tube heat exchanger with one shell and any multiple of two tube passes (two, four, etc., tube passes).

(d) Effectiveness of a shell-and-tube heat exchanger with two shell passes and any multiple of four tube passes (four, eight, etc., tube passes).

(e) Heat exchanger effectiveness for cross-flow with both fluids unmixed.

(f) Heat exchanger effectiveness for cross-flow with one fluid mixed, the other unmixed. Where $C_{mixed}/C_{unmixed} > 1$, NTU_{max} is based on $C_{unmixed}$.

Figure 2-8 Heat exchanger effectiveness for various arrangements. (Used with permission, from D. R. Pitts and L. E. Sissom, *Schaum's Outline of Theory and Problems of Heat Transfer*, McGraw-Hill Book Company, 1977.)

$C_{unmixed}$. The last two entries of Table 2-1 give closed-form expressions for the two cases shown in Fig. 2-8(f). The C within these expressions has the usual meaning (i.e., $C = C_{min}/C_{max}$).

The NTU method can be used for heat exchangers with phase change (condensers or evaporators) by noting that the specific heat for the fluid undergoing the phase change is essentially unbounded since the temperature remains constant. Thus, for condensers or evaporators, $C = 0$ and all the expressions in Table 2-1 reduce to a much simpler form.

Example 2-1

A heat exchanger with one shell pass and eight tube passes raises 100,000 lbm/h of water from 180° to 300°F. The tube-side fluid is air ($C_p = 0.24$ Btu/lbm °F) and enters at 650°F and exits at 350°F. If $U = 5$ Btu/h ft^2 °F, find the surface area required.

Solution. This is a design problem, since a geometrical attribute, A, is to be calculated. The temperature–area diagram for this situation is given in Fig. 2-9.

$$\text{LMTD} = \frac{350 - 170}{\ln 350/170} = 249.3\text{°F}$$

The total energy transfer is

$$q = \dot{m}C_p\,\Delta T$$
$$= 100{,}000\,\frac{\text{lbm}}{\text{h}}\,1\,\frac{\text{Btu}}{\text{lbm °F}}\,(300 - 180)\text{°F}$$
$$= 12 \times 10^6 \text{ Btu/h}$$

and

$$P = \frac{T_{t_{out}} - T_{t_{in}}}{T_{s_{in}} - T_{t_{in}}} = \frac{350 - 650}{180 - 650} = 0.638$$

$$Z = \frac{T_{s_{in}} - T_{s_{out}}}{T_{t_{out}} - T_{t_{in}}} = \frac{180 - 300}{350 - 650} = 0.4$$

from whence F is found from Fig. 2-3(b) to be

$$F = 0.88$$

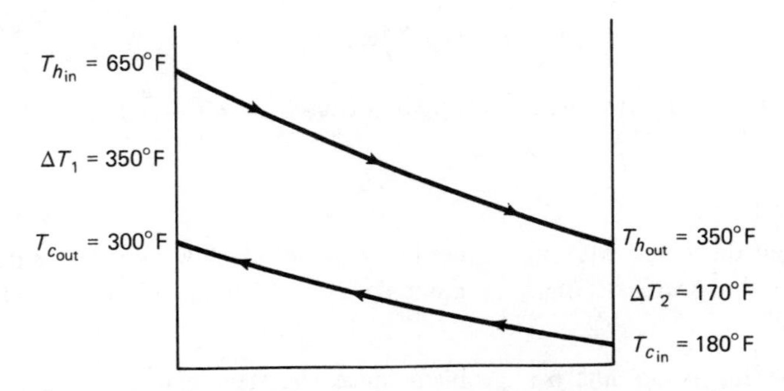

Figure 2-9 Temperature–area diagram for Example 2-1.

But

$$12 \times 10^6 \frac{\text{Btu}}{\text{h}} = 5 \frac{\text{Btu}}{\text{h ft}^2\ ^\circ\text{F}} A(249.3^\circ\text{F})(0.88)$$

from which

$$A = 10{,}940 \text{ ft}^2$$

Alternate Method

The NTU approach can also be used. Assuming negligible energy lost to the surroundings,

$$\Delta T_h \left(\dot{m}C_p \right)_h = \Delta T_c \left(\dot{m}C_p \right)_c$$

$$\dot{m}_{\text{air}} 0.24 \frac{\text{Btu}}{\text{lbm }^\circ\text{F}} (650 - 350)^\circ\text{F} = 100{,}000 \frac{\text{lbm}}{\text{h}} 1 \frac{\text{Btu}}{\text{lbm }^\circ\text{F}} (300 - 180)^\circ\text{F}$$

$$\dot{m}_{\text{air}} = 166{,}667 \frac{\text{lbm}}{\text{h}}$$

The capacities are:

$$\left(\dot{m}C_p \right)_h = 166{,}667 \frac{\text{lbm}}{\text{h}} 0.24 \frac{\text{Btu}}{\text{lbm }^\circ\text{F}} = 40{,}000 \frac{\text{Btu}}{\text{h }^\circ\text{F}}$$

$$\left(\dot{m}C_p \right)_c = 100{,}000 \frac{\text{lbm}}{\text{h}} 1 \frac{\text{Btu}}{\text{lbm }^\circ\text{F}} = 100{,}000 \frac{\text{Btu}}{\text{h }^\circ\text{F}}$$

Thus $(\dot{m}c_p)_h$ is minimum and

$$C = \frac{C_{\text{min}}}{C_{\text{max}}} = 0.40$$

and

$$\xi = \frac{q}{q_{\text{max}}} = \frac{12 \times 10^6}{C_{\text{min}}(650 - 180)} = 0.6383$$

Using Fig. 2-8(c),

$$\text{NTU} = \frac{UA}{C_{\text{min}}} = 1.36$$

$$A = 1.36 \frac{C_{\text{min}}}{U} = 10{,}880 \text{ ft}^2$$

The difference in the two areas computed is caused by errors on reading the various figures.

Example 2-2

What would the water exit temperature be if the hot air flow rate were reduced by 30 percent in the heat exchanger in Example 2-1, assuming all other quantities are unchanged?

Solution. This is an analysis problem since the heat exchanger area is known and an exit temperature is to be computed. If the LMTD method is used, the solution is iterative, since P and Z are functions of the unknown exit temperature of the water. The NTU method, however, can be used to directly compute the water exit temperature.

The new air flow rate is $0.7\dot{m}_{air}$ and is 116,667 lbm/h. Then

$$(\dot{m}C_p)_h = 116{,}667\frac{\text{lbm}}{\text{h}}0.24\frac{\text{Btu}}{\text{lbm °F}} = 28{,}000\frac{\text{Btu}}{\text{h °F}}$$

and

$$C = \frac{28{,}000}{100{,}000} = 0.28$$

Here we know the NTU, since

$$\text{NTU} = \frac{UA}{C_{min}} = 5\frac{\text{Btu}}{\text{h ft}^2\text{ °F}}10{,}940\text{ ft}^2\frac{\text{h °F}}{28{,}000\text{ Btu}} = 1.95$$

from which Fig. 2-8(c) yields

$$\xi = 0.76$$

and

$$\begin{aligned} q &= \xi C_{min}(T_{h_{in}} - T_{c_{in}}) \\ &= (0.76)28{,}000\frac{\text{Btu}}{\text{h °F}}(650 - 180)\text{°F} \\ &= 10 \times 10^6\frac{\text{Btu}}{\text{h}} \end{aligned}$$

Writing an energy balance on the cold fluid,

$$q = (\dot{m}C_p)_c\Delta T_c = (\dot{m}C_p)_c(T_{c_{out}} - T_{c_{in}})$$

$$\begin{aligned} T_{c_{out}} &= T_{c_{in}} + \frac{q}{C_c} \\ &= 180\text{°F} + 100\text{°F} = 280\text{°F} \end{aligned}$$

Thus the exit temperature of the water is 280°F. The gas exit temperature for this case is 293°F. Reducing the mass flow rate of air results in a lower exit temperature for the air but less total energy transfer.

Special case: Heat transfer from a pipe immersed in an ambient environment. A recurring situation for many energy systems is a fluid-conveying pipe or duct exposed to an ambient environment. This situation, illustrated in Fig. 2-10, occurs in situations such as chilled or hot water heating, ventilating, and air-conditioning (HVAC) systems and in many industrial processes where hot or cold compounds must be piped from one location to another. For the situation illustrated in Fig. 2-10, the driving potential for the heat transfer process traditionally has been taken as the difference between T_∞ and the average of T_{in} and T_{out}; that is,

$$\overline{\Delta T} = T_\infty - \tfrac{1}{2}(T_{in} + T_{out}) \qquad (2\text{-}36)$$

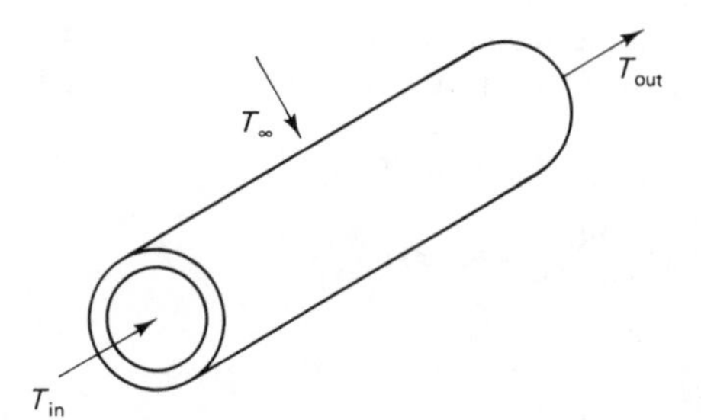

Figure 2-10 Pipe in an ambient environment.

so that

$$q = UA\,\overline{\Delta T} \tag{2-1}$$

However, as pointed out by Wolf (4) and as discussed by Suryanarayana (7), when $UA/C_{\min}$ is greater than 2.0, the second law of thermodynamics would be violated since $(T_{in} - T_\infty)$ and $(T_{out} - T_\infty)$ would have different signs. The reason for this anomaly is that the situation illustrated in Fig. 2-10 has a temperature–area diagram similar to that shown in Fig. 2-2(c) and (d) so that a simple arithmetic average is not accurate or meaningful for all $(T_{in} - T_\infty)$ and $(T_{out} - T_\infty)$ values.

The pipe in an ambient environment is thus a heat exchanger and should be analyzed as such. Since the change in the reference temperature of the ambient environment is zero or near zero, the capacity $C_{\max}$ is effectively a very large number, and $C_{\min}/C_{\max}$ is near zero. Thus, the appropriate temperature difference to use in Eq. (2-1) is the log mean temperature difference, LMTD, as given by Eq. (2-10). For analysis problems, Eqs. (2-28), (2-29), and (2-33) to (2-35) all reduce to

$$\xi = 1 - \exp(-\text{NTU}) \tag{2-37}$$

and are available for use. Using either the LMTD or the NTU–ξ method, no second-law violation occurs as might in the arithmetical average approach.

Example 2-3

Chilled water, initially at 50°F, flows through a 4-in. ID uninsulated pipe 1000 ft long. The overall convective heat transfer coefficient is estimated as 350 Btu/ft^2 h °F. The pipe is exposed to air at 100°F. What is the exit temperature of the chilled water if the average water velocity is 5 ft/s?

Solution. This is an analysis problem so the NTU $-$ ξ method is appropriate for use. The capacity $C_{\min}$ is

$$\begin{aligned} C_{\min} &= C_p \dot{m} \\ &= 1.002 \frac{\text{Btu}}{\text{lbm °F}} 62.38 \frac{\text{lbm}}{\text{ft}^3} \frac{\pi}{4}\left(\frac{4}{12}\right)^2 \text{ft}^2\, 5\frac{\text{ft}}{\text{s}} \\ &= 27.273 \text{ Btu/s °F} \end{aligned}$$

and

$$\text{NTU} = \frac{UA}{C_{\min}} = 350 \frac{\text{Btu}}{\text{h ft}^2\ {}^\circ\text{F}} \pi \left(\frac{4}{12}\right) \text{ft } 1000 \text{ ft} \frac{\text{s }{}^\circ\text{F}}{27.273 \text{ Btu}} \frac{\text{h}}{3600 \text{ s}}$$

$$= 3.73$$

Then the effectiveness is

$$\xi = 1 - \exp(-3.73) = 0.976$$

and

$$q = (0.976)27.273 \frac{\text{Btu}}{\text{s }{}^\circ\text{F}} (100 - 50)^\circ\text{F}$$

$$= 1330.9 \text{ Btu/s}$$

An energy balance on the chilled water yields

$$1330.9 \frac{\text{Btu}}{\text{s}} = 27.273 \frac{\text{Btu}}{\text{s }{}^\circ\text{F}} (T_{\text{exit}} - 50)$$

$$T_{\text{exit}} = 98.8^\circ\text{F}$$

which illustrates both the principle involved in the calculation and the necessity for using insulation on chilled water systems for hot climates, although the specified value of U is unrealistically large for such systems. Had $\overline{\Delta T}$ been approximated as an arithmetical average, T_{exit} would have been computed as 115.1°F, an obvious violation of the second law of thermodynamics.

Examples 2-1 through 2-3 illustrate typical uses of the LMTD and NTU approaches. In any reasonable problem, we cannot expect U, the overall heat transfer coefficient, to be given. The next section contains a brief review of heat transfer and presents a rather complete set of convective heat transfer coefficient correlations.

2-3 HEAT TRANSFER FUNDAMENTALS

Composite Walls in the Steady State

The average convective heat transfer coefficient $\bar{h}_c$ for a wall at temperature T_s and a fluid at bulk temperature T_b is defined by

$$\frac{q}{A} = \bar{h}_c(T_s - T_b) \tag{2-38}$$

This expression can be used to compute the heat transfer given $\bar{h}_c$, T_s, and T_b or to evaluate $\bar{h}_c$ given q/A, T_s, and T_b. In heat exchangers, as in most heat transfer situations of interest, the surface temperature T_s is seldom known, and conduction in series with convection takes place across one or more thicknesses of material.

The electrical resistance analogy between steady-state heat flow and dc circuits, as discussed in most heat transfer textbooks, can be used to establish

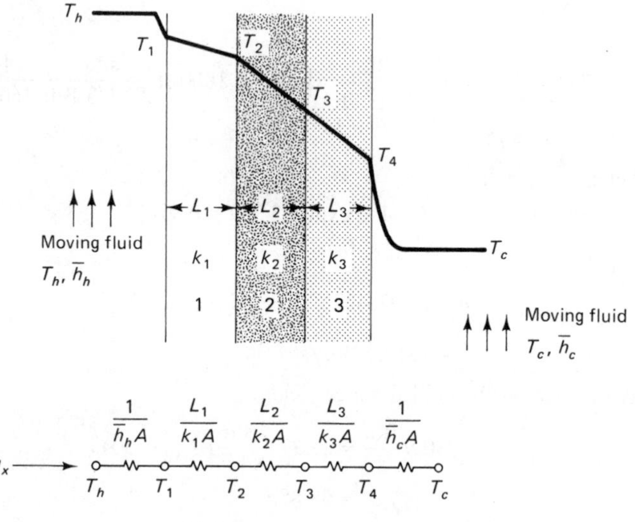

(a) Planar composite wall.

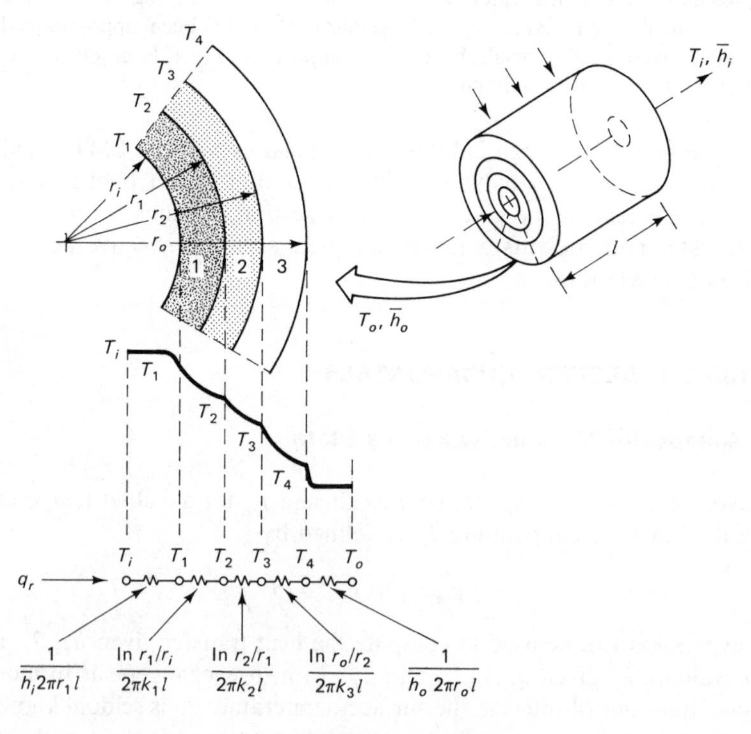

(b) Cylindrical composite wall.

Figure 2-11 Composite wall schematic. (Used with permission, from F. P. Incropera and D. P. DeWitt, *Fundamentals of Heat Transfer*, John Wiley & Sons, Inc., 1981.)

the overall heat transfer rate in many situations. The temperature profile and convective heat transfer resistances for a planar composite wall of surface area A separating two fluids at temperatures T_h and T_c are shown in Fig. 2-11(a). The steady state heat transfer rate can be determined by

$$q = \frac{\Delta T}{\sum_{i=1} R_i} \tag{2-39}$$

where the planar convective resistances are

$$R_{\text{conv}} = \frac{1}{\bar{h}_i A} \tag{2-40}$$

and the planar conductive resistances are

$$R_{\text{cond}} = \frac{L_i}{k_i A} \tag{2-41}$$

So, for the situation sketched in Fig. 2-11(a), Eq. (2-40) becomes

$$q = \frac{T_h - T_c}{\dfrac{1}{\bar{h}_h A} + \dfrac{L_1}{k_1 A} + \dfrac{L_2}{k_2 A} + \dfrac{L_3}{k_3 A} + \dfrac{1}{\bar{h}_c A}} \tag{2-42}$$

An overall heat transfer coefficient U is defined such that

$$q = UA(T_h - T_c) \tag{2-1}$$

This overall heat transfer coefficient for the planar wall is

$$U = \frac{1}{\dfrac{1}{\bar{h}_h} + \dfrac{L_1}{k_1} + \dfrac{L_2}{k_2} + \dfrac{L_3}{k_3} + \dfrac{1}{\bar{h}_c}} \tag{2-43}$$

The cylindrical composite wall follows the same sequence with the convective resistance as

$$R_{\text{conv}} = \frac{1}{\bar{h}_i 2\pi r_i l} \tag{2-44a}$$

and the cylindrical conductive resistance as

$$R_{\text{cond}} = \frac{\ln r_{i+1}/r_i}{2\pi k_i l} \tag{2-44b}$$

For the cylindrical segment (of length l) example shown in Fig. 2-11(b), the heat transfer rate becomes

$$q = \frac{T_o - T_i}{\dfrac{1}{\bar{h}_i 2\pi r_i l} + \dfrac{\ln r_1/r_i}{2\pi k_1 l} + \dfrac{\ln r_2/r_1}{2\pi k_2 l} + \dfrac{\ln r_o/r_2}{2\pi k_3 l} + \dfrac{1}{\bar{h}_o 2\pi r_o l}} \tag{2-45}$$

The overall heat transfer coefficient for the cylindrical case, unlike the planar case, is dependent on the area to which it is referenced. The product UA is the important quantity, and U can be expressed based on an inside area as $U_i A_i$ or an outside area as $U_o A_o$. The outside area is the most common, and for the situation portrayed in Fig. 2-11(b), U_o becomes

$$U_o = \frac{1}{\dfrac{r_o}{r_i \bar{h}_i} + \dfrac{r_o \ln r_1/r_i}{k_1} + \dfrac{r_o \ln r_2/r_1}{k_2} + \dfrac{r_o \ln r_o/r_2}{k_3} + \dfrac{1}{\bar{h}_o}} \tag{2-46}$$

where $A_o = 2\pi r_o l$.

The concept of resistance to heat transfer is very useful for many one-dimensional heat transfer calculations; and, in the case of the planar or cylindrical composite wall, allows us to very simply evaluate U or U_o for any number of composite layers. The fact that the resistances add also permits the rapid assessment of which resistance, if any, is controlling the overall heat transfer process. The larger the heat transfer resistance term in comparison to the other heat transfer resistances, the more controlling the larger term. Example 2-4 illustrates this.

Example 2-4

A nominal 1-in. schedule 40 steel pipe has steam condensing on the outside at 212°F and water flowing on the inside at 70°F. If $\bar{h}_o$ is 2000 Btu/h ft^2 °F and $\bar{h}_i$ is 100 Btu/h ft^2 °F, find the controlling resistance.

Solution. At the temperatures of interest

$$k_{\text{steel}} = 26 \text{ Btu/h ft °F}$$

and for 1-in. schedule 40 pipe

$$\text{OD} = 1.315 \text{ in.} \quad \text{or} \quad r_o = 0.0548 \text{ ft}$$

$$\text{ID} = 1.049 \text{ in.} \quad \text{or} \quad r_i = 0.0437 \text{ ft}$$

Then U_o becomes

$$U_o = \frac{1}{\dfrac{r_o}{r_i \bar{h}_i} + \dfrac{r_o \ln r_i/r_o}{k} + \dfrac{1}{\bar{h}_o}}$$

$$= \frac{1}{\dfrac{0.0548}{0.0437(100)} + \dfrac{0.0548 \ln (0.0548/0.0437)}{26} + \dfrac{1}{2000}}$$

$$= \frac{1}{\underset{\text{inner convective}}{0.0128} + \underset{\text{wall}}{0.00048} + \underset{\text{outer convective}}{0.0005}} \; \frac{\text{Btu}}{\text{h ft}^2 \text{ °F}}$$

$$= 72.57 \text{ Btu/h ft}^2 \text{ °F}$$

In this case the inner convective resistance is controlling, and if $\bar{h}_o$ was increased to

4000 Btu/h ft^2 °F, the effect on U_o would be less than a 2 percent change. Thus the inside convective resistance limits the heat transfer rate, and changing the other two resistances would have very little effect on the overall heat transfer rate.

Convective Heat Transfer Correlations

For almost all preliminary design work and for many final design calculations, the convective heat transfer coefficients are calculated using previously established and verified correlations based on experimental data. Depending on the fluids and velocities involved, the convective heat transfer coefficient can attain magnitudes between 1 Btu/h ft^2 °F and 20,000 Btu/h ft^2 °F. Typical values are given in Table 2-2. The range of values for different convection regimes is significant.

Convection is an energy transport process due to random molecular motion and the macroscopic motion of the fluid. Viscous effects play a large role in determining the convective heat transfer rate. At a local position, the heat transfer is

$$dq_{\text{local}} = h_c(T_s - T_\infty)\, dA \tag{2-47a}$$

where h_c is the *local convective heat transfer coefficient*. Since flow conditions normally vary from point to point on a surface, the local heat transfer rate and, hence, the local convective heat transfer coefficient vary from point to point on the surface. The total heat transfer rate over the entire surface is

$$q = \int_A h_c(T_s - T_\infty)\, dA \tag{2-47b}$$

and if T_s and T_∞ are constant,

$$q = (T_s - T_\infty)\int_A h_c\, dA \tag{2-48}$$

If we define the *average heat transfer coefficient* such that

$$q = \bar{h}_c A(T_s - T_\infty) \tag{2-49}$$

TABLE 2-2 Typical Values of Convective Heat Transfer Coefficients

	Btu/h ft^2 °F	w/m^{2}K
Air, free convection	1–5	6–30
Superheated steam or air, forced convection	5–50	30–300
Oil, forced convection	10–300	60–1,800
Water, forced convection	50–2,000	300–6,000
Water, boiling	500–10,000	3,000–60,000
Steam, condensing	1,000–20,000	6,000–120,000

Used with permission, from F. Kreith, *Principles of Heat Transfer*, 3rd ed., Harper & Row, 1973.

TABLE 2-3 Frequently Used Dimensionless Groups

Group	Definition	Interpretation
Biot number (Bi)	$\frac{hL}{k_s}$	Ratio of the internal thermal resistance of a solid to the boundary layer thermal resistance.
Mass transfer Biot number (Bi_m)	$\frac{h_m L}{D_{AB}}$	Ratio of the internal species transfer resistance to the boundary layer species transfer resistance.
Coefficient of friction (C_f)	$\frac{\tau_s}{\rho u_\infty^2/2}$	Dimensionless surface shear stress.
Eckert number (Ec)	$\frac{u_\infty^2}{c_p(T_s - T_\infty)}$	Kinetic energy of the flow relative to the boundary layer enthalpy difference.
Fourier number (Fo)	$\frac{\alpha t}{L^2}$	Ratio of the heat conduction rate to the rate of thermal energy storage in a solid. Dimensionless time.
Mass transfer Fourier number (Fo_m)	$\frac{D_{AB} t}{L^2}$	Ratio of the species diffusion rate to the rate of species storage. Dimensionless time.
Friction factor (f)	$\frac{\Delta p}{(L/D)(\rho u_m^2/2)}$	Dimensionless pressure drop for internal flow.
Grashof number (Gr_L)	$\frac{g\beta(T_s - T_\infty)L^3}{\upsilon^2}$	Ratio of buoyancy to viscous forces.
Colburn j factor (j_H)	$StPr^{2/3}$	Dimensionless heat transfer coefficient.
Colburn j factor (j_m)	$St_m Sc^{2/3}$	Dimensionless mass transfer coefficient.
Lewis number (Le)	$\frac{\alpha}{D_{AB}}$	Ratio of the molecular thermal and mass diffusivities.
Nusselt number (Nu_L)	$\frac{hL}{k_f}$	Ratio of convection heat transfer to conduction in a fluid slab of thickness L.
Peclet number (Pe_L)	$\frac{u_\infty L}{\alpha} = Re_L Pr$	Dimensionless independent heat transfer parameter.
Prandtl number (Pr)	$\frac{C_p \mu}{k} = \frac{\nu}{x}$	Ratio of the molecular momentum and thermal diffusivities.
Reynolds number (Re_L)	$\frac{u_\infty L}{\nu}$	Ratio of the inertia and viscous forces.
Schmidt number (Sc)	$\frac{\nu}{D_{AB}}$	Ratio of the molecular momentum and mass diffusivities.
Sherwood number (Sh_L)	$\frac{h_m L}{D_{AB}}$	Ratio of convection mass transfer to diffusion in a slab of thickness L.
Stanton number (St)	$\frac{h}{\rho u_\infty c_p} = \frac{Nu_L}{Re_L Pr}$	Dimensionless heat transfer coefficient.

Used with permission, from F. P. Incropera and D. P. DeWitt, *Fundamentals of Heat Transfer*, John Wiley & Sons, Inc., 1981.

then

$$\bar{h}_c = \frac{1}{A}\int_A h_c \, dA \tag{2-50}$$

For heat exchanger design as viewed in this chapter, we shall be concerned predominately with average convective heat transfer coefficients.

Virtually all engineering data are reduced or correlated in terms of dimensionless parameters. Not only is such representation convenient in terms of data presentation, but a deeper understanding of the underlying physical processes often results. The more frequently used and referenced dimensionless parameters are given in Table 2-3. In most of these parameters containing L, the proper interpretation of L is as a characteristic length. Pipe flow, for example, would be described or correlated by $\mathrm{Re}_D = \rho VD/\mu$, whereas L is given in Table 2-3. The proper characteristic length, L, for a pipe is the diameter D. Parameters of particular interest to use in this chapter are the friction factor f, the Grashof number Gr_L, the Nusselt number Nu_L, and Prandtl number Pr, the Reynolds number Re_L, and the Stanton number St. The Stanton number is a dimensionless heat transfer coefficient, which can be expressed as the quotient ($\mathrm{Nu}_L/\mathrm{Re}_L\mathrm{Pr} = \mathrm{St}$) of the Reynolds, Prandtl, and Nusselt numbers.

Generally, we expect the local Nusselt number Nu_x to functionally become

$$\mathrm{Nu}_x = f(x, \mathrm{Re}_x, \mathrm{Pr}) \tag{2-51}$$

and the average Nusselt number to be functionally represented as

$$\overline{\mathrm{Nu}}_L = f(\mathrm{Re}_L, \mathrm{Pr}) \tag{2-52}$$

A large body of experimental data suggests

$$\overline{\mathrm{Nu}}_L = C\,\mathrm{Re}_L^m \mathrm{Pr}^n \tag{2-53}$$

or, in the case of free convection,

$$\overline{\mathrm{Nu}}_L = C(\mathrm{Gr}_L \mathrm{Pr})^n \tag{2-54}$$

Equations (2-53) and (2-54) are often developed assuming constant fluid properties, but we know that the fluid temperature and hence properties vary across the boundary layer. Correlations such as these forms are usually evaluated at (1) a temperature called the film temperature T_f, where

$$T_f = \frac{T_s + T_m}{2} \tag{2-55}$$

with T_s the surface temperature and T_m the mean fluid temperature, or (2) at $\bar{T}_m$, the average of the inlet and outlet mean temperature, or (3) at T_m with property variations across the boundary layer accounted for by multiplying the expression by property ratios such as

$$\left(\frac{\mathrm{Pr}_m}{\mathrm{Pr}_s}\right)^r \quad \text{or} \quad \left(\frac{\mu_m}{\mu_s}\right)^r \tag{2-56}$$

It is very important that the correct method for accounting for property variations be established prior to using any convective heat transfer correlation equation, as significant errors can result if properties are improperly evaluated.

Most convective heat transfer data and resulting correlations are broadly grouped into internal flows (such as pipe flow) and external flows (such as flow over a cylinder). The flow over a flat plate that has a zero pressure gradient is a rather special case of an external flow and is normally treated separately. Whether the boundary layer is laminar or turbulent and, if turbulent, the point of transition, makes a significant difference in both the average skin friction and the average convective heat transfer.

The flat plate will be examined first since it offers a simple flow that we can use to delineate several concepts of interest. Figure 2-12 schematically indicates important features of incompressible flow across a flat plate. The flow regimes are classified as laminar or turbulent with the process of transition bridging the gap between the two regimes. Transitional flows start out, near the laminar range, with surface measurable attributes of laminar flow and end up, near the turbulent range, with surface measurable attributes of turbulent flow. Transition is an exceedingly complex process, influenced by many factors, and is really not amenable to an exact analysis. Typically, and accurately enough for most purposes, a *transition* Reynolds number $\mathrm{Re}_{x,\mathrm{crit}}$ is specified and denotes the location at which laminar flow ends and at which turbulent flow begins. For situations of engineering interest, $\mathrm{Re}_{x,\mathrm{crit}}$ is usually taken to be 500,000. In this representation the transition region is collapsed to a single point.

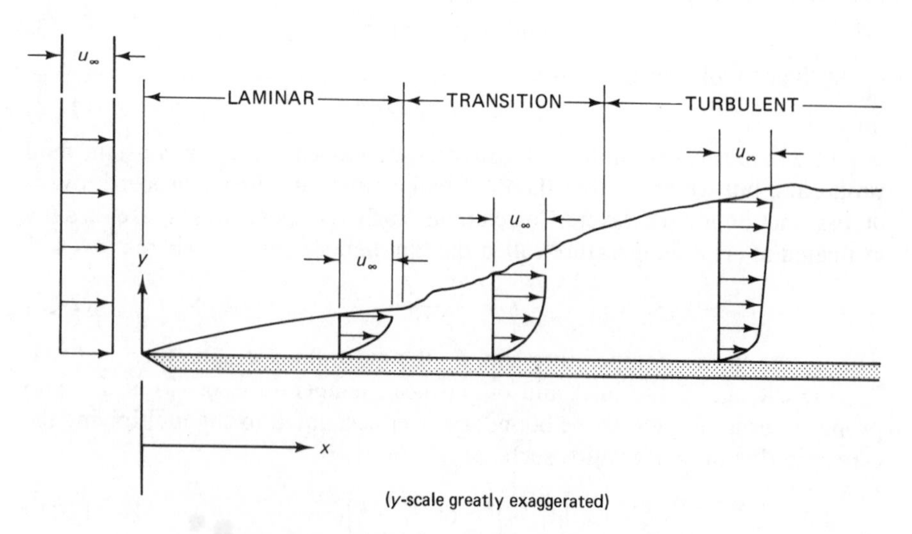

Figure 2-12 Flat-plate boundary layer development.

The local skin friction coefficient and the local Nusselt number are strong functions of the regime of flow. Table 2-4 provides a useful and convenient summary of local and average skin friction and Nusselt number relationships for both laminar and turbulent flow and for flows with $\mathrm{Re}_{x,\mathrm{crit}}$ of 5×10^5. All physical properties must be evaluated at the mean film temperature.

Nusselt number and skin friction (or friction factor) data and correlations for other more complex external flows and for internal flows are presented in much the same fashion. We shall examine each of these categories in turn.

Internal flows constitute one of the two major categories into which heat transfer correlations are grouped. Chapter 1 dealt at length with the calculation of wall friction losses in pipes and ducts. As with flat plate flows, internal flows may be classed as laminar or turbulent; and, as with flat plate flows, the convective heat transfer correlations will be markedly different for laminar or turbulent flows. Unlike flat plate flows, the Reynolds number is based on the characteristic diameter or hydraulic diameter, rather than a length; and, unlike flat plate flows, the critical Reynolds number for turbulent flow is normally taken as 2600, based on the average velocity and the pipe or hydraulic diameter. Thus, depending on the value of $\mathrm{Re}_D = \rho \bar{V} D/\mu$, the entire pipe is considered to be in either laminar or turbulent flow. The flow in a pipe or duct is said to be fully developed if the wall viscous effects extend all the way out to the center line. The length, normally expressed in terms of L/D, required for this condition to develop is called the *hydrodynamic entry length*. This concept is quite important in heat exchanger design as some types of heat exchangers have low values of L/D, and the resulting flow is never fully developed. Whether or not a flow is fully developed also has an important effect on the local convective heat transfer and thus on the average convective heat transfer coefficient.

Table 2-5 presents a summary of correlations for skin friction (friction factor) and heat transfer in a pipe. These equations possess sufficient accuracy for most engineering purposes; they are, however, approximate and may possess errors of as much as 25 percent. These equations should be examined carefully as they cover a wide range of developing or fully developed laminar or turbulent flows inside pipes or ducts. The friction factor correlations are expressed in terms of the Darcy–Weisbach friction factor.

For fully developed laminar flow in tubes, the Nusselt number is constant, but its value depends on whether a constant heat flux (q_s'') or a constant wall temperature (T_s) is prescribed. Table 2-6 gives the constant Nusselt numbers for fully developed laminar flow in tubes of different cross section.

There are many different turbulent heat transfer correlations (and sources) for internal flows in pipes and ducts. All correlations are approximate, with the degree of accuracy varying. Nonetheless, for specific applications some correlations can be found that are much more accurate than others, especially

TABLE 2-4 Useful Equations for Flat Plate Flows[a]

Laminar Flow		
Local friction coefficient	$C_{fx} = 0.664\mathrm{Re}_x^{-0.5}$	$\mathrm{Re}_x < 5 \times 10^5$
Local Nusselt number at distance x from leading edge	$\mathrm{Nu}_x = 0.332\mathrm{Re}_x^{0.5}\mathrm{Pr}^{0.33}$	$\mathrm{Pr} > 0.1$, $\mathrm{Re}_x > 5 \times 10^5$
	$\mathrm{Nu}_x = 0.565(\mathrm{Re}_x\mathrm{Pr})^{0.5}$	$\mathrm{Pr} < 0.1$, $\mathrm{Re}_x < 5 \times 10^5$
Average friction coefficient	$\overline{C}_f = 1.33\mathrm{Re}_L^{-0.5}$	$\mathrm{Re}_L < 5 \times 10^5$
Average Nusselt number between $x = 0$ and $x = L$	$\overline{\mathrm{Nu}}_L = 0.664\mathrm{Re}_L^{0.5}\mathrm{Pr}^{0.33}$	$\mathrm{Pr} > 0.1$, $\mathrm{Re}_L < 5 \times 10^5$
	$\overline{\mathrm{Nu}}_L = 1.13(\mathrm{Re}_L\mathrm{Pr})^{0.5}$	$\mathrm{Pr} < 0.1$, $\mathrm{Re}_L < 5 \times 10^5$
Turbulent Flow		
Local friction coefficient	$C_{fx} = 0.0576\mathrm{Re}_x^{-0.2}$	$\mathrm{Re}_x > 5 \times 10^5$, $\mathrm{Pr} > 0.5$
Local Nusselt number at distance x from leading edge	$\mathrm{Nu}_x = 0.0288\mathrm{Re}_x^{0.8}\mathrm{Pr}^{0.3}$	
Average friction coefficient	$\overline{C}_f = 0.072[\mathrm{Re}_L^{-0.2} - 0.0464(x_{\mathrm{crit}}/L)]$	$\mathrm{Re}_L > 5 \times 10^5$, $\mathrm{Pr} > 0.5$
Average Nusselt number between $x = 0$ and $x = L$ with transition at $\mathrm{Re}_{x,\mathrm{crit}} = 5 \times 10^5$	$\overline{\mathrm{Nu}}_L = 0.036\mathrm{Pr}^{0.33}(\mathrm{Re}_L^{0.8} - 23{,}200)$	

[a] Applicable to low-speed flow (Mach number < 0.5) of gases and liquids if all physical properties at the mean film temperature $T_f = (T_s + T_\infty)/2$.

$$C_{fx} = \tau_s/(\rho u_\infty^2/2g_c) \qquad \overline{C}_f = (1/L)\int_0^L C_{fx}\,dx$$

$$\mathrm{Nu}_x = h_c x/k \qquad \overline{\mathrm{Nu}} = \bar{h}_c L/k \qquad \bar{h}_c = (1/L)\int_0^L h_c(x)\,dx$$

$$\mathrm{Re}_x = \rho u_\infty x/\mu \qquad \mathrm{Re}_L = \rho u_\infty L/\mu \qquad \mathrm{Pr} = c_p\mu/k$$

Used with permission, from F. Kreith, *Principles of Heat Transfer*, 3rd ed., Harper & Row, 1973.

TABLE 2-5 Internal Flow Convective Heat Transfer Correlations

Correlation	Eq.	Conditions
$f = 64/\mathrm{Re}_D$	2-57	Laminar, fully developed
$\mathrm{Nu}_D = 4.36$	2-58	Laminar, fully developed, constant q_s'', $\mathrm{Pr} \gtrsim 0.6$
$\mathrm{Nu}_D = 3.66$	2-59	Laminar, fully developed, constant T_s, $\mathrm{Pr} \gtrsim 0.6$
$\overline{\mathrm{Nu}}_D = 3.66 + \dfrac{0.0668(D/L)\mathrm{Re}_D\mathrm{Pr}}{1 + 0.04[(D/L)\mathrm{Re}_D\mathrm{Pr}]^{2/3}}$	2-60	Laminar, thermal entry length ($\mathrm{Pr} \gg 1$ or an unheated starting length), constant T_s
or		
$\overline{\mathrm{Nu}}_D = 1.86\left(\dfrac{\mathrm{Re}_D\mathrm{Pr}}{L/D}\right)^{1/3}(\mu/\mu_s)^{0.14}$	2-61	Laminar, $\left[(\mathrm{Re}_D Pr/L/D)^{1/3}(\mu/\mu_s)^{0.14}\right] \gtrsim 2$, *constant* T_s, $0.48 < Pr < 16{,}700$, $0.0044 < (\mu/\mu_s) < 9.75$
$f = 0.316\mathrm{Re}_D^{-1/4}$ [a]	2-62	Turbulent, fully developed, $\mathrm{Re}_D \lesssim 2 \times 10^4$
$f = 0.184\mathrm{Re}_D^{-1/5}$ [a]	2-63	Turbulent, fully developed, $\mathrm{Re}_D \gtrsim 2 \times 10^4$
$\mathrm{Nu}_D = 0.023\mathrm{Re}_D^{4/5}\mathrm{Pr}^{1/3}$	2-64	Turbulent, fully developed, $0.6 \leq \mathrm{Pr} \leq 160$, $\mathrm{Re}_D \gtrsim 10{,}000$, $L/D \gtrsim 60$
or		
$\mathrm{Nu}_D = 0.023\mathrm{Re}_D^{4/5}\mathrm{Pr}'^{n}$	2-65	Turbulent, fully developed, $0.6 \leq \mathrm{Pr} \leq 160$, $\mathrm{Re}_D \geq 10{,}000$, $L/D \gtrsim 60$, $n = 0.4$ for $T_s > T_m$ and $n = 0.3$ for $T_s < T_m$
or		
$\mathrm{Nu}_D = 0.027\mathrm{Re}_D^{4/5}\mathrm{Pr}^{1/3}(\mu/\mu_s)^{0.14}$	2-66	Turbulent, fully developed, $0.7 \leq \mathrm{Pr} \leq 16{,}700$, $\mathrm{Re}_D \gtrsim 10{,}000$, $L/D \gtrsim 60$
$\mathrm{Nu}_D = 4.82 + 0.0185(\mathrm{Re}_D\mathrm{Pr})^{0.827}$	2-67	Liquid metals, turbulent, fully developed, constant q_s'', $3.6 \times 10^3 < \mathrm{Re}_D < 9.05 \times 10^5$, $10^2 < \mathrm{Pe}_d < 10^4$
$\mathrm{Nu}_D = 5.0 + 0.025(\mathrm{Re}_D\mathrm{Pr})^{0.8}$	2-68	Liquid metals, turbulent, fully developed, constant T_s, $\mathrm{Pe}_D > 100$

[a] Smooth tubes only.

Properties in Eqs. (2-58), (2-59) and (2-64) through (2-68) are evaluated at T_m. Properties in Eqs. (2-57), (2-62), and (2-63) are evaluated at T_f. Properties in Eqs. (2-60) and (2-61) are based on the mean $\overline{T}_m$.)

Used with permission, from F. P. Incropera and D. P. DeWitt, *Fundamentals of Heat Transfer*, John Wiley & Sons, Inc., 1981.

TABLE 2-6 Fully Developed Laminar Flow Nusselt Numbers

Cross Section	$\frac{b}{a}$	$Nu_D \equiv hD_h/k$ (Constant q_s'')	(Constant T_s)
 Circle	—	4.36	3.66
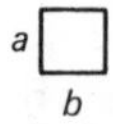 Rectangle a × b	1.0	3.63	2.98
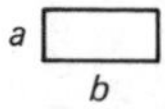 Rectangle a × b	1.4	3.78	—
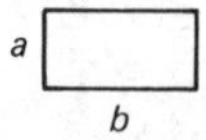 Rectangle a × b	2.0	4.11	3.39
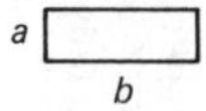 Rectangle a × b	3.0	4.77	—
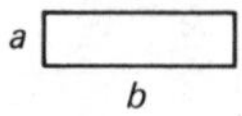 Rectangle a × b	4.0	5.35	4.44
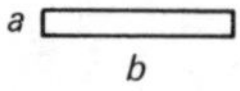 Rectangle a × b	8.0	6.60	5.95
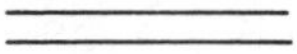 Parallel plates	∞	8.23	7.54
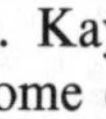 Triangle	—	3.00	2.35

Used with permission, from W. M. Kays and M. Crawford, *Convective Heat and Mass Transfer*, 2nd ed., McGraw-Hill Book Company, 1980.

for internal flows. Kays and Crawford (8) offer alternative correlation equations, which in some cases may offer greater accuracy than those presented here.

Historically, two equations are worthy of note: (1) the Colburn equation

$$\overline{Nu}_D = 0.023 Re_D^{0.8} Pr^{0.333} \tag{2-69}$$

and (2) the Dittus–Boelter equation

$$\overline{Nu}_D = 0.023 Re_D^{0.8} Pr^{n} \tag{2-70a}$$

where

$$n = 0.4 \qquad T_s > T_m \tag{2-70b}$$

$$n = 0.3 \qquad T_s < T_m \tag{2-70c}$$

The preceding are valid for $0.7 \leq Pr \leq 160$, $Re_D \geq 10{,}000$, and $L/D \geq 60$. Equations (2-69) and (2-70a) are valid only for moderate temperature differences with all properties evaluated at T_m, the mean fluid temperature.

Where property variations are large, the following expression is recommended:

$$\overline{\mathrm{Nu}}_D = 0.027\mathrm{Re}_D^{0.8}\mathrm{Pr}^{0.333}\left(\frac{\mu}{\mu_s}\right)^{0.14} \tag{2-71}$$

where all properties except μ_s are evaluated at T_m. In situations where T_m varies significantly, an arithmetical mean should be used for property variation. Care should be exercised in selecting an equation from Table 2-5 as the appropriate ranges of validity must be observed.

External flows are the remaining class for which we shall need forced convective heat transfer correlations. Obviously, we could discuss local and average Nusselt number correlations for an open-ended range of shapes in external flow. This discussion will be limited to laminar or turbulent flows over single cylinders or to bundles of cylinders and to some special cases of single noncircular cross sections in a cross flow. Since the tubes in most shell-and-tube heat exchangers are circular in cross section, the correlations provided here will be sufficient for use in design and analysis of most shell-and-tube heat exchangers.

Several correlations are available for the average Nusselt number over a single circular cylinder in a cross flow. Incropera and DeWitt recommend the Churchill–Bernstein equation:

$$\overline{\mathrm{Nu}}_D = 0.3 + \frac{0.62\mathrm{Re}_D^{1/2}\mathrm{Pr}^{1/3}}{\left[1+\left(\frac{0.4}{\mathrm{Pr}}\right)^{2/3}\right]^{1/4}}\left[1+\left(\frac{\mathrm{Re}_D}{28200}\right)^{5/8}\right]^{4/5} \tag{2-72}$$

Equation (2-72) is valid for $\mathrm{Re}_D\mathrm{Pr} > 0.2$; properties should be evaluated at the film temperature T_f.

Average Nusselt number correlations for noncircular cylinders in a cross flow are usually expressed as

$$\overline{\mathrm{Nu}}_D = C\mathrm{Re}_D^m\mathrm{Pr}^{1/3} \tag{2-73}$$

where Table 2-7 presents typical values of C and m for several orientations and shapes of noncircular cylinders in a cross flow. In Eq. (2-73), properties are again evaluated at T_f.

Tube bundles are used in many different heat transfer applications. One fluid flows over the exterior of the tubes, while another at a different temperature flows inside the tubes. Figure 2-13 schematically illustrates such an arrangement. When compared with the heat transfer from a single tube, two additional parameters, tube arrangement and tube rows, become important. The two typical arrangements, as shown in Fig. 2-14, are aligned or staggered. For each arrangement, geometrical parameters are indicated and labeled. Except for the first few rows, tubes are immersed in the very turbulent wakes of preceding tubes, and the average heat transfer coefficients associated with these inner row tubes tend to remain constant. The dual effects of increased

TABLE 2-7 Constants for Noncircular Correlations

Geometry	Re_D	C	m
Square (diamond, V →, D)	5×10^3–10^5	0.246	0.588
Square (V →, D)	5×10^3–10^5	0.102	0.675
Hexagon (V →, D)	5×10^3–1.95×10^4	0.160	0.638
	1.95×10^4–10^5	0.0385	0.782
Hexagon (V →, D)	5×10^3–10^5	0.153	0.638
Vertical plate (V →, D)	4×10^3–1.5×10^4	0.228	0.731

Used with permission, from F. P. Incropera and D. P. DeWitt, *Fundamentals of Heat Transfer*, John Wiley & Sons, Inc., 1981.

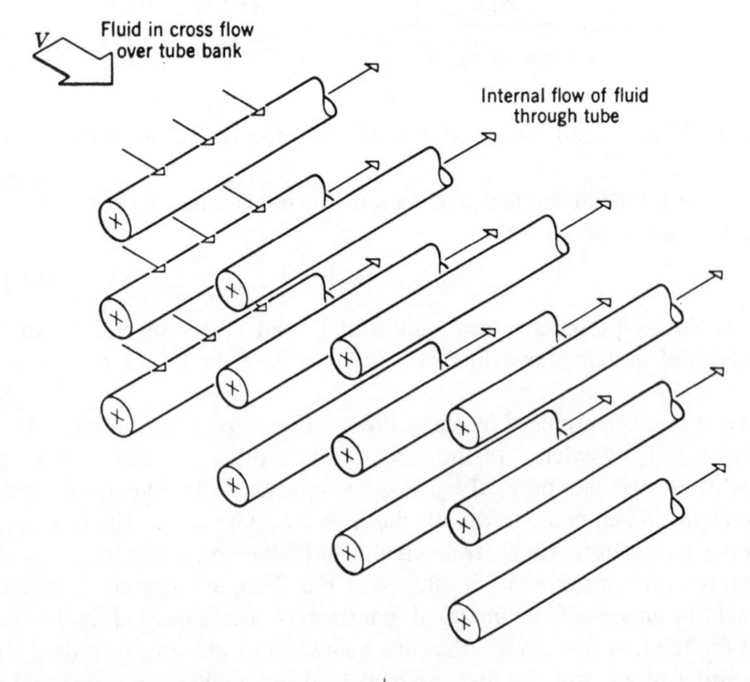

Figure 2-13 Tube bank schematic. (Used with permission, from F. P. Incropera and D. P. DeWitt, *Fundamentals of Heat Transfer*, John Wiley & Sons, Inc., 1981.)

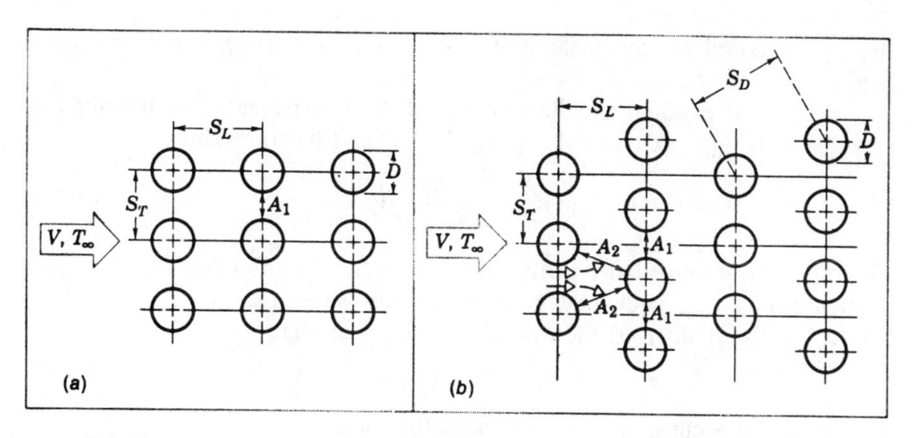

Figure 2-14 Tube bank nomenclature. (Used with permission, from F. P. Incropera and D. P. DeWitt, *Fundamentals of Heat Transfer*, John Wiley & Sons, Inc., 1981.)

turbulence and a higher effective velocity enhance the average heat transfer coefficient for these inner row tubes when compared to a single tube in a cross flow. Since for heat exchanger design we are interested in the bundle, not the performance of discrete tubes, the correlations are usually for the average heat transfer coefficient over the entire bundle.

As with the previous correlations presented, many different formulations have been suggested for correlating heat transfer for tube bundles. The most widely used correlation, that of Zhukaukas, has the form

$$\overline{\mathrm{Nu}}_D = C\mathrm{Re}_{D,\max}^m \mathrm{Pr}^{0.36}(\mathrm{Pr}_\infty/\mathrm{Pr}_s)^{1/4} \tag{2-74}$$

which is valid for N (number of tubes) ≥ 20, $0.7 \leq \mathrm{Pr} \leq 500$, and $1000 < \mathrm{Re}_{D,\max} < 2 \times 10^6$. In Eq. (2-74) all properties except Pr_s are evaluated at T_∞. The constants, C and m, are listed in Table 2-8, and the Reynolds number

TABLE 2-8 Tube Bundle Correlation Constants

Configuration	$\mathrm{Re}_{D,\max}$	C	m
Aligned	10^3–2×10^5	0.27	0.63
Staggered ($S_T/S_L < 2$)	10^3–2×10^5	$0.35(S_T/S_L)^{1.5}$	0.60
Staggered ($S_T/S_L > 2$)	10^3–2×10^5	0.40	0.60
Aligned	2×10^5–2×10^6	0.021	0.84
Staggered	2×10^5–2×10^6	0.022	0.84

Used with permission, from F. P. Incropera and D. P. DeWitt, *Fundamentals of Heat Transfer*, John Wiley & Sons, Inc., 1981.

$Re_{D,max}$ is based on the maximum velocity V_{max} occurring within the tube bank.

V_{max} is dependent on the tube size and arrangement. For the aligned arrangement, V_{max} occurs at A_1 of Fig. 2-14(a) and has the value

$$V_{max} = \frac{S_T}{S_T - D} V \tag{2-75}$$

where V is the unimpeded velocity approaching the tube bundle. The staggered configuration's V_{max} is dependent on which transverse area, A_1 or A_2 [see Fig. 2-14(b)], is the smaller. If the rows are spaced such that

$$2(S_D - D) < (S_T - D) \tag{2-76}$$

then V_{max} will occur at A_2 and will have the value

$$V_{max} = \frac{S_T}{2(S_D - D)} V \tag{2-77}$$

For the case where V_{max} occurs at A_1, Eq. (2-75) may be used.

All the heat transfer correlations examined so far in this section have been for forced convection over flat plates or for internal or external flows. In contrast to forced convection characterized by an external forcing condition, free convection has no forced velocity but is driven by convective currents induced by buoyancy forces. As we saw in Eq. (2-54), the appropriate dimensionless parameter is the Grashof number–Prandtl number product ($Gr_L Pr$). This product appears so frequently in free convection that it has been given the name Rayleigh number Ra_L:

$$Ra_L = Gr_L Pr = \frac{g\beta(T_s - T_\infty)L^3}{\nu\alpha} \tag{2-78}$$

Most free convection correlations take the form

$$\overline{Nu}_L = C Ra_L^n \tag{2-79}$$

where $n \sim \frac{1}{4}$ for laminar flow and $n \sim \frac{1}{3}$ for turbulent flow. Since $Ra_L \propto L^3$ and $\overline{Nu}_L \propto Ra_L^{1/3}$, it follows that for turbulent flow $\overline{Nu}_L$ is independent of L, the characteristic length of the geometry. Generally, all properties are evaluated at the film temperature $T_f = (T_s + T_\infty)/2$.

Virtually any heat transfer text book presents a plethora of free convective heat transfer correlations for a number of situations and geometrical arrangements. The information given herein appeared in Incropera and DeWitt (2). Figure 2-15 presents a sketch of the geometrical arrangement, as well as appropriate Nusselt number–Rayleigh number correlations for a number of arrangements of interest.

Free convective Nusselt numbers are smaller than forced convective Nusselt numbers, so the corresponding free convection heat transfer coefficients are smaller than the forced convective coefficients. Table 2-2 provides a

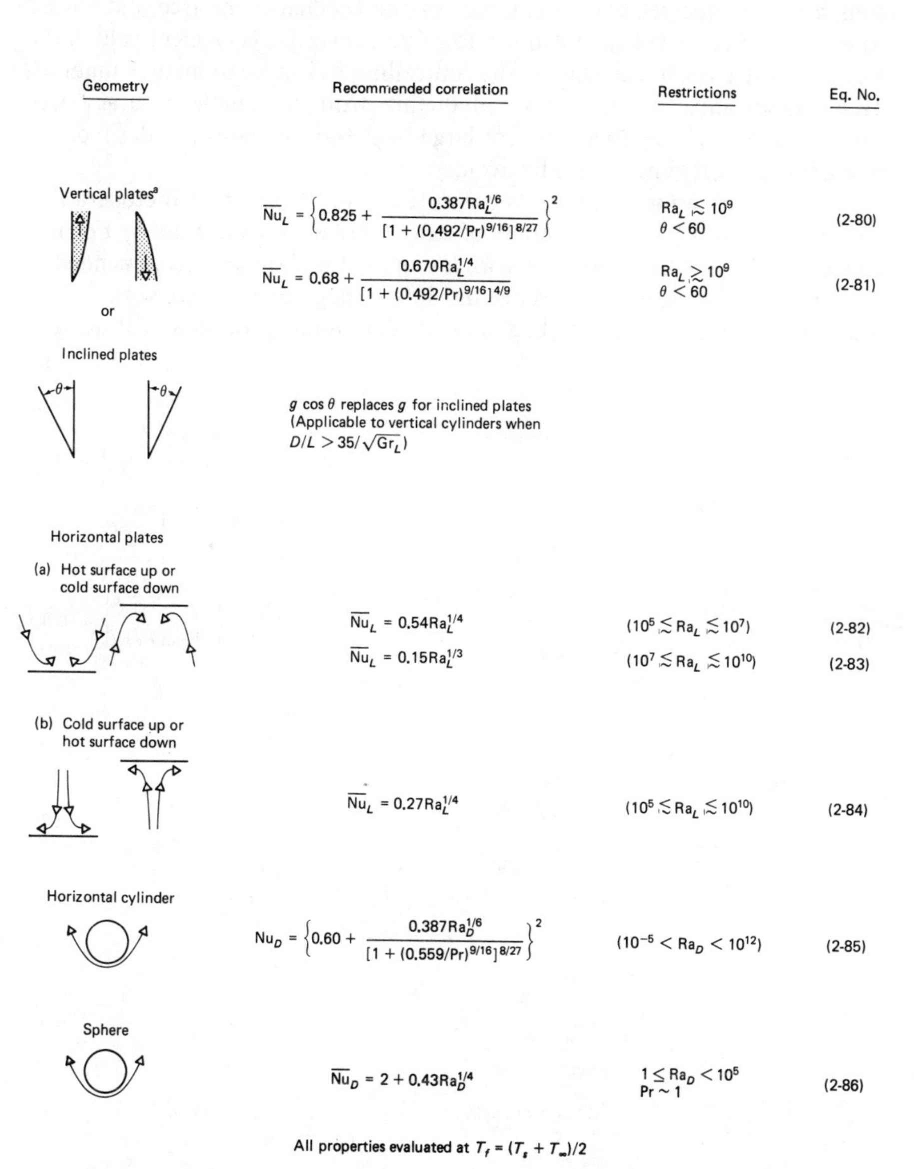

Geometry	Recommended correlation	Restrictions	Eq. No.
Vertical plates[a]	$\overline{Nu}_L = \left\{0.825 + \dfrac{0.387 Ra_L^{1/6}}{[1 + (0.492/Pr)^{9/16}]^{8/27}}\right\}^2$	$Ra_L \lesssim 10^9$, $\theta < 60$	(2-80)
	$\overline{Nu}_L = 0.68 + \dfrac{0.670 Ra_L^{1/4}}{[1 + (0.492/Pr)^{9/16}]^{4/9}}$	$Ra_L \gtrsim 10^9$, $\theta < 60$	(2-81)
or Inclined plates	$g \cos \theta$ replaces g for inclined plates (Applicable to vertical cylinders when $D/L > 35/\sqrt{Gr_L}$)		
Horizontal plates (a) Hot surface up or cold surface down	$\overline{Nu}_L = 0.54 Ra_L^{1/4}$	$(10^5 \lesssim Ra_L \lesssim 10^7)$	(2-82)
	$\overline{Nu}_L = 0.15 Ra_L^{1/3}$	$(10^7 \lesssim Ra_L \lesssim 10^{10})$	(2-83)
(b) Cold surface up or hot surface down	$\overline{Nu}_L = 0.27 Ra_L^{1/4}$	$(10^5 \lesssim Ra_L \lesssim 10^{10})$	(2-84)
Horizontal cylinder	$Nu_D = \left\{0.60 + \dfrac{0.387 Ra_D^{1/6}}{[1 + (0.559/Pr)^{9/16}]^{8/27}}\right\}^2$	$(10^{-5} < Ra_D < 10^{12})$	(2-85)
Sphere	$\overline{Nu}_D = 2 + 0.43 Ra_D^{1/4}$	$1 \leq Ra_D < 10^5$, $Pr \sim 1$	(2-86)

All properties evaluated at $T_f = (T_s + T_\infty)/2$

Figure 2-15 Free convection empirical correlations for external geometries. (Used with permission, from F. P. Incropera and D. P. DeWitt, *Fundamentals of Heat Transfer*, John Wiley & Sons, Inc., 1981.)

comparison of the relative magnitudes of the coefficient for free and forced convection. Because of the relatively low free convective coefficient values, the free convective resistance may be the controlling resistance in many situations. Thus, importance is attached to predicting with acceptable accuracy free convective Nusselt numbers. Where large heat transfer rates are desired, the free convection regime should be avoided.

The information correlations presented in Fig. 2-15 (from Incropera and DeWitt) are valid for a wide variety of external free convective flows. For free convective flows within ducts, the correlations of Fig. 2-16 are recommended.

If free convection represents the lower magnitude of convective heat transfer coefficients, then boiling and condensation, processes that involve

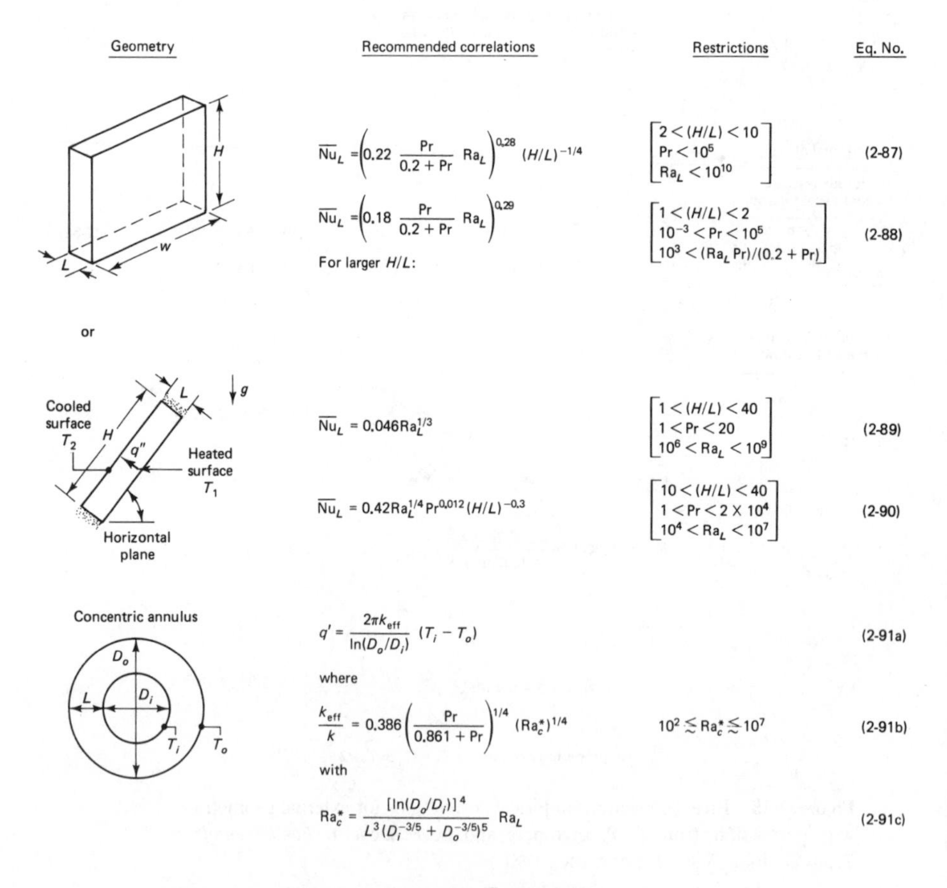

Figure 2-16 Free convection empirical correlations for internal geometries. (Used with permission, from F. P. Incropera and D. P. DeWitt, *Fundamentals of Heat Transfer*, John Wiley & Sons, Inc., 1981.)

phase change, represent the larger magnitudes. The significant latent heat effects associated with the phase change make possible heat transfer rates that are generally much larger than those of forced convection heat transfer in the absence of phase change. Since the phase change can result in energy transfer without a fluid temperature change, high heat transfer rates may be attained with small temperature differences through boiling or condensation. As opposed to free convection, whose low convective heat transfer coefficients are often the controlling resistance, phase change resistances are usually small when compared to other conductive or convective resistances in a composite situation.

Boiling occurs when evaporation takes place at a solid–liquid interface. Condensation occurs when the temperature of a vapor is reduced below its saturation temperature. We shall examine boiling first. The phenomenological aspects of boiling can best be explored by referring to Fig. 2-17, a typical representation of surface heat flux q'' as a function of excess temperature ΔT_e. The excess temperature is the difference between the surface temperature T_s and the saturation temperature T_{sat}. This figure is for water and shows several different regimes as ΔT_e is increased from 1 to 10,000°F. These data were generated by a heated platinum wire immersed in water. Other fluids show a similar response.

When ΔT_e is less than 10°F, the heat transfer is by free convection. The superheated liquid rises because of the buoyant forces. As the excess tempera-

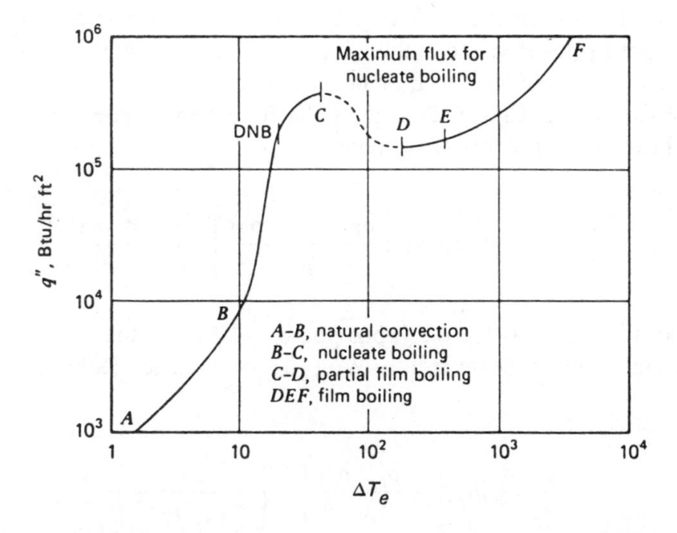

Figure 2-17 Pool boiling heat flux as a function of temperature excess. (Used with permission, from W. H. McAdams, *Heat Transmission*, 3rd ed., McGraw-Hill Book Company, 1954.)

ture is increased from near 10° to about 90°F, free convection ceases and bubble nucleation takes place. The bubbles form and rise to the surface. This is generally called nucleate boiling. In this regime the surface becomes densely populated with bubbles. The subsequent separation and rise induces additional fluid mixing near the surface. Much of the energy exchange in nucleate boiling results from direct transfer to the liquid at the surface and not through the rising bubbles.

As ΔT_e continue to increase, the point labeled DNB, *departure from nucleate boiling*, is reached. The point is also called the critical heat flux or the critical excess temperature or the burnout point. [Some authors, Burmeister (11), for example, say that the DNB point occurs a few degrees of excess before burnout is reached.] As ΔT_e is increased to DNB, the wire's surface becomes increasingly covered with bubbles, and finally at DNB regular bubble formation ceases.

Beyond DNB, the heat flux actually decreases as the surface becomes momentarily covered with patches of vapor whose insulating effect inhibits heat transfer. The dashed region of Fig. 2-17, unstable film boiling, is as its name implies an unstable situation, difficult to obtain experimentally. As ΔT_e is increased beyond the unstable film boiling regime, a region of stable film boiling is reached. For the case of water as shown in Fig. 2-17, stable film boiling is reached at a ΔT_e of near 1000°F. The region of stable film boiling generally requires surface temperatures that exceed the melting point of the solid surface. Because of the unstable film boiling region, increases in the unit heat flux beyond DNB initiate a jump in ΔT_e to the stable film boiling region; hence, the label *burnout* for the DNB point.

With the aforementioned catastrophic results occurring for the film boiling regions (unstable and stable), we should not operate or design devices to operate in excess of the DNB or burnout point.

Rohsenow and Choi (12) suggest the following equation for predicting the unit heat transfer at the DNB point:

$$\left(\frac{q}{A}\right)_{\text{DNB}} = 0.18\rho_v h_{fg}\left[\frac{\sigma g g_c(\rho_l - \rho_v)}{\rho_v^2}\right]^{1/4}\left(\frac{\rho_l}{\rho_l - \rho_v}\right)^{1/2} \tag{2-92}$$

where the subscript v is for the vapor phase and l is for the liquid phase. For pool boiling in the nucleate region, Wolf (3) recommends Rohsenow's expression

$$\frac{q}{A} = \mu_l h_{fg}\left[\frac{g(\rho_l - \rho_v)}{g_c \sigma}\right]^{1/2}\left(\frac{C_{pl}\Delta T_e}{C_{sf} h_{fg} \text{Pr}^{1.7}}\right)^3 \tag{2-93}$$

where C_{sf} values are given for various liquid–surface combinations in Table

TABLE 2-9 C_{sf} Values for Various Liquid–Metal Combinations

Liquid–Surface Material	C_{sf}
Isopropyl alcohol–copper	0.0025
Ethyl alcohol–chromium	0.0027
50 percent K_2CO_3–copper	0.0027
35 percent K_2CO_3–copper	0.0054
n-Butyl alcohol–copper	0.0030
Water–brass	0.006
Water–nickel	0.006
Benzene–chromium	0.010
Water–platinum	0.013
Water–copper	0.013
Carbon tetrachloride–copper	0.013
N-pentane-chromium	0.015

Used with permission, from W. M. Rohsenow and H. Y. Choi, *Heat, Mass, and Momentum Transfer*, Prentice-Hall, 1961.

2-9. As before, subscript l represents the liquid values and subscript v the vapor, with σ the surface tension.

Condensation represents the other phase change phenomenon of interest. This process occurs when vapor comes in contact with a surface below the saturation temperature of the vapor. Condensation occurs in two modes: (1) film condensation on clean, uncontaminated surfaces, and (2) dropwise condensation on surfaces that inhibit wetting. In both modes the condensate flows down the surface because of gravity, and the condensate, either film or drops, provides a resistance to heat transfer between the vapor and the surface. As this resistance increases with condensate thickness, short vertical surfaces or horizontal cylinders are usually employed in practice. The convective heat transfer coefficient for dropwise condensation is higher than for film condensation, although the convective coefficients of both modes are much higher than conventional forced convection. (See Table 2-2.) Over an extended period of time, most heat transfer condensation surfaces permit too much wetting for dropwise condensation to be maintained. Hence, film condensation is typically used as the basis for the design of condensers. Griffith (13) and Burmeister (11) present additional information on dropwise condensation.

We shall examine here only film condensation. For film condensation on a vertical plate, Nusselt devised the following expression and others later empirically adjusted the constant to provide better agreement with data:

$$\bar{h}_L = 1.13\left[\frac{g\rho_l(\rho_l - \rho_v)k_l^3 h_{fg}}{\mu_l(T_{\text{sat}} - T_s)L}\right]^{1/4} \tag{2-94}$$

All properties should be evaluated at $T_f = (T_{\text{sat}} + T_s)/2$, and h_{fg} should be evaluated at T_{sat}. The total condensation is

$$\dot{m} = \frac{\bar{h}_L A(T_{sat} - T_s)}{h_{fg}} \tag{2-95}$$

In film condensation on a vertical surface, transition is indicated when the Reynolds number

$$\mathrm{Re}_\delta = \frac{\rho_l u_m 4\delta}{\mu_l} \tag{2-96}$$

(where u_m is the average film velocity and δ is the condensation film thickness) exceeds 1800. For values of the Reynolds number, as defined by Eq. (2-96), greater than 1800, the following expression for the convective heat transfer coefficient is recommended:

$$\bar{h}_L = 0.0077\mathrm{Re}_\delta^{0.4}\left[\frac{g\rho_l(\rho_l - \rho_v)k_l^3}{\mu_l^2}\right]^{1/3} \tag{2-97}$$

Equation (2-95) can again be used to calculate the total condensation rate. Equation (2-97) must be solved by iteration, since δ depends on $\dot{m}$, and $\dot{m}$ depends on $\bar{h}_L$ [via Eq. (2-95)]. Wolf reminds us that the $\bar{h}_L$ from Eq. (2-97) is developed from experimental data that include a laminar portion; so within the accuracy of the correlation, laminar and transition effects are already included in the $\bar{h}_L$.

The average convective coefficient for a single horizontal tube is given by

$$\bar{h}_D = 0.728\left[\frac{g\rho_l(\rho_l - \rho_v)k_l^3 h_{fg}}{\mu_l(T_{sat} - T_s)D}\right]^{1/4} \tag{2-98}$$

For a vertical tier of N horizontal tubes, an approximate value for the average convective heat transfer coefficient over the N tubes is

$$\bar{h}_N = \frac{\bar{h}_D}{\sqrt[4]{N}} \tag{2-99}$$

where $\bar{h}_D$ is evaluated from Eq. (2-98).

We have examined in this section a number of correlations for many different situations. The correlations presented here are all widely used and accepted, but they are certainly not the only accepted, or even the ultimate, correlations. Incropera and DeWitt suggest the following methodology for selection and application of convective correlations for any general flow situation:

1. Become immediately cognizant of the flow geometry. Does the problem involve flow over a flat plate, a sphere, a cylinder? The specific form of the convection correlation depends, of course, on the geometry.
2. Specify the appropriate reference temperature and then evaluate the pertinent fluid properties at that temperature. For moderate boundary

layer temperature differences, it has been found that the film temperature may be used for this purpose. However, there are correlations that require property evaluation at the free-stream temperature and include a property ratio to account for the nonconstant property effect.

3. Determine whether the flow is laminar or turbulent. This determination is made by calculating the appropriate Reynolds number and comparing the value with the appropriate transition criterion. For example, if a problem involves parallel flow over a flat plate for which the Reynolds number is $\mathrm{Re}_L = 10^6$ and the transition criterion is $\mathrm{Re}_{x,c} = 5 \times 10^5$, it is obvious that a mixed boundary layer condition exists.

Fouling Resistance

The composite expressions presented earlier were derived under the assumption that only conductive and convective resistances were present. When a heat exchanger is new, that is a reasonable assumption; but as the exchanger ages, the deposition of a film or scale on the surfaces can greatly increase the resistance to heat transfer between the fluids. The effect of this additional resistance, called *fouling*, can be estimated by adding a total fouling resistance to the usual composite expression such that

$$U_o = \frac{1}{\dfrac{1}{\bar{h}_o} + R_{f_o} + \dfrac{r_o}{k}\ln\dfrac{r_o}{r_i} + \dfrac{r_o}{r_i}R_{f_i} + \dfrac{r_o}{r_i}\dfrac{1}{\bar{h}_i}} \tag{2-100}$$

where R_{f_o} and R_{f_i} are the fouling factors. Figure 2-18 delineates the usually accepted values of fouling factors for various circumstances. Fouling is a very complex subject and many references and expressions are available in the literature that allow prediction of the growth rate of scales on tube surfaces. Typical practice is to design the heat exchanger in such a manner that when new it is oversized and that just before cleaning the design rating is reached.

Finned Surfaces

An examination of any of the previously stated composite wall expressions [Eq. (2-45), for example] reveals that the overall heat transfer rate is dependent on the various convective, conductive, and fouling resistances. The heat transfer rate can only be increased (or decreased) by decreasing (or increasing) the sum of the resistances. One effective technique for enhancing the heat transfer rate for the lower end of the convective heat coefficient spectrum is to add extended surfaces or *fins*. It is appropriate for us to examine fins and their effects in some detail since many heat exchangers, especially the cross-flow types, utilize finned surfaces to achieve increases in the heat transfer rates.

Fluid 1	Fluid 2	Total Fouling Resistance hr ft^2°F/Btu
Water	Water	0.0015
"	Gas, about 10psig	0.001
"	Gas, about 100psig	0.001
"	Gas, about 1000psig	0.001
"	Light organic liquids	0.0015
"	Medium organic liquids	0.002
"	Heavy organic liquids	0.0025
"	Very heavy organic liquids Heating Cooling	0.004
Steam	Gas, about 10psig	0.0005
"	Gas, about 100psig	0.0005
"	Gas, about 1000psig	0.0005
"	Light organic liquids	0.001
"	Medium organic liquids	0.0015
"	Heavy organic liquids	0.002
"	Very heavy organic liquids	0.0035
" (no non-condensables)	Water	0.001
Light organic liquids	Light organic liquids	0.002
"	Medium organic liquids	0.0025
"	Heavy organic liquids Heating Cooling	0.003

Figure 2-18 Selected fouling factors for heat exchangers (units are h ft^2 °F/Btu). (Used with permission, from S. Kakac, A. E. Bergles, and F. Mayinger, eds., *Heat Exchangers: Thermal–Hydraulic Fundamentals and Design*, Hemisphere Publishing Corporation, 1981.)

Light organic liquids	Very heavy organic liquids Heating Cooling	0.004
Medium organic liquids	Medium organic liquids	0.003
"	Heavy organic liquids Heating Cooling	0.0035
"	Very heavy organic liquids Heating Cooling	0.0045
Heavy organic liquids	Heavy organic liquids	0.005
"	Very heavy organic liquids	0.006
Gas, about 10psig	Gas, about 10psig	0
"	Gas, about 100psig	0
"	Gas, about 1000psig	0
Gas, about 100psig	Gas, about 100psig	0
"	Gas, about 1000psig	0
Gas, about 1000psig	Gas, about 1000psig	0
Water	Condensing light organic vapors, pure component	0.001
"	Condensing medium organic vapors, pure compenent	0.001
"	Condensing heavy organic vapors, pure component	0.002

Figure 2-18 (continued)

General Notes

1. The total fouling resistance and the overall heat transfer coefficient are based on the total outside tube area.

2. Allowable pressure drops on each side are assumed to be about 10psi except for (a) low pressure gas and condensing vapor where the pressure drop is assumed to be about 5 per cent of the absolute pressure, and (b) heavy organics where the allowable pressure drop is assumed to be about 20 to 30 psi.

3. Aqueous solutions give approximately the same coefficients as water.

4. Liquid ammonia gives about the same results as water.

5. "Light organic liquids" include liquids with viscosities less than about 0.5 cp, such as hydrocarbons through C_8, gasoline, light alcohols and ketones, etc.

6. "Medium organic liquids" include liquids with viscosities between about 0.5 cp and 1.5 cp, such as kerosene, straw oil, hot gas oil, absorber oil, and light crudes.

7. "Heavy organic liquids" include liquids with viscosities greater than 1.5 cp, but not over 50 cp, such as cold gas oil, lube oils, fuel oils, and heavy and reduced crudes.

8. "Very heavy organic liquids" include tars, asphalts, polymer melts, greases, etc., having viscosities greater than about 50 cp. Estimation of coefficients for these materials is very uncertain.

9. These values may be used for vapor mixtures when the condensing range of the vapor is less than half of the temperature difference between the outlet coolant and the vapor. If the condensing range is greater than this, or if there are significant amounts of non-condensable gas present, the coefficient should be reduced towards the values shown for gas cooling; in these cases, the accuracy of the estimation is very uncertain.

Figure 2-18 (continued)

We shall start our examination by recalling the ordinary differential equation governing the temperature distributions in fins,

$$\frac{d^2\theta}{dx^2} - m^2\theta = 0 \tag{2-101}$$

where $m^2 = \bar{h}_c P/kA_c$ and $\theta = T - T_\infty$, the temperature excess. Equation (2-101) was derived under the following assumptions:

1. Constant cross-sectional area.
2. One-dimensional conditions in the x direction.
3. Constant material thermal conductivity.
4. Steady state.
5. Constant value of $\bar{h}_c$.

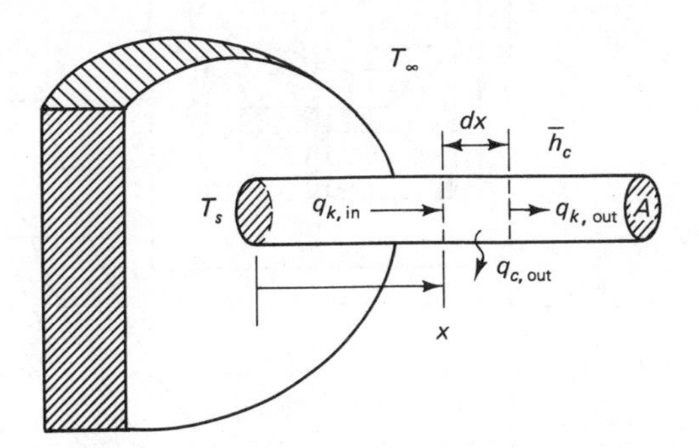

Figure 2-19 Fin nomenclature.

The nomenclature is prescribed by Figure 2-19.

Within the assumptions given in the previous paragraph, three cases are of interest: (1) the infinitely long fin, (2) the finite fin with an insulated end, and (3) the finite fin with an end convective heat loss. The appropriate solutions are as follows:

1. Infinitely long fin, as $x \to \infty$, $T \to T_\infty$:

$$T = T_\infty + (T_s - T_\infty)e^{-mx}$$
$$q_{\text{fin}} = \sqrt{\bar{h}_c PkA}\,(T_s - T_\infty) \tag{2-102}$$

2. Finite fin with an insulated end:

$$\frac{T - T_\infty}{T_s - T_\infty} = \frac{\cosh m(L - x)}{\cosh mL}$$
$$q_{\text{fin}} = \sqrt{\bar{h}_c PkA}\,(T_s - T_\infty)\tanh(mL) \tag{2-103}$$

3. Finite fin with end loss by convection:

$$\frac{T - T}{T_s - T_\infty} = \frac{\cosh m(L - x) + (\bar{h}_L/mk)\sinh m(L - x)}{\cosh mL + (\bar{h}_L/mk)\sinh mL}$$
$$q_{\text{fin}} = \sqrt{\bar{h}_c PkA}\,(T_s - T_\infty)\frac{\sinh mL + (\bar{h}_L/mk)\cosh mL}{\cosh mL + (\bar{h}_L/mk)\sinh mL} \tag{2-104}$$

where $\bar{h}_L$ is the average convective heat transfer coefficient of the exposed fin's end.

Equations (2-102) to (2-104) present both the temperature distribution down the fin, $(T - T_\infty)/(T_s - T_\infty)$, and the heat transfer at the fin's base, q_{fin}. In all expressions, T_s is the surface temperature at the fin's base. Such expressions

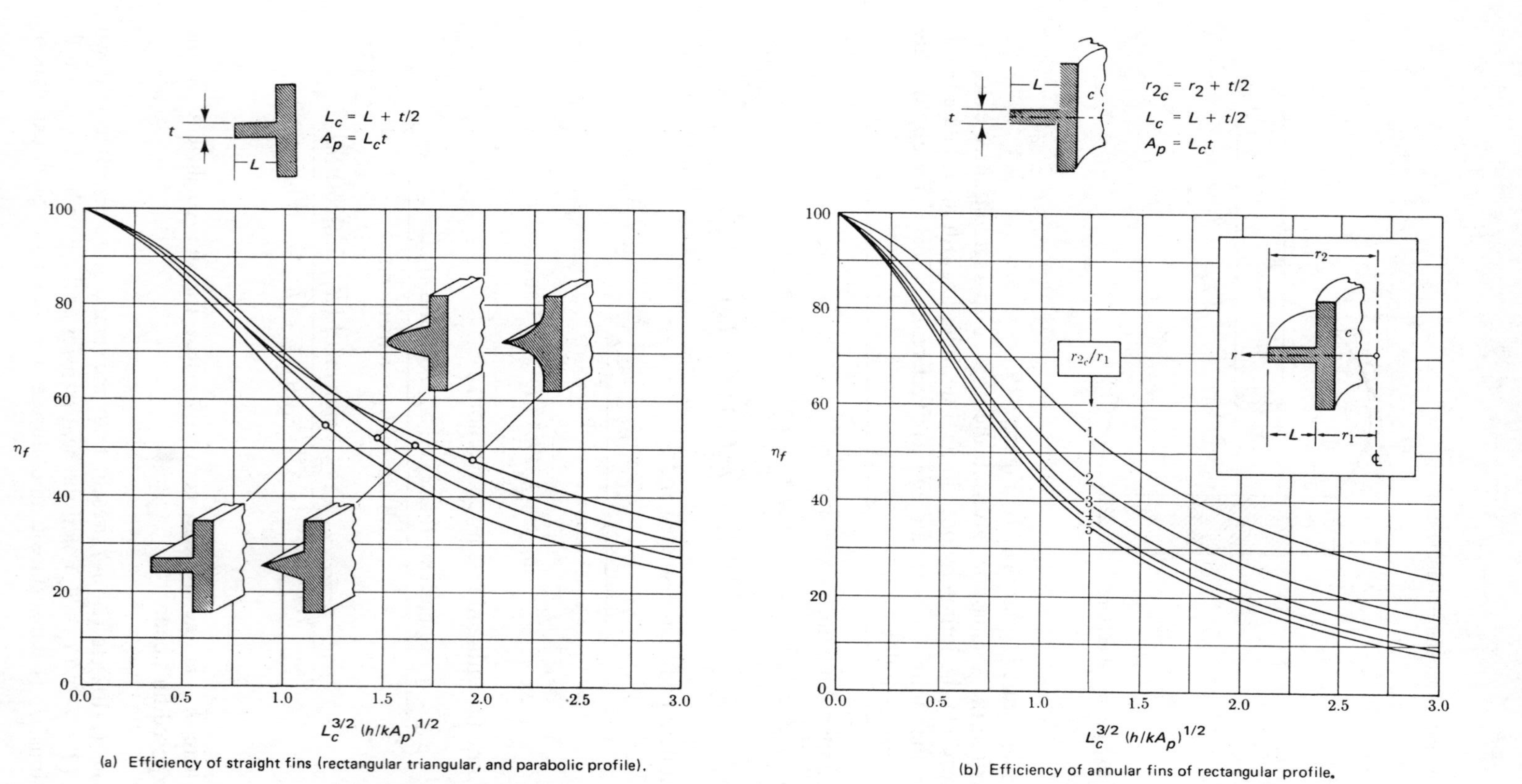

(a) Efficiency of straight fins (rectangular triangular, and parabolic profile).

(b) Efficiency of annular fins of rectangular profile.

Figure 2-20 Fin efficiencies for several fin geometries. (Used with permission, from P. J. Schneider, *Conduction Heat Transfer*, Addison–Wesley, 1955.)

are awkward to use in situations where multiple fins are present and where T_s is not known.

A more fruitful use of Eqs. (2-102) to (2-104) or, for that matter, any other such expression for fin performance is the fin efficiency. The fin efficiency, η_f, is defined as

$$\eta_f = \frac{q_{\text{fin}}}{q \text{ if entire fin were at } T_s} \tag{2-105}$$

For example, for the circular pin fin of diameter D and length L with an

Type	Fin profile	Efficiency	Area / Volume
Straight fins	① Rectangular, $y = \delta$	$e = \frac{1}{ml}\tanh ml, \quad m = \sqrt{h/k\delta}$	$A = 2\delta l$
Straight fins	② Parabolic, $y = \delta_1(1 - x/l)^{\frac{1}{2}}$	$e = \frac{1}{m_1 l}\frac{I_{\frac{2}{3}}(\frac{4}{3}m_1 l)}{I_{-\frac{1}{3}}(\frac{4}{3}m_1 l)}, \quad m_1 = \sqrt{h/k\delta_1}$	$A = \frac{4}{3}\delta_1 l$
Straight fins	③ Triangular, $y = \delta_1(1 - x/l)$	$e = \frac{1}{m_1 l}\frac{I_1(2m_1 l)}{I_0(2m_1 l)}$	$A = \delta_1 l$
Straight fins	④ Parabolic, $y = \delta_1(1 - x/l)^2$	$e = \frac{2}{\sqrt{4(m_1 l)^2 + 1} + 1}$	$A = \frac{2}{3}\delta_1 l$
Circular fins	⑤ Rectangular, $y = \delta$	$e = \frac{-2}{ml(r_t/r_1 + 1)}\left[\frac{I_1(mr_1)K_1(mr_t) - I_1(mr_t)K_1(mr_1)}{I_0(mr_1)K_1(mr_t) + I_1(mr_t)K_0(mr_1)}\right]$	$A = 2\delta(r_t - r_1)$
Circular fins	⑥ Hyperbolic, $y = \delta_1(r_1/r)$	$e = \frac{2}{m_1 l(r_t/r_1 + 1)}\left[\frac{I_{\frac{2}{3}}(\frac{2}{3}m_1 r_1)I_{-\frac{2}{3}}(\frac{2}{3}m_1 r_t\sqrt{r_t/r_1}) - I_{\frac{2}{3}}(\frac{2}{3}m_1 r_t\sqrt{r_t/r_1})I_{-\frac{2}{3}}(\frac{2}{3}m_1 r_1)}{I_{\frac{1}{3}}(\frac{2}{3}m_1 r_1)I_{\frac{2}{3}}(\frac{2}{3}m_1 r_t\sqrt{r_t/r_1}) - I_{-\frac{2}{3}}(\frac{2}{3}m_1 r_t\sqrt{r_t/r_1})I_{-\frac{1}{3}}(\frac{2}{3}m_1 r_1)}\right]$	$A = 2\delta_1 r_1 \ln\frac{r_t}{r_1}$
Spines (circular cross-section)	⑦ Rectangular, $y = \delta$	$e = \frac{1}{\sqrt{2}ml}\tanh\sqrt{2}ml$	$V = \pi\delta^2 l$
Spines (circular cross-section)	⑧ Parabolic, $y = \delta_1(1 - x/l)^{\frac{1}{2}}$	$e = \frac{2}{(\frac{4}{3}\sqrt{2}m_1 l)}\frac{I_1(\frac{4}{3}\sqrt{2}m_1 l)}{I_0(\frac{4}{3}\sqrt{2}m_1 l)}$	$V = \frac{\pi}{2}\delta_1^2 l$
Spines (circular cross-section)	⑨ Triangular, $y = \delta_1(1 - x/l)$	$e = \frac{4}{(2\sqrt{2}m_1 l)}\frac{I_2(2\sqrt{2}m_1 l)}{I_1(2\sqrt{2}m_1 l)}$	$V = \frac{\pi}{3}\delta_1^2 l$
Spines (circular cross-section)	⑩ Parabolic, $y = \delta_1(1 - x/l)^2$	$e = \frac{2}{\sqrt{\frac{8}{9}(m_1 l)^2 + 1} + 1}$	$V = \frac{\pi}{5}\delta_1^2 l$

Figure 2-21 Fin efficiencies in closed form. For this figure use $e = \eta_f$. (Used with permission, from W. M. Rohsenow and J. P. Hartnett, *Handbook of Heat Transfer*, McGraw-Hill Book Company, 1973.)

insulated end, Eqs. (2-103) and (2-105) can be used to find

$$\eta_f = \frac{\tanh\sqrt{4L^2\bar{h}_c/kD}}{\sqrt{4L^2\bar{h}_c/kD}} \tag{2-106}$$

Once the fin efficiency η_f is known, the heat transfer of a fin can be easily evaluated. Figure 2-20 presents fin efficiencies versus the fin parameter $L_c^{3/2}(\bar{h}_c/kA_p)^{1/2}$ for several of the more commonly used fin configurations. Figure 2-21 gives the fin efficiencies in closed form for a rather wide variety of fins and spines. The equation representations are convenient for computer program applications.

Consider, as illustrated by Fig. 2-22, the case of a wall with fins and the following definitions:

1. A, total heat transfer area.
2. A_f, total exposed fin area.
3. η_f, fin efficiency.
4. η_t, total surface effectiveness.

Then the total heat transfer from the finned wall is

$$q_{\text{total}} = q_{\text{fin}} + q_{\text{nonfin}} \tag{2-107}$$

Using the total surface effectiveness as

$$q_{\text{total}} = A\eta_t\bar{h}_c(T_s - T_\infty) \tag{2-108}$$

then

$$A\eta_t\bar{h}_c(T_s - T_\infty) = A_f\eta_f\bar{h}_c(T_s - T_\infty) + (A - A_f)\bar{h}_c(T_s - T_\infty) \tag{2-109}$$

assuming the fin and wall $\bar{h}_c$ to be equal. Solving Eq. (2-109) yields

$$\eta_t = 1 - \frac{A_f}{A}(1 - \eta_f) \tag{2-110}$$

and, rewriting Eq. (2-108),

$$q = \frac{T_s - T_\infty}{1/(A\eta_t\bar{h}_c)} \tag{2-111}$$

Thus the resistance for a finned wall can be easily expressed as

$$R = \frac{1}{A\eta_t\bar{h}_c} \tag{2-112}$$

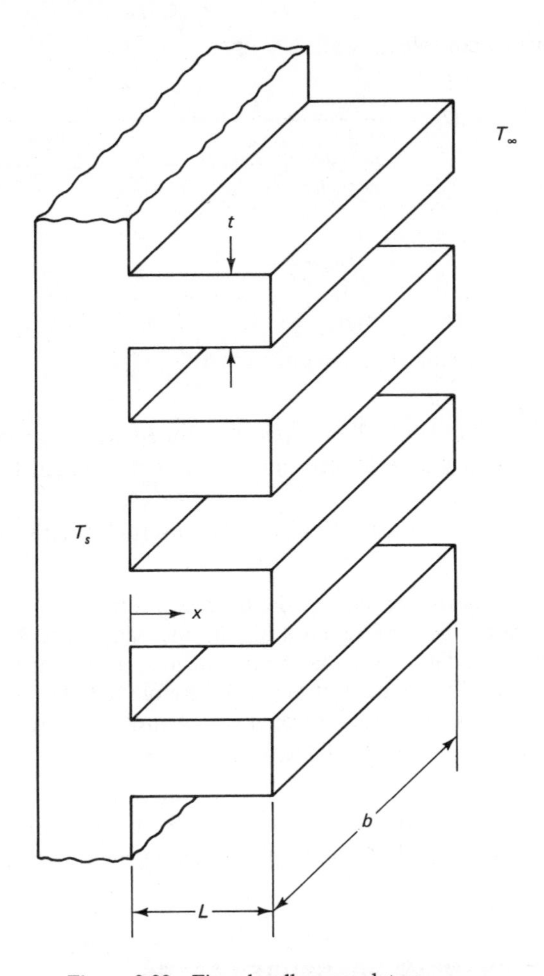

Figure 2-22 Finned wall nomenclature.

The results of the previous paragraph in conjunction with the previously developed composite wall expressions yield the following:

1. Planar composite wall:

$$UA = \frac{1}{\dfrac{1}{\eta_{t_i} A_i \bar{h}_i} + \dfrac{R_{f_i}}{A_i} + \sum_{j=1}^{J} \dfrac{l_j}{A_j k_j} + \dfrac{R_{f_o}}{A_o} + \dfrac{1}{\eta_{t_o} A_o \bar{h}_o}} \tag{2-113}$$

where J is the number of composite layers and A_i and A_o are the total inside and outside areas.

2. Cylindrical composite wall:

$$U_{\text{ref}} = \frac{1}{\dfrac{A_{\text{ref}}}{A_i \eta_{t_i} \bar{h}_i} + \dfrac{A_{\text{ref}}}{A_i} R_{f_i} + \sum_{j=1}^{J} A_{\text{ref}} \dfrac{\ln r_{j+1}/r_j}{2\pi kj} + \dfrac{A_{\text{ref}}}{A_o} R_{f_o} + \dfrac{A_{\text{ref}}}{A_o \eta_{t_o} \bar{h}_o}} \tag{2-114}$$

where J, A_o, and A_i are as in item 1.

The use of Eq. (2-113) or (2-114) permits finned wall heat exchanger calculations to proceed in a fashion not much more complicated than for nonfinned walls.

It is reasonable to ask, When are fins useful? Kreith discusses that with

> For a plane surface of area A, the thermal resistance is $1/\bar{h}A$. The addition of fins increases the surface area but at the same time it also introduces a conductive resistance over that portion of the original surface at which the fins are attached. The addition of fins will therefore not always increase the rate of heat transfer. In practice, the addition of fins is hardly ever justified unless $\bar{h}A/Pk$ is considerably less than unity.

Thus, given the choice of whether to place the fins on the high $\bar{h}_c$ side or on the low $\bar{h}_c$ side, the choice is to place fins on the low $\bar{h}_c$ side.

2-4 HEAT EXCHANGER PRESSURE LOSSES

The preceding topics in this chapter have examined various aspects of heat exchangers from a thermal viewpoint; nothing has been said about the equally important problem of fluid dynamic losses. The losses in pressure experienced by each fluid's path through the heat exchanger determine the pumping power required. The energy required to overcome the fluid pressure losses represents an "inefficiency" for the heat exchanger and, if large enough, can abrogate the requirement for a heat exchanger. Heat exchangers used for waste heat recovery, for example, can cost more energy than they recover if the pumps' energy input is larger than the recovered energy. Because of the importance of the pressure drop characteristics of various candidate heat exchanger cores, friction factor as well as Nusselt number data are usually taken in heat exchanger test rigs and are usually presented as design data.

The most commonly used expression for a shell-and-tube core pressure drop is that of Kays and London (3).

$$\frac{\Delta P}{P_1} = \frac{G^2}{2g_c}\frac{v_1}{P_1}\left[\underbrace{(K_c + 1 - \sigma^2)}_{\text{entrance effect}} + \underbrace{2\left(\frac{v_2}{v_1} - 1\right)}_{\text{flow acc.}} + \underbrace{f_F\frac{A}{A_c}\frac{v_m}{v_1}}_{\text{core friction}} - \underbrace{(1 - \sigma^2 - K_e)\frac{v_2}{v_1}}_{\text{exit effect}}\right] \tag{2-115}$$

where $G = \rho_1 V_1$, mass velocity based on free-flow area
v_1 = entering specific volume
v_2 = exiting specific volume
v_m = mean specific volume
P_1 = inlet pressure
ΔP = core pressure drop
K_c = entrance loss coefficient
K_e = exit loss coefficient
σ = free-flow area to frontal area ratio
A = total heat transfer area
A_c = free-flow area
f = core Fanning friction factor

Equation (2-115) is valid for situations as shown in Fig. 2-23, a core with a sudden contraction from the frontal area A_{fr} to the free-flow area A_c and a sudden expansion from the free-flow area A_c to the exit area A_{fr}, with a total surface area A of *interior* heat transfer surface. For flows normal to tube

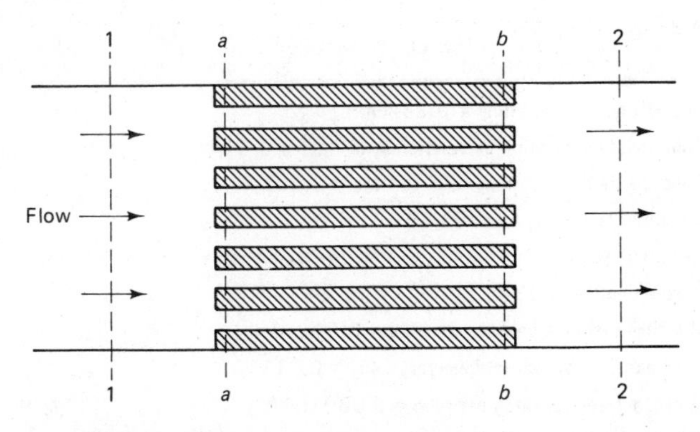

Figure 2-23 Heat exchanger pressure drop schematic. (Used with permission, from W. M. Kays and A. L. London, *Compact Heat Exchangers*, 2nd ed., McGraw-Hill Book Company, 1964.)

banks, the entrance and exit losses are accounted for in the friction factor and the appropriate equation is

$$\frac{\Delta P}{P_1} = \frac{G^2}{2g_c}\frac{v_1}{P_1}\Bigg[\underbrace{(1+\sigma^2)\left(\frac{v_2}{v_1}-1\right)}_{\text{flow acceleration}} + \underbrace{f_F\frac{A}{A_c}\frac{v_m}{v_1}}_{\text{core friction}}\Bigg] \tag{2-116}$$

Equations (2-115) and (2-116) also show why pressure drop for a given mass flow rate of a gas is generally much larger than the pressure drop for the same mass flow rate of a liquid. The entering specific volume v_1 appears in

Flow inside dimpled flattened tubes, surface FTD-1.

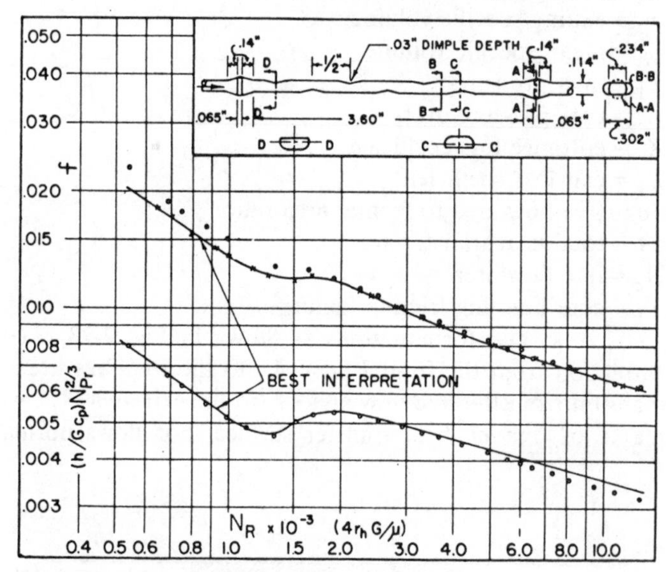

Tube ID before flattening = 0.234 in.

Tube inside dimension perpendicular to flats = 0.114 in.

Tube inside dimension parallel to flats = 0.302 in.

Length of flat along tube = 3.60 in.

Length of section from flat-to-circular-to-flat cross section = 0.345 in.

Dimple depth = 0.03 in.

Minimum distance between dimples = 0.5 in.

Flow passage hydraulic diameter, $4r_h = 0.01116$ ft

Minimum free-flow area per tube = 0.000169 ft^2

Figure 2-24 Example of Kays and London's data presentation. In this figure $N_R = \text{Re}_{D_H}$, $N_{Pr} = \text{Pr}$, $h = h_c$, and $f = f_F$, Fanning friction factor. (Used with permission, from W. M. Kays and A. L. London, *Compact Heat Exchangers*, 2nd ed., McGraw-Hill Book Company, 1964.)

the numerator of the leading term. Thus in liquid-to-liquid heat exchangers the friction pressure losses are relatively unimportant, but in heat exchanger loops with gases the higher specific volume of the gas greatly multiplies the losses.

The correct mean specific volume to use is

$$v_m = \frac{1}{A}\int_0^A v\,dA \tag{2-117}$$

For nonparallel flow exchangers with $C_{\min}/C_{\max}$ near unity, Kays and London suggest

$$v_m = \tfrac{1}{2}(v_1 + v_2) \tag{2-118}$$

If one temperature T_{const} is essentially uniform as in a condenser, evaporator, or intercooler, then they suggest

$$\frac{v_m}{v_1} = \frac{P_1}{P_{\text{avg}}}\frac{T_{lma}}{T_1} \tag{2-119}$$

where

$$T_{lma} = T_{\text{const}} \pm \Delta T_{lma} \tag{2-120}$$

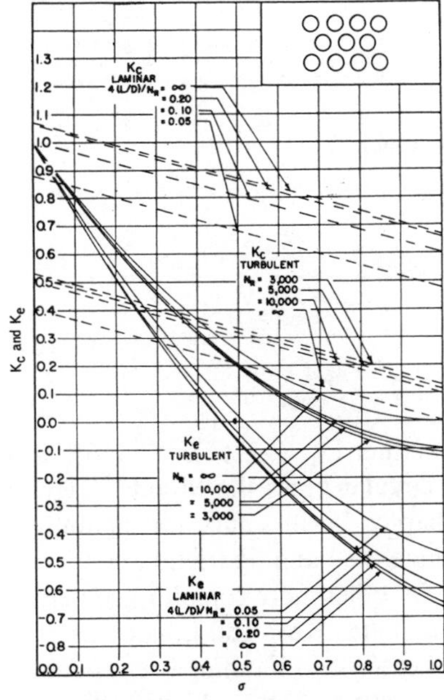

(a) Entrance and exit pressure loss coefficients for a multiple-circular-tube heat exchanger core with abrupt-contraction entrance and abrupt-expansion exit.

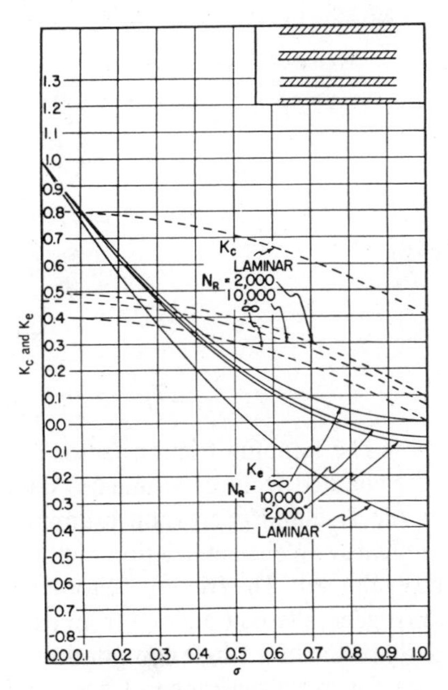

(b) Entrance and exit pressure loss coefficients for a multiple-tube flat-duct heat exchanger core with abrupt-contraction entrance and abrupt-expansion exit.

Figure 2-25 Recommended values of K_c and K_e for different geometries. (Used with permission, from W. M. Kays and A. L. London, *Compact Heat Exchangers*, 2nd ed., McGraw-Hill Book Company, 1964.)

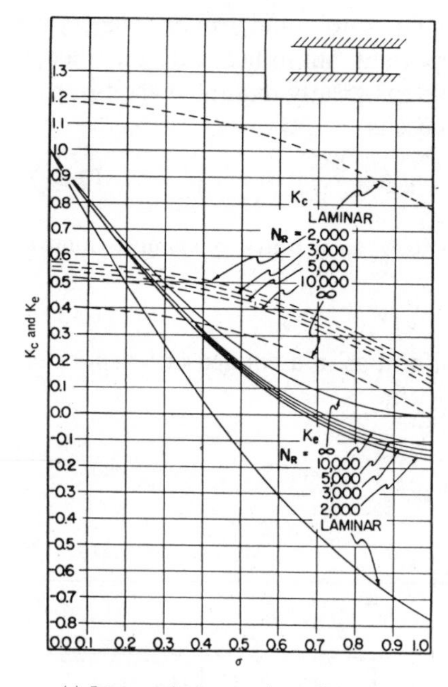

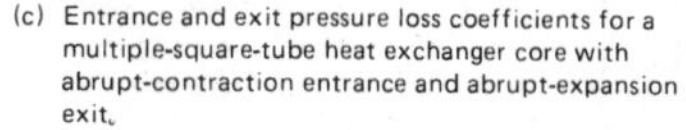

(c) Entrance and exit pressure loss coefficients for a multiple-square-tube heat exchanger core with abrupt-contraction entrance and abrupt-expansion exit.

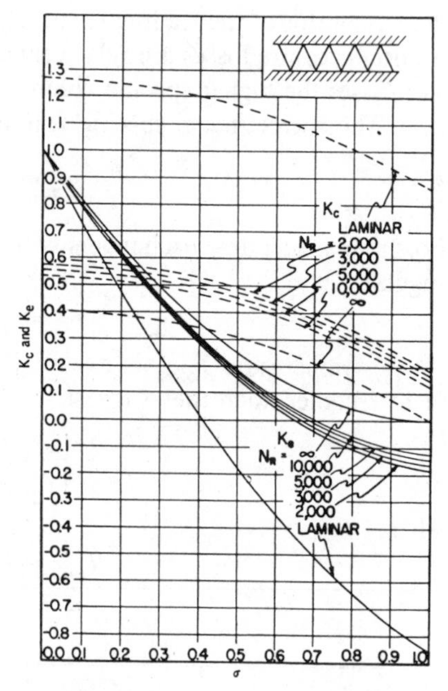

(d) Entrance and exit pressure loss coefficients for a multiple-triangular-tube heat exchanger core with abrupt-contraction entrance and abrupt-expansion exit.

Figure 2-25 (continued)

with

$$\Delta T_{lma} = \frac{T_{h_{\text{in}}} - T_{h_{\text{out}}}}{\text{NTU}} \qquad T_c = T_{\text{const}}$$
$$\Delta T_{lma} = \frac{T_{c_{\text{out}}} - T_{c_{\text{in}}}}{\text{NTU}} \qquad T_h = T_{\text{const}} \tag{2-121}$$

The Fanning friction factor for use in either Eq. (2-115) or (2-116) must be evaluated from experimental data. For circular pipes the Moody diagram or the Haaland expression examined in Chapter 1 will suffice for computing the Darcy–Weisbach friction factor from which the Fanning friction factor can be evaluated. The friction factor evaluation for other more complex geometries must generally be obtained from experimental data. Kays and London present perhaps the best single compilation of heat transfer and Fanning friction factor data for a diverse range of core surface geometries. Figure 2-24 illustrates a typical data presentation from Kays and London. This data is for flow inside a dimpled flattened tube, a geometry for which we would be at a loss to make any estimates concerning friction loss or heat transfer characteristics.

The entrance and exit loss coefficients, K_c and K_e, can be estimated using the results of Chapter 1 or taken from Kays and London's recommendation as reproduced in Fig. 2-25. Typically, we would not expect the entrance and exit losses to be dominant.

Equation (2-115) or (2-116), in conjunction with appropriate experimental data or theoretical calculations, is sufficient for estimating the fluid dynamics losses in either fluid's loop of a heat exchanger.

REFERENCES

1. Kreith, F., *Principles of Heat Transfer*, 3rd ed. New York: Harper & Row, 1973.
2. Incropera, F. P., and DeWitt, D. P., *Fundamentals of Heat Transfer*. New York: Wiley, 1981.
3. Kays, W. M., and London, A. L., *Compact Heat Exchangers*, 2nd ed. New York: McGraw-Hill, 1964.
4. Wolf, H., *Heat Transfer*. New York: Harper & Row, 1983.
5. *Standards of the Tubular Exchanger Manufacturers Association*, 6th ed. New York: TEMA, 1978.
6. Pitts, D. R., and Sissom, L. E., *Schaum's Outline Series*: *Heat Transfer*, New York: McGraw-Hill, 1977.
7. Suryanarayana, N. V., "Some Remarks on the Use of Convective Heat Transfer Coefficients and Integral Methods for Natural Convection on Vertical Surfaces," *Mechanical Engineering News*, Vol. 20, No. 2, May 1983, pp. 8–10.
8. Kays, W. M., and Crawford, M., *Convective Heat and Mass Transfer*, 2nd ed. New York: McGraw-Hill, 1980.
9. Rohsenow, W. M., and Hartnett, J. P., *Handbook of Heat Transfer*. New York: McGraw-Hill, 1973.
10. Shah, R. K., "Compact Heat Exchangers" and "Compact Heat Exchanger Design Procedures," in Kakac, Bergles, and Mayinger, eds. *Thermal-Hydraulic Fundamentals and Design*. New York: Hemisphere, 1981.
11. Burmeister, L. C., *Convective Heat Transfer*. New York: Wiley-Interscience, 1983.
12. Rohsenow, W. M., and Choi, H. Y., *Heat, Mass, and Momentum Transfer*. Englewood Cliffs, N.J.: Prentice-Hall, 1961.
13. Griffith, P., "Dropwise Condensation," Chapter 12B, *Handbook of Heat Transfer*, Rohsenow and Harnett, eds. New York: McGraw-Hill, 1973.

REVIEW QUESTIONS

1. Why is the effectiveness, ξ, of a heat exchanger a good measure of the amount of energy that could be exchanged from two streams at different temperatures?
2. Are fins constructed from a low k or a high k material more effective? Explain.
3. Are fins more effective when added to a high $\bar{h}_c$ or a low $\bar{h}_c$ wall? Explain.
4. Is a parallel or counterflow heat exchanger more effective for a given area? Why?
5. What is the difference in *accuracy* between the LMTD method and the NTU method?

6. C_{min}/C_{max} for a condenser is ______________________

7. In the composite wall expression

$$U = \frac{1}{\underbrace{0.0015}_{\substack{\text{inner}\\\text{convective}}} + \underbrace{0.0001}_{\text{conductive}} + \underbrace{0.15}_{\substack{\text{outer}\\\text{convective}}}} \frac{\text{BTU}}{\text{h ft}^2\ {}^\circ\text{F}}$$

which resistance is controlling? Why?

8. In a counterflow heat exchanger of very large area, the inlet temperatures are $T_{h_{in}} = 500°R$ and $T_{c_{in}} = 100°R$. If $\dot{m}_h c_{p_h} < \dot{m}_c c_{p_c}$, what is the outlet temperature?

9. Why is the NTU method easier to use than the LMTD method for an *analysis* problem?

10. For a bank of aligned tubes in a cross flow with $Re_D = 1 \times 10^5$ (based upon V_{max}) and $Pr = Pr_\infty = Pr_s = 1.0$, what is the Nusselt number?

11. What is burnout?

PROBLEMS

1. A heat exchanger tube constructed of copper has an OD = 1.256 in. and an ID = 0.975 in. If $\bar{h}_o = 2000$ Btu/h ft^2 °F and $\bar{h}_i = 5$ Btu/hr ft^2 °F, what is the dominant mechanism controlling the heat transfer?

2. The NTU-effectiveness for a single-pass counterflow heat exchanger is given by

$$\xi = \frac{1 - \exp[-\text{NTU}(1 - C)]}{1 - C\exp[-\text{NTU}(1 - C)]}$$

$$C = C_{min}/C_{max}$$

and is indeterminate when $C = 1$. Using l'Hôpital's rule, verify the results given in Figure 2.4b for $C_{min}/C_{max} = 1$.

3. Consider a cross-flow heat recovery unit comprised of a number of tubes that have engine exhaust flowing normal to the tubes in mixed flow and water inside the tubes in unmixed flow. The hot exhaust gas enters at 1200°F at a flow rate of 2000 lbm/h, and the water enters at 40°F at 1800 lbm/h. Under these conditions the exchanger has an overall heat transfer coefficient of 55.7 Btu/h ft^2 °F for the tube area of 27.5 ft^2. At what pressure must the water be kept to prevent boiling in the tubes? What is the rate of heat transfer? What are the exit temperatures?

4. An oil is being cooled by water in a double-pipe parallel-flow heat exchanger. The water enters the center pipe at a temperature of 50°F and is heated to 120°F. The oil that flows in the annulus is cooled from 270° to 150°F. It is proposed to cool the oil to a lower final temperature by increasing the length of the exchanger. Neglecting external heat loss from the exchanger, determine (a) the minimum temperature to which the oil may be cooled; (b) the exit-oil temperature as a function of the fractional increase in the exchanger length; (c) the exit temperature of each stream if the existing exchanger were switched to counterflow operation;

(d) the lowest temperature to which the oil could be cooled with counterflow operation.

5. In a single-pass counterflow heat exchanger, 10,000 lbm/h of water ($C = 1.0$ Btu/lbm °F) enters at 60°F and cools 20,000 lbm/h of oil ($C = 0.5$ Btu/lbm °F), which enters at 200°F. For the exchanger $UA = 5500$ Btu/h °F. Determine the fluid outlet temperatures.

6. Compressed air is used in a heat-pump system to heat water that is subsequently used to warm a house. The house demand is 95,000 Btu/h. Air enters the exchanger at 200°F and leaves at 120°F; water enters and leaves the exchanger at 90°F and 125°F, respectively. Choose from the following alternatives the unit that is most compact.

(a) A counterflow surface with $U = 30$ Btu/h ft^2 °F and a surface-to-volume ratio of 130 ft^2/ft^3.

(b) A cross-flow configuration with the water unmixed and air mixed having $U = 40$ Btu/h ft^2 °F and a surface-to-volume ratio of 100 ft^2/ft^3.

(c) A cross-flow unit with both fluids unmixed with $U = 50$ Btu/h ft^2 °F and surface-to-volume ratio of 90 ft^2/ft^3.

7. Water enters a counterflow, double-pipe heat exchanger at a rate of 150 lbm/min and is heated from 60° to 140°F by an oil with a specific heat of 0.45 Btu/lbm °F. The oil enters at 240°F and leaves at 80°F. The overall heat transfer coefficient is 50 Btu/h ft^2 °F.

(a) What heat transfer area is required?

(b) What area is required if all conditions remain the same except that a shell-and-tube heat exchanger is used, with the water making one shell pass and the oil making two tube passes?

(c) What exit water temperature would result if, for the exchanger of part (a), the water flow rate were decreased to 120 lbm/min?

8. Water at 300°F and 8 ft/s enters a 50-ft-long schedule 40 1-in. cast-iron pipe. If air at 32°F flows perpendicular to the pipe at 40 ft/s, determine (a) the outlet temperature of the water, (b) the outlet water temperature if the air velocity is increased to 160 ft/s, (c) the water outlet temperature if the air velocity is 40 ft/s and the water velocity is 40 ft/s. (d) Discuss the implications of (a), (b), and (c).

9. A double-pipe heat exchanger with the dimensions shown in Fig. P2-9 is used to heat water ($T_{in} = 50$°F). Steam at 250°F is condensing on the outside with an

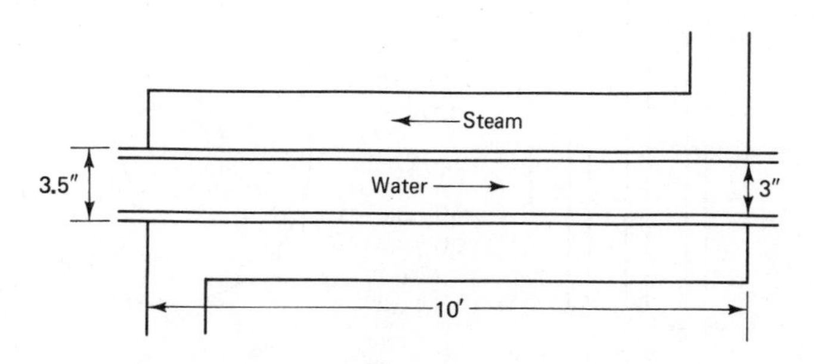

Figure P2-9

$\bar{h}_o = 2500$ Btu/ft^2 h °F. Sufficient steam is available. If the water flow rate is 3 ft/s, find the outlet temperature of the water. Neglect the thermal resistance of the pipe.

10. Heat is to be transferred from an unknown fluid to air through an aluminum wall. It is proposed to add rectangular fins 0.05 in. thick and 0.25 in. long spaced 0.08 in. apart to the surface to increase the heat transfer. The wall is 0.125 in. thick and the air side convective heat transfer coefficient is 8 Btu/h ft^2 °F.

(a) If the fins are made from aluminum and the unknown fluid's $\bar{h}_c$ is 100 Btu/h ft^2 °F, then:

(1) Find the percent of increase in q if the fins are added to the air side.

(2) Find the percent of increase in q if the fins are added to the other fluid's side.

(3) Find the increase in heat transfer if fins are added to both sides.

(b) Rework part (a) assuming the fins are made of copper.

(c) Rework part (a) if $\bar{h}_c$ of both sides are changed to 10,000 Btu/h ft^2 °F.

(d) Discuss the implications of the answers from parts (a), (b), and (c).

11. A pipe with circumferential fins has the dimensions given in Fig. P2-11. On the outside a gas with the following properties is flowing:

$\bar{h}_c = 5$ Btu/ft^2 h °F $k = 0.0140$ Btu/h ft °F
$\rho = 0.81$ lbm/ft^3 $\Pr = 0.72$
$C_p = 0.24$ Btu/lbm °F $T = 32$°F
$\mu = 1.165 \times 10^{-5}$ lbm/ft s

On the inside a liquid has the following properties:

$\bar{V}$(average velocity) $= 5$ ft/s $k_m = 0.394$ Btu/h ft °F
$\rho = 60.1$ lbm/ft^3 $\Pr_m = 1.89$
$C_{p_m} = 1.0$ Btu/lbm °F $T_{in} = 200$°F
$\mu_m = 0.205 \times 10^{-3}$ lbm/ft s Pipe and fins are copper.
$\nu_m = 0.341 \times 10^{-5}$ ft^2/s $k = 226$ Btu/h ft °F

Compute the pipe length required to cool the liquid to 180°F.

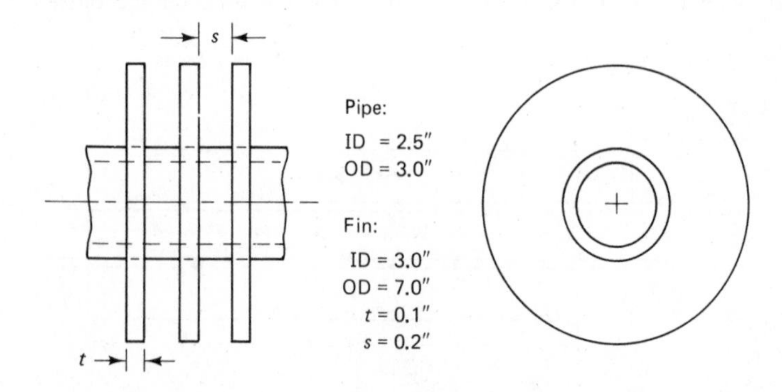

Figure P2-11

12. Heat is to be transferred from water to air through an aluminum wall. It is proposed to add rectangular fins 0.05 in. thick and $\frac{1}{4}$ in. long spaced 0.08 in. apart to the aluminum surface to aid in transferring heat. The heat transfer coefficients on the air and water sides are 3 Btu/h ft^2 °F and 25 Btu/h ft^2 °F, respectively. Evaluate the percent of increase in heat transfer if these fins are added to (a) the airside, (b) the waterside, or (c) both sides. What conclusions may be reached regarding this result?

13. **(a)** A 1 in.-OD steel tube has its outside wall surface maintained at 250°F. It is proposed to increase the rate of heat transfer by adding fins of $\frac{1}{12}$ in. thickness and $\frac{1}{4}$ in. long to the outside tube surface. Compare the increase in heat transfer achieved by adding 12 longitudinal straight fins or circular fins with the same total surface area as the 12 longitudinal fins. The surrounding air is at 80°F and the convective heat transfer coefficient is 6 Btu/h ft^2 °F.

(b) Solve the previous problem if the convective heat transfer coefficient is increased to 60 Btu/h ft^2 °F by forcing air past the tube surface.

14. Saturated steam at 10 psia condenses on a 2-ft-tall vertical plate. The 175°F surface temperature can be considered uniform. Compute the average convective heat transfer coefficient, the condensate thickness, and the condensation rate.

3

HEAT EXCHANGERS II

3-1 INTRODUCTION

In Chapter 2 we developed all the analysis techniques that we shall require for the analysis or design of shell-and-tube and cross-flow heat exchangers. Included in that chapter are sufficient heat transfer correlations and fundamentals to allow us to model a wide range of physically realistic conditions.

This chapter uses these fundamental concepts to develop and explain methodologies required for the analysis and design of a wide variety of heat exchangers. Extensive use is made of the existing data base for various heat transfer studies, and at least some consideration is given to various codes and standards. At the conclusion of this chapter, we should have a rather comprehensive understanding of the thermohydraulic aspects of heat exchangers. As with the previous chapters, the material presented within this chapter is not original. It has been assembled from a number of sources, which are cited in the References.

3-2 TEMPERATURE-DEPENDENT FLUID PROPERTIES

The operating conditions for many heat exchangers are such that the fluids involved undergo *large* changes in temperature. Because of the nonnegligible dependence of many of the thermophysical properties on temperature, considerable errors can be present in calculations made assuming no such dependences. Basically, two approaches are used to account for these effects: (1)

evaluation of certain or all properties at some specific *reference temperature* involving the wall (T_s) and the mean fluid temperature (T_m), or (2) evaluation of all properties at the *mean fluid temperature* with a correction applied to account for wall and mean temperature property differences. Generally, *external* flow convective relations use method 1 while *internal* flow convective relations use method 2.

The need for proper evaluation of fluid properties was noted in Chapter 2 [Eqs. (2-55) and (2-56)]; indeed, the appropriate temperature for property evaluation was duly noted in each expression or table. This section constitutes both a reminder of the importance of property evaluation and a compact source of the recommended heat exchanger design procedures.

Problems arise when attempting to account for thermophysical property variations in data taken from experimental setups in which the temperature differences, the driving forces for heat transfer, and, hence, the physical property variations are small. Much heat transfer data are acquired in this manner. The approach presented is from Kays and London (1) and differs slightly from the recommendations of Shah (2).

Most heat transfer data are presented as St $\mathrm{Pr}^{2/3}$ and f as a function of Reynolds number. These heat transfer data are reduced to the form St $\mathrm{Pr}^{2/3}$

TABLE 3-1 Exponents for Property Corrections, Gas

	n	m	Basis
Laminar boundary layer on flat plate:			
Gas heating	−0.08	−0.08	Analytic solution
Gas cooling	−0.045	−0.045	Analytic solution
Flow normal to circular tube or bank of circular tubes:			
Gas heating	0.0	0.0	Heat transfer experiments
Gas cooling	0.0	0.0	Heat transfer experiments
Fully developed laminar flow in circular tube:			
Gas heating	0.0	—	Experiments
		+0.45[a]	Analytic solution
Gas cooling	0.0	—	Experiments
		+0.30[a]	Analytic solution
Fully developed turbulent flow in circular tube:			
Gas heating	−0.5	—	Experiments, analysis
		−0.1	Experiments, analysis
Gas cooling	0.0	—	Experiments
		0.0	Analytic solution

[a]These may have a large error band.

Used with permission, from W. M. Kays and A. L. London, *Compact Heat Exchangers*, 2nd ed., McGraw-Hill Book Company, 1964.

by assuming constant properties. For *gas flows*, the temperature ratio, T_s/T_m, where T_s is the surface temperature and T_m the mean fluid temperature, provides the basis for accounting for property variations. The variable property quantities Nu and f are related to Nu_m and f_m, which are evaluated at T_m, by

$$\frac{\overline{\text{Nu}}}{\overline{\text{Nu}}_m} = \left(\frac{T_s}{T_m}\right)^n \tag{3-1}$$

and

$$\frac{f}{f_m} = \left(\frac{T_s}{T_m}\right)^m \tag{3-2}$$

where Table 3-1 provides values of m and n for various situations. In a similar fashion, the appropriate variable property corrections for *liquids* are

$$\frac{\overline{\text{Nu}}}{\overline{\text{Nu}}_m} = \left(\frac{\mu_s}{\mu_m}\right)^n \tag{3-3}$$

and

$$\frac{f}{f_m} = \left(\frac{\mu_s}{\mu_m}\right)^m \tag{3-4}$$

where Table 3-2 presents values of the exponents for various situations.

We need to be cognizant of the property evaluation stipulation of each friction or convective heat transfer correlation that we use. Generally, external flow correlations that are evaluated at the film temperature

$$T_f = \tfrac{1}{2}(T_s + T_\infty) \tag{2-55}$$

TABLE 3-2 Exponents for Property Corrections, Liquids

	n		m	
Pr	Heating	Cooling	Heating	Cooling
Laminar	−0.11	−0.11	0.58	0.50
1	−0.20	−0.19	0.09	0.12
3	−0.27	−0.21	0.06	0.09
10	−0.36	−0.22	0.03	0.05
30	−0.39	−0.21	0.00	0.03
100	−0.42	−0.20	−0.04	0.01
1,000	−0.46	−0.20	−0.12	−0.02

Used with permission, from W. M. Kays and A. L. London, *Compact Heat Exchangers*, 2nd ed., McGraw-Hill Book Company, 1964.

do not require the use of corrections such as Eqs. (3-1) to (3-4). The property variations that "feed" in through the film temperature are adequate to account for variable property effects. The entries in Table 2-4 are good examples of correlations for which the film temperature concept is acceptable.

Many of the *internal* heat transfer correlations of Chapter 2 are evaluated at the mean temperature T_m. Where the surface temperature results in thermophysical properties markedly different from those at T_m, the correction of Eqs. (3-1) to (3-4) are recommended. For example, the Colburn equation, Eq. (2-69), and the Dittus–Boelter equation, Eq. (2-70), should have their properties evaluated at T_m. Both of these equations are valid only for moderate temperature differences. For extensions to large temperature differences, Eqs. (3-1) and (3-2) or Eqs. (3-3) and (3-4) should be used. Equation (2-71) was thus modified from the Colburn equation by the addition of Eq. (3-3).

3-3 SHELL-AND-TUBE HEAT EXCHANGERS

We have now developed sufficient background material to consider engineering aspects of heat exchangers in some detail. Because of the differences between shell-and-tube and cross-flow heat exchangers, each class will be examined in turn. The shell-and-tube category will be covered first. These have traditionally been extensively used in the petrochemical, power, and process industries and are manufactured in a wide range of ratings and sizes.

TEMA Standards

The most compact available document discussing the effects of various standards on the design of heat exchangers resulted from a 1979 ASME session entitled "The Interrelationships between Codes, Standards, and Customer Specifications for Process Heat Transfer Equipment" (3). The session chairman and proceedings editor, F. L. Rubin, limited material to shell-and-tube heat exchangers; therefore, that content is particularly appropriate for inclusion in this section. Of the two usually cited standards for shell-and-tube exchangers, the TEMA (Tubular Exchanger Manufacturers Association) *Standards* (4) and the *API* (American Petroleum Institute) *Standard 660*, the TEMA standards enjoy the most widespread acceptance. The only code applicable (in the United States) to heat exchangers is Section VII, Division One of the ASME Boiler and Pressure Vessel Code—Unfired Pressure Vessels. The ASME code is generally intended to apply only to the pressure-containing envelope of a shell-and-tube heat exchanger.

We shall examine in some detail the TEMA standards. Copyright laws and length make it impractical to reproduce the entire TEMA standards, but some pertinent requirements will be examined. The standards deal with essentially four areas:

1. Heat exchanger nomenclature
2. Fabrication tolerances
3. Mechanical standards
4. Recommended good practices

A brief survey of heat exchanger nomenclature is needed prior to any heavy involvement with the standards. Heat exchanger sizes and types are designated by *numbers* describing shell diameters and tube lengths and by *letters* describing the stationary head, shell, and rear head. The TEMA standards specify that the nominal diameter is the inside diameter, rounded off to the nearest inch. For kettle reboilers, the nominal diameter is specified to be the port diameter followed by the shell diameter, both rounded off to the nearest inch. The nominal tube length is taken as the actual overall length for straight tubes and as the straight length from the end of the tube to the bend tangent for U-tube arrangements. The type designation is specified by letters describing the head, shell, and rear head, in that order, as indicated by Fig. 3-1.

TEMA general nomenclature for heat exchanger items and components is given in Table 3-3.

TABLE 3-3 TEMA Component Nomenclature

1. Stationary head—channel	20. Slip-on backing flange
2. Stationary head—bonnet	21. Floating head cover—external
3. Stationary head flange—channel or bonnet	22. Floating tubesheet skirt
4. Channel cover	23. Packing box flange
5. Stationary head nozzle	24. Packing
6. Stationary tubesheet	25. Packing follower ring
7. Tubes	26. Lantern ring
8. Shell	27. Tie rods and spacers
9. Shell cover	28. Transverse baffles or support plates
10. Shell flange—stationary head end	29. Impingement baffle
11. Shell flange—rear head end	30. Longitudinal baffle
12. Shell nozzle	31. Pass partition
13. Shell cover flange	32. Vent connection
14. Expansion joint	33. Drain connection
15. Floating tubesheet	34. Instrument connection
16. Floating head cover	35. Support saddle
17. Floating head flange	36. Lifting lug
18. Floating head backing device	37. Support bracket
19. Split shear ring	38. Weir
	39. Liquid level connection

Used with permission, from the *TEMA Standards*, TEMA, 1978.

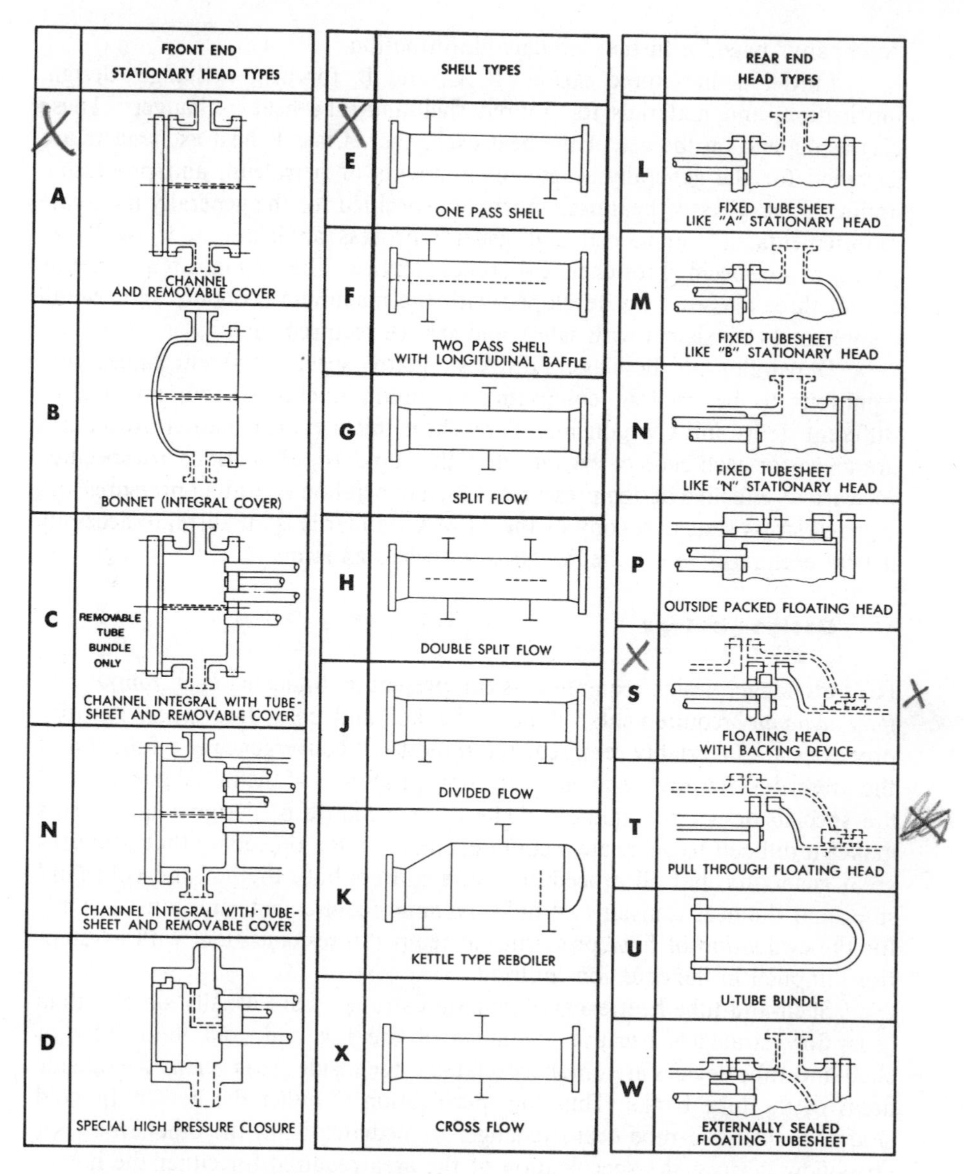

Figure 3-1 Letter type designations for TEMA nomenclature. (Used with permission, from *TEMA Standards*, TEMA, 1978.)

An example, taken from the TEMA standards, of size and type designations, as well as identification of components, is presented in Fig. 3-2. Other combinations are possible depending on the specifications and requirements.

A standard *heat exchanger specification sheet* has also been developed by TEMA. This sheet, shown in Fig. 3-3, is widely used throughout the shell-and-tube manufacturing and using industries. Many firms utilize a second "company" page for further tabulated information.

TEMA defines three *classes*, R, C, and B, for heat exchanger design, fabrication, and materials for unfired shell-and-tube heat exchangers. These classes depend on the use of the heat exchanger. Class R heat exchangers are specified for the generally severe requirements of petroleum and processing applications. Class C heat exchangers are specified for the generally moderate requirements of commercial and general process applications. Class B exchangers are specified for chemical process service. The standard requirements for all three classes were developed with the maximum economy and overall compactness consistent with safety and service requirements.

Depending on the class, TEMA standards specify different values (and equations to be used in computing minimum thickness requirements) for different items and components. Tube sheet thickness calculation procedures are given for each class of exchanger in the standards. Equations are specified for various effective design pressures, which depend on operating pressures and thermal stress states. A copy of the TEMA standards is an absolute necessity if heat exchangers are to be designed to these standards.

Design Strategy

Heat exchanger design requirements can present multifaceted *heat transfer* and *fluid dynamic* requirements on both the hot and cold fluid sides. Iterative processes are invariably required; the rapidity of convergence is a function of the prescribed iterative sequence, the appropriateness of the initial guesses, and the specific design requirements. The diverse nature of design requirements makes it difficult to delineate a comprehensive design procedure that possesses great generality, but all procedures must address both the hot and cold fluid sides and the heat transfer and fluid dynamic requirements and must provide for the evaluation of fluid properties at temperatures consistent with assumptions implicit in the equations utilized.

Shell-and-tube heat exchanger design strategies are usually simpler than cross-flow strategies because the areas of the hot and cold fluid sides of shell-and-tube devices are uniquely related—the inside areas versus the outside areas of the tube banks. Thus the specification of either the hot or the cold side of a shell-and-tube heat exchanger immediately fixes the other side. For cross-flow devices, the specification of the area required for either the hot or the cold fluid side does not specify the remaining side's geometry.

Heat exchanger specifications are normally centered about the *rating* and the allowable *pressure drop*. The rating refers to the capacity of the device and is usually given in Btu/h (or kilowatts). The pressure drop can be a design constraint on either or both the hot and cold fluids. Any successful design strategy must result in the satisfaction of both the rating and the pressure drop constraints. Two categories of procedures exist: (1) enforce the rating requirement and vary the geometry until the pressure drop constraint is satisfied, and (2) enforce the pressure drop requirement and vary the geometry until the rating is achieved.

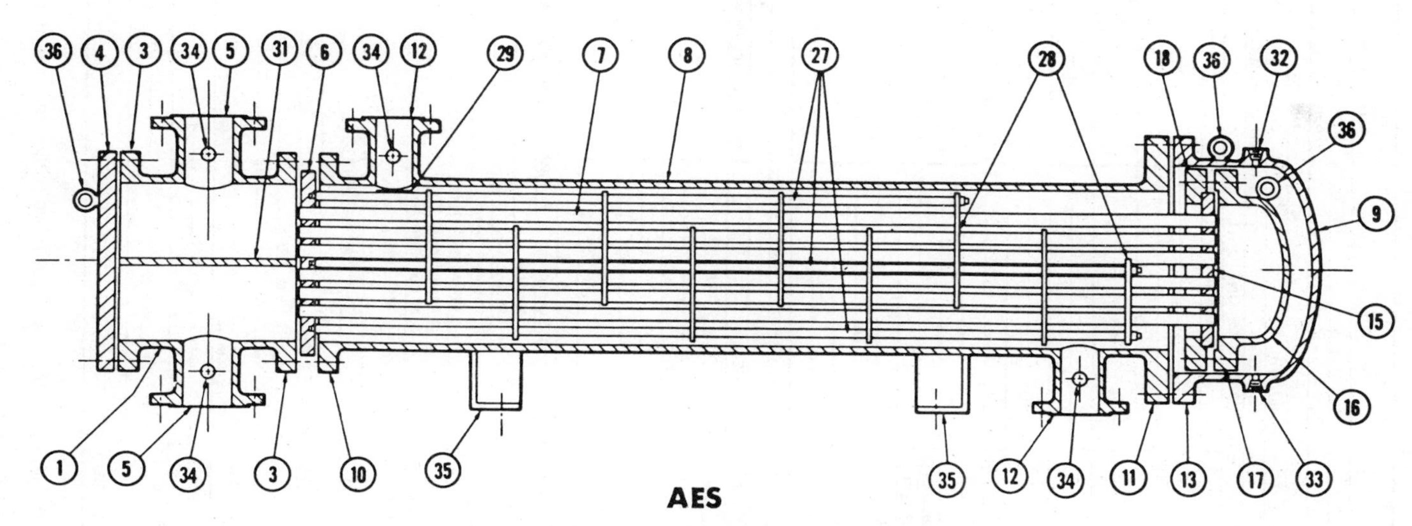

Figure 3-2 TEMA designation example. (Used with permission, from *TEMA Standards*, TEMA, 1978.)

1					Job No.	
2	Customer				Reference No.	
3	Address				Proposal No.	
4	Plant Location				Date	Rev.
5	Service of Unit				Item No.	
6	Size	Type	(Hor/Vert)		Connected In	Parallel Series
7	Surf/Unit (Gross/Eff.)	Sq Ft; Shells/Unit			Surf/Shell (Gross/Eff.)	Sq Ft
8	PERFORMANCE OF ONE UNIT					
9	Fluid Allocation		Shell Side		Tube Side	
10	Fluid Name					
11	Fluid Quantity, Total	Lb/Hr				
12	Vapor (In/Out)					
13	Liquid					
14	Steam					
15	Water					
16	Noncondensable					
17	Temperature (In/Out)	°F				
18	Specific Gravity					
19	Viscosity, Liquid	Cp				
20	Molecular Weight, Vapor					
21	Molecular Weight, Noncondensable					
22	Specific Heat	Btu/Lb °F				
23	Thermal Conductivity	Btu Ft/Hr Sq Ft °F				
24	Latent Heat	Btu/Lb @ °F				
25	Inlet Pressure	Psig				
26	Velocity	Ft/S				
27	Pressure Drop, Allow./Calc.	Psi	/		/	
28	Fouling Resistance (Min.)					
29	Heat Exchanged		Btu/Hr; MTD (Corrected)			°F
30	Transfer Rate, Service		Clean			Btu/Hr Sq Ft °F
31	CONSTRUCTION OF ONE SHELL				Sketch (Bundle/Nozzle Orientation)	
32			Shell Side	Tube Side		
33	Design/Test Pressure	Psig	/	/		
34	Design Temperature	°F				
35	No. Passes per Shell					
36	Corrosion Allowance	In.				
37	Connections	In				
38	Size &	Out				
39	Rating	Intermediate				
40	Tube No.	OD	In.;Thk (Min/Avg)	In.; Length	Ft; Pitch	In. ⊲ 30 △ 60 ⊟ 90 ◇ 45
41	Tube Type			Material		
42	Shell	ID	OD In.	Shell Cover		(Integ.) (Remov.)
43	Channel or Bonnet			Channel Cover		
44	Tubesheet-Stationary			Tubesheet-Floating		
45	Floating Head Cover			Impingement Protection		
46	Baffles-Cross	Type		% Cut (Diam/Area)	Spacing: c/c	Inlet In.
47	Baffles-Long			Seal Type		
48	Supports-Tube		U-Bend		Type	
49	Bypass Seal Arrangement			Tube-Tubesheet Joint		
50	Expansion Joint			Type		
51	ρv^2-Inlet Nozzle		Bundle Entrance		Bundle Exit	
52	Gaskets-Shell Side			Tube Side		
53	-Floating Head					
54	Code Requirements				TEMA Class	
55	Weight/Shell		Filled with Water		Bundle	Lb
56	Remarks					
57						
58						
59						
60						
61						

Figure 3-3 TEMA heat exchanger specification sheet. (Used with permission, from *TEMA Standards*, TEMA, 1978.)

Procedures that belong to category 1 are generally preferred, but more understanding of the interrelationships that exist between the various parameters of a shell-and-tube heat exchanger is possible with procedures that belong to category 2. Consider the following problem:

Water is to be heated from 50° to 90°F at the rate of 300,000 gal/h in a single-pass shell-and-tube heat exchanger made of 1-in. schedule 40 steel pipe. The convective heat transfer coefficient on the shell side (steam is the fluid) is taken to be 20,000 Btu/h ft^2 °F. Water is supplied to the tubes at 100 psia and must exit at not less than 85 psia. The steam is saturated at 30 psia. Calculate the number of tubes and the tube length required to fulfill the requirements.

A category 2 procedure to calculate the number of tubes and the tube length required is as follows:

1. Assume N, the number of tubes.
2. Estimate the appropriate temperatures for all property evaluations.
3. Evaluate the fluid properties at the estimated temperatures.
4. Compute Re_D and obtain the friction factor for the tube.
5. Compute the tube length required for $\Delta p = 15$ psia. Use Eq. (2-115).
6. Compute the tube side heat transfer coefficient $\bar{h}$ and the overall heat transfer coefficient U.
7. Using the results of step 6, evaluate the surface temperatures.
8. Compute the area based on steps 5 and 1.
9. Compute the heat exchanger rating, q, based on the area obtained in step 8.
10. Compare q with q_{required}.
11. Repeat steps 1 to 10 until $q = q_{\text{required}}$. After the first pass, use the results of steps 7 and 3.

In this procedure, the tube length is always constrained to satisfy the 15-psia pressure drop (step 5), and the resulting rating is then computed (step 9). The number of tubes is varied until the rating specification is satisfied. Qualitatively, Fig. 3-4 illustrates the behavior.

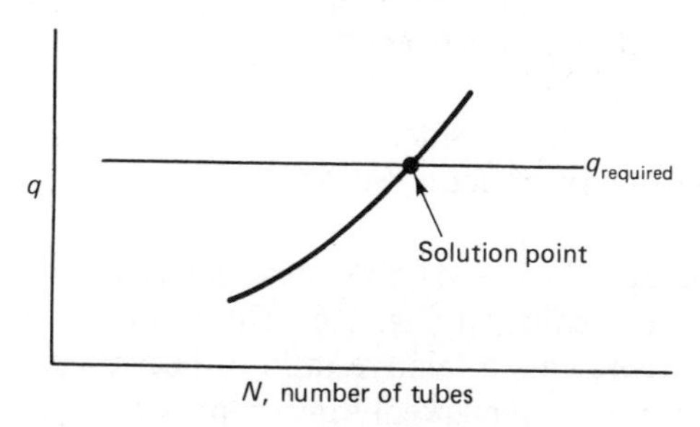

Figure 3-4 Solution method illustration.

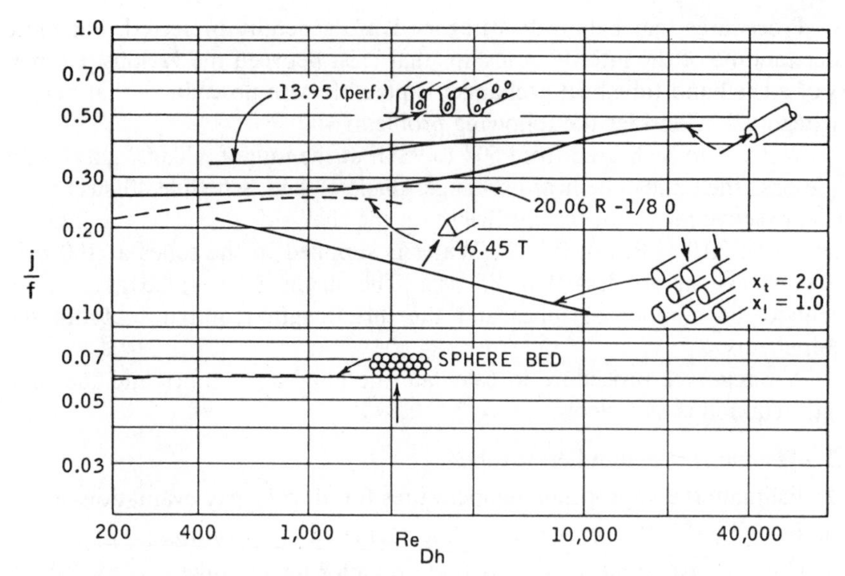

Figure 3-5 Ratio of j/f for diverse surfaces. (Used with permission, from W. M. Kays and A. L. London, *Compact Heat Exchangers*, 2nd ed., McGraw-Hill Book Company, 1964.)

Although the preceding approach shows the effect of increasing the number of tubes until the solution point is reached, no guidance is available for choosing a starting point. Shah (2) provides a procedure developed around an expression for the reasonable initial estimate of the mass velocity ($G = \rho V$) of the tube bank.

By starting with Eq. (2-115), making reasonable assumptions on the importance of each term, and using j/f (the Colburn j factor), Shah was able to develop a simple expression for estimating G. In particular, Shah's expression is

$$G = \left(\frac{2 g_c \eta_t}{v_m \mathrm{Pr}^{2/3}} \frac{\Delta P}{\mathrm{NTU}} \frac{j}{f} \right)^{1/2} \tag{3-5}$$

where η_t = total area effectiveness (no fins, $\eta_t = 1$; fins, $\eta_t = 0.8$ for first estimate)

NTU = NTU for one side

j = Colburn j factor, St $\mathrm{Pr}^{2/3}$

The usefulness of Eq. (3-5) is visibly enhanced by the results of j/f (for a series of surfaces) presented in Fig. 3-5. The ratio j/f or St $\mathrm{Pr}^{2/3}/f$, "the goodness factor," represents a relative ratio of heat transfer to friction losses and is, as Fig. 3-6 shows, remarkedly invariant with Reynolds number for a given surface, and exhibits only a sixfold range for markedly different surfaces.

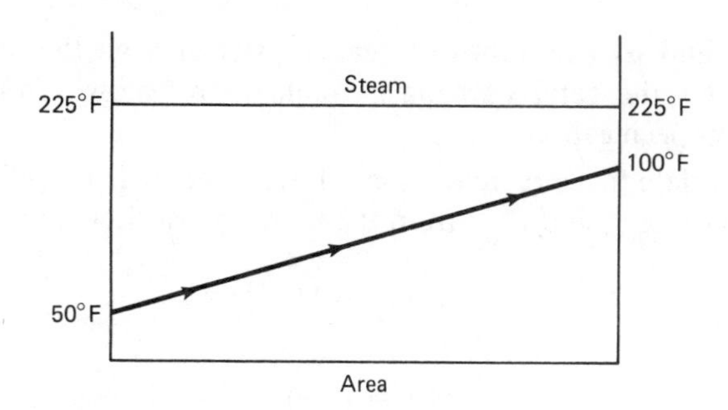

Figure 3-6 Temperature–area diagram for Example 3-1.

The procedure recommended by Shah for given ξ, ΔP_h, and ΔP_c design requirements is as follows:

1. Determine the fluid outlet temperatures from the specified exchanger effectiveness. Obtain the fluid mean temperature on each side and evaluate the fluid physical properties ρ_i, ρ_o, $(1/\rho)_m$, c_p, μ, k, and Pr.
2. Calculate C_{min}/C_{max} and ξ, and determine NTU for the exchanger for the selected flow arrangement. The influence of longitudinal heat conduction, if any, is ignored in the first approximate calculation.
3. Determine NTU on one side by the approximations

$$\text{NTU}_1 = 2\ \text{NTU} \quad \text{for both fluids with the same state}$$

$$\text{NTU}_{\text{gas side}} = 1.1\ \text{NTU} \quad \text{for liquid/gas arrangements}$$

4. For the selected surface on the critical side, plot the j/f versus Re curve and obtain a ball-park value of j/f. If fins are employed on this side, also assume $\eta_t = 0.80$.
5. Evaluate G from Eq. (3-5) using the information from steps 1 to 4 and input the value of $\Delta P/P$.
6. Calculate Reynolds number Re, and determine $\bar{h}_c$ and f for this Re.
7. Compute $\bar{h}$, η_f and η_t. To determine U_1, either lump the fouling resistances with the convective film resistances on the respective sides or ignore them. Similarly, either divide and lump the wall thermal resistance with the respective hot and cold side resistances or ignore it. U_1, after replacing the subscript $\bar{h}$ and c with 1 and 2, is

$$\frac{1}{U_1} = \frac{1}{\eta_{t_1}\bar{h}_1} + \frac{A_1/A_2}{\eta_{t_2}\bar{h}_2} \tag{3-6}$$

where

$$\frac{A_1}{A_2} = \frac{\alpha_1}{\alpha_2} \tag{3-7}$$

Here α_1 and α_2 (the ratios of heat transfer area on the respective side divided by the total exchanger volume) are known from the input geometry specifications.

8. Now calculate the core dimensions. Determine A_1 from NTU for this U_1 and known $C_{\min}$, and $A_{c,1}$ from the known mass flow rate and G:

$$A_1 = \text{NTU}\frac{C_{\min}}{U_1} \tag{3-8}$$

$$A_{c,1} = \left(\frac{\dot{m}}{G}\right)_1 \tag{3-9}$$

so that the frontal area A_{fr} becomes

$$A_{fr} = \frac{A_{c,1}}{\sigma_1} \tag{3-10}$$

Finally, evaluate the core length L_1 from the known $A_{c,1}$, A_1, and D_{h_1} as

$$L_1 = \frac{D_{h_1} A_1}{4A_{c,1}} \tag{3-11}$$

Once the frontal area is calculated, the core dimensions are chosen in reasonable size such that it will meet the pressure drop on the other side, too.

9. Now correct the f factor for the variable property effects and calculate $\Delta P/P$ [Eq. (2-115)].

10. If the calculated value of $(\Delta P/P)$ is within the input specification, the approximate solution to the sizing problem is finished. Finer refinements in the core dimensions may be carried out at this time.

    If the calculated value of $\Delta P/P$ is larger than the input value, compute an improved value of G using the specified $\Delta P/P$ and the values of f and L from the preceding step. With this new value of G, repeat steps 6 to 10 until both heat transfer and pressure drops are met as specified.

We should pay particular attention to the details of this procedure for ensuring the correct evaluation of all properties for each equation. Virtually any iterative design scheme must incorporate these steps, albeit not in this same order.

Step 8 in this procedure determines the strategy implemented. In step 8, the heat transfer surface area A_1 is computed by requiring satisfaction of the NTU calculated in step 2. Once the heat transfer surface geometry (A_1, L_1, A_{fr}) has been computed, the resulting pressure drop is calculated in step 9 using Eq. (2-115). The iteration proceeds by recomputing G using the latest values of f, $\overline{\text{Nu}}$, K_e, K_c, A_1, and L_1 and repeating steps 6 to 10. This procedure thus forces satisfaction of the heat transfer requirement (ξ–NTU) at each iterate while searching for the geometry that will simultaneously satisfy the pressure drop requirement.

The category 1 procedure of Shah is recommended over the category 2 procedure previously examined.

Once a solution is reached that satisfies the heat transfer and fluid dynamic requirements, strength of material considerations as well as code* and standards come into play. Even in the earlier heat transfer and fluid dynamic analyses, standards like the TEMA standards can affect these aspects of design since tube array *arrangement* and tube bundle baffling *limits* are prescribed by the standard. The satisfaction of all design requirements, code requirements, and standards specifications completes the heat exchanger design. Design requirements can be so flexible that an optimum design may exist. Within the design specification envelope, an optimum can be found once a value or judgment function is specified. In many energy systems today, cost is the value function chosen to define the optimum. Usually a very special cost, the life-cycle cost, is the cost of interest. Life-cycle costing forces the designer and/or customer into an engineering economic analysis where the cost of the unit over its entire operating life must be estimated.

Example 3-1

A heat exchanger to be used as a converter is to be designed. Water, 250,000 gal/h, is to be heated from 50° to 100°F with a pressure drop not to exceed 16 psia (2304 psf). The exchanger is to use 1-in. schedule 40 commercial steel pipes, which are to be placed so that 50 percent of the tube sheet frontal area is available for flow; that is, $\sigma = A_{c,1}/A_{fr} = 0.5$. Steam is supplied at 225°F, and the steam side heat transfer coefficient can be taken as 12,000 Btu/ft^2 h °F.

Solution. Shah's procedure will be followed. Schematically, the area–temperature relationship appears as in Fig. 3-6.

Since $\bar{h}_o$ is specified and since the Δp constraint is on the water side, only the water side calculations need be considered.

$$T_m = \tfrac{1}{2}(100 + 50) = 75°\text{F}$$

for which

$$\rho_m = 62.25 \text{ lbm/ft}^3$$

$$C_{p_m} = 0.998 \text{ Btu/lbm °F}$$

$$\mu_m = 0.618 \times 10^{-3} \text{ lbm/ft s}$$

$$k_m = 0.350 \text{ Btu/h ft °F}$$

$$\text{Pr}_m = 6.355$$

*ASME Unfired Pressure Vessel Code.

The heat transfer rating is

$$q = \dot{m} C_p \, \Delta T$$

$$= 250{,}000 \frac{\text{gal}}{\text{h}} \frac{\text{ft}^3}{7.481 \text{ gal}} 62.25 \frac{\text{lbm}}{\text{ft}^3} 0.998 \frac{\text{Btu}}{\text{lbm °F}} (100 - 50)\text{°F}$$

$$= 103.81 \times 10^6 \text{ Btu/h}$$

$C_{max} \sim \infty$ since the steam is condensing

$C_{min} = (\dot{m} c_p)_c = 2.076 \times 10^6$ Btu/h °F

Thus $C_{min}/C_{max} \sim 0$ and

$$\xi = \frac{q}{C_{min}\left(T_{h_{in}} - T_{c_{in}}\right)} = \frac{103.81 \times 10^6}{2.076 \times 10^6 (225 - 50)} = 0.2857$$

For $C_{min}/C_{max} = 0$, $\xi = 1 - e^{-\text{NTU}}$, for which

$$\text{NTU} = -\ln(1 - \xi) = -\ln(1 - 0.2857) = 0.33649$$

Using Eq. (3-5), the initial mass velocity, G, estimate is 1648. lbm/ft^2 s. If steps 6 to 10 of Shah's procedure are followed, Table 3-4 results. The converged solution requires 82 tubes each of length 23.4 ft to meet the rating and pressure drop constraints. The computer program block diagram is given in Fig. 3-7.

The dependence of the thermophysical properties on temperature is significant for this problem. Had Eqs. (3-3) and (3-4) not been used to account for these effects, the calculations would have estimated 85 tubes (instead of 82) of length 25.3 ft (instead of 23.4). As a result, an increase of 12 percent in the heat transfer area would have been predicted.

From Table 3-4 it is apparent why iteration using the mass velocity G is preferred to other variables. The initial guess on G, while in error by a significant amount (nearly 30 percent), is still an excellent starting point, an advantage not shared by some other iterative sequences. The approach to convergence of Shah's procedure is satisfying as the number of tubes required overshoots and then undershoots the converged results for alternate iterations.

TABLE 3-4 Heat Exchanger Design Problem Approach to Convergence

Iteration	G (lbm/ft^2 s)	f	j/f	N	L (ft)	ΔP (lbf/ft^2)
Initial	1648.	0.00569	0.432	59	29.5	5512.
2	1066.	0.00578	0.464	91	21.8	1762.
3	1218.	0.00575	0.454	79	23.9	2492.
4	1172.	0.00576	0.457	82	23.3	2251.
5	1185.	0.00576	0.457	81	23.4	2320.
6	1181.	0.00576	0.457	82	23.4	2299.
Converged	1182.	0.00576	0.456	82	23.4	2305.

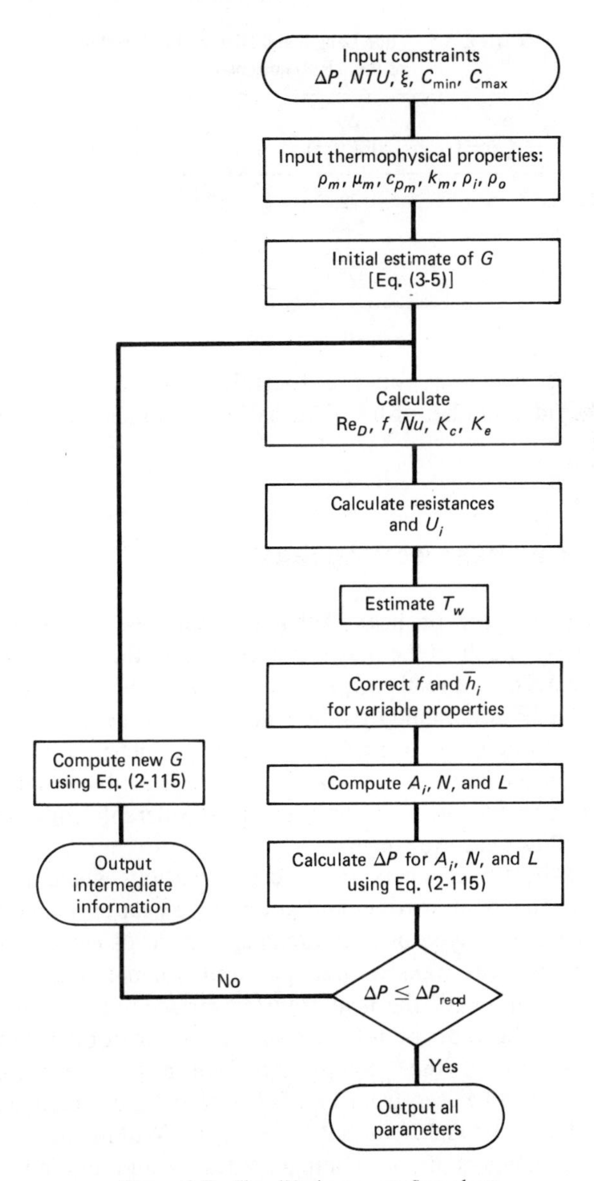

Figure 3-7 Simplified program flow chart.

An interesting study can be made using the computer program developed for Example 3-1. The heat exchanger geometrical requirements (number and length of tubes required) are dependent on the allowable pressure drop. Table 3-5 was generated by specifying various allowable pressure drops while keeping all other requirements constant. As the allowable pressure drop is increased, the number of tubes required goes down while the tube lengths increase. The

TABLE 3-5 Tube Length and Diameter a Function of the ΔP Requirement

ΔP (lbf/in.^{2})	ΔP (lbf/ft^2)	N	L (ft)
5	720	129	17.5
10	1440	98	20.8
20	2880	75	24.8
25	3600	69	26.3

relationships are *nonlinear*, since a fourfold increase in ΔP results in a decrease in the tube number of less than half and an increase in tube length of about a third.

3-4 CROSS-FLOW HEAT EXCHANGERS

The remaining category of heat exchangers that we will examine is the cross-flow variety. Much of the material presented in the section on shell-and-tube heat exchangers is directly applicable to cross-flow heat exchangers. The equivalent of the TEMA standards does not exist for general cross-flow heat exchangers. The wide range of surface and core geometries precludes describing the heat transfer and pressure drop characteristics by simple correlation expressions, so the existing data base is of considerable importance in the design of cross-flow heat exchangers.

Cross-flow heat exchangers are used in situations where small size is a virtue; many aeronautical, marine, and automotive heat exchangers are of the cross-flow variety. This type of heat exchanger achieves *compactness* by providing much more heat transfer area per unit volume than shell-and-tube exchangers. The symbol β, the heat transfer surface area density, is usually used to designate the ratio of total transport area to total volume, since it serves as a measure of the "compactness" of a given surface geometry configuration. As can be seen from Fig. 3-8, surfaces are classified as compact when β exceeds 700 m^2/m^3 or about 200 ft^2/ft^3. Within this scheme, shell-and-tube heat exchangers are not compact surfaces, and automotive radiators are in the low range of compact surface. The surface area density of the human lung generally exceeds that of any realizable heat exchanger surface geometry. The extreme range that β, the heat transfer surface area density, spans precludes wide use of simple correlations.

Heat exchanger surfaces, planar or tubular, finned or unfinned, come in a wide range of sizes and shapes. *Fins* on tubes cover a spectrum from the usually expected circumferential annular fins to helical fins with cuts and/or slots. Figure 3-9 illustrates variations that external circumferential and helical

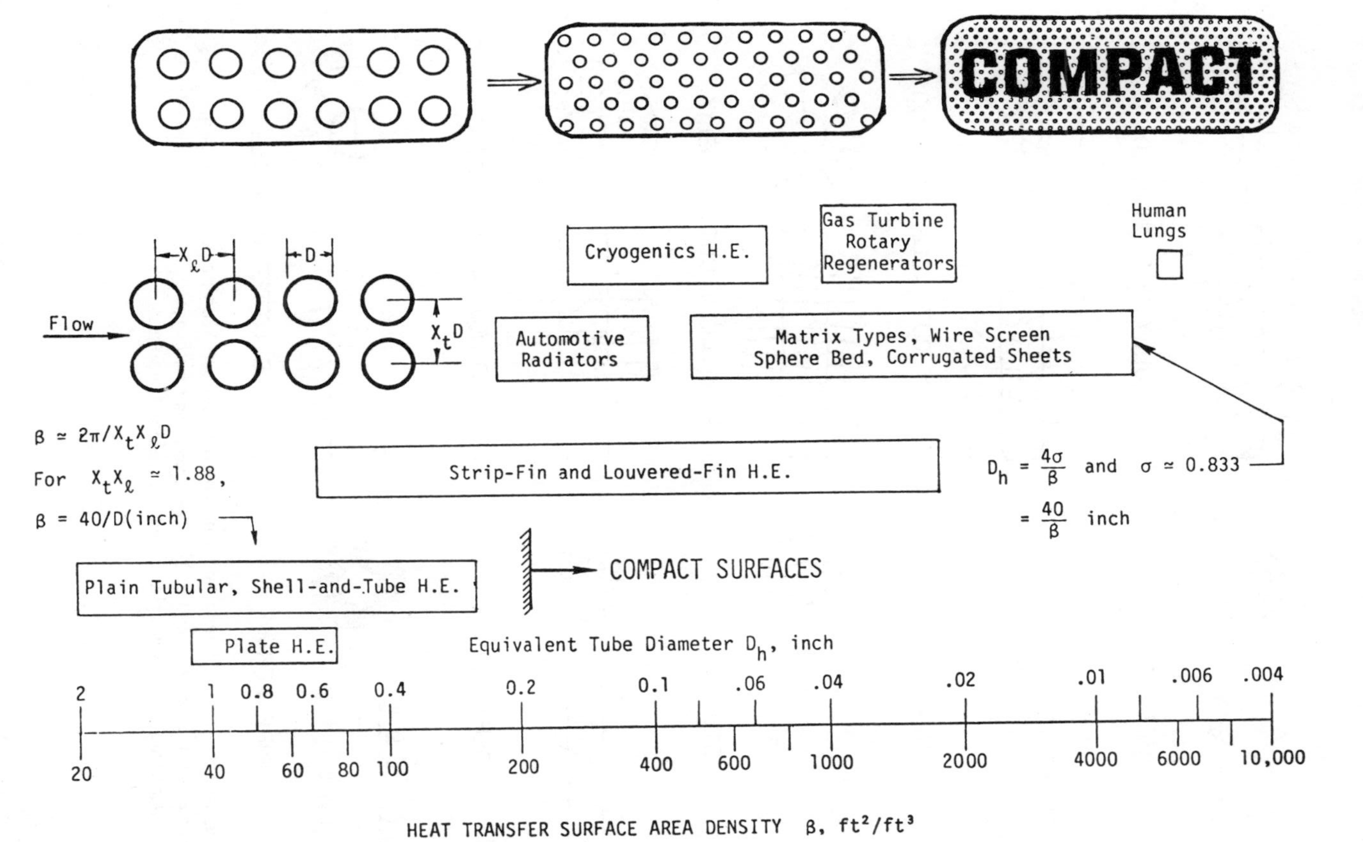

Figure 3-8 Heat transfer surface area density spectrum of exchanger surfaces. (Used with permission, from A. L. London, "A Brief History of Compact Heat Exchanger Technology," in *Compact Heat Exchangers—History, Technological Advancement and Mechanical Design Problems*, ASME HTD-10, 1981.)

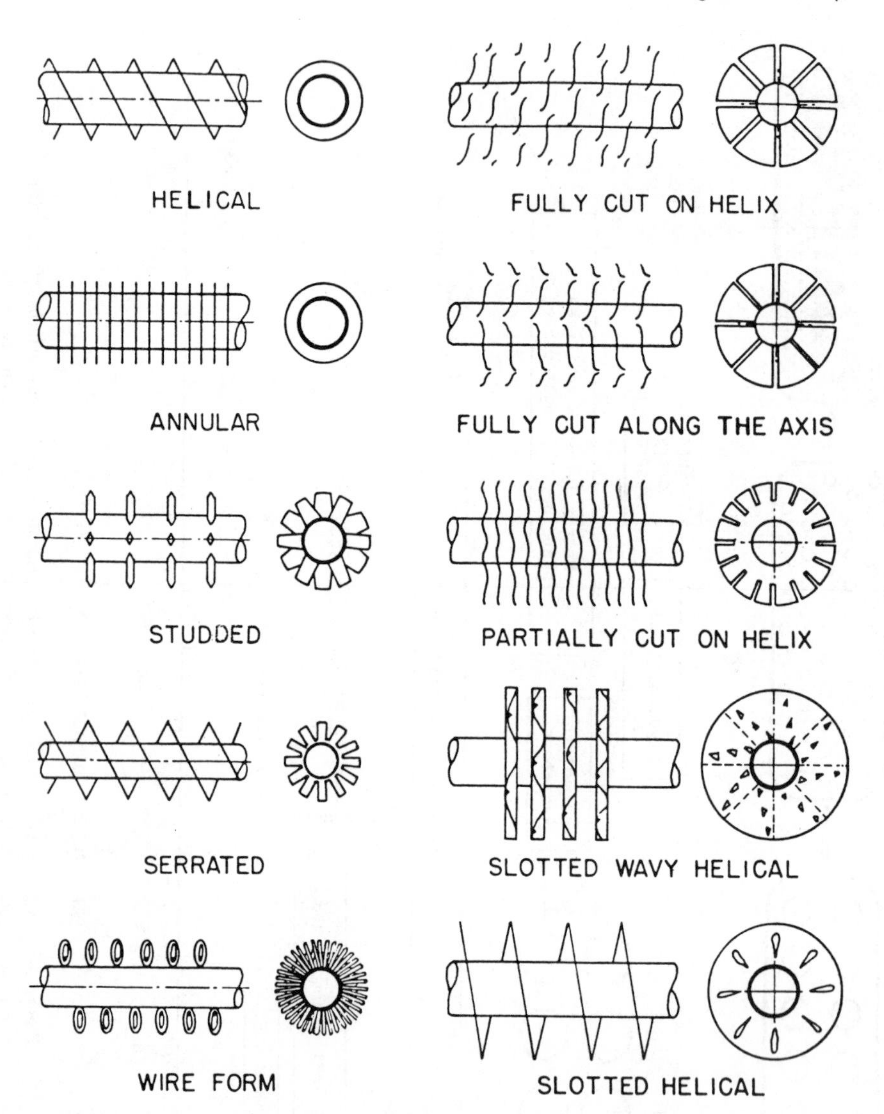

Figure 3-9 Circumferential and helical fin geometries. (Used with permission, from S. Kakac, A. E. Bergles, and F. Mayinger, eds., *Heat Exchangers: Thermal-Hydraulic Fundamentals and Design*, Hemisphere Publishing Corporation, 1981.)

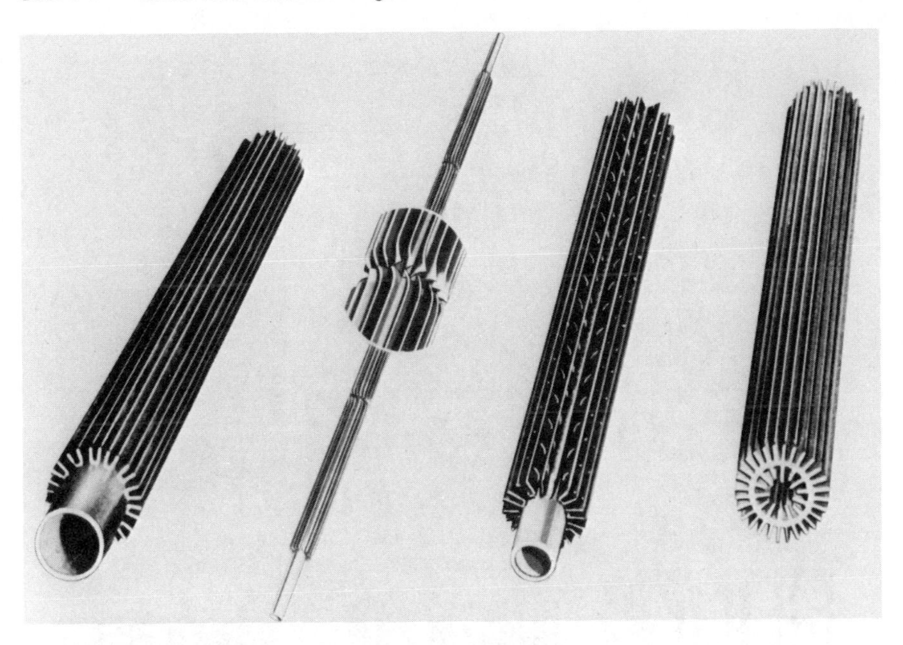

Figure 3-10 Typical longitudinal fins. (Used with permission, from Brown Fintube Co., Houston, Texas.)

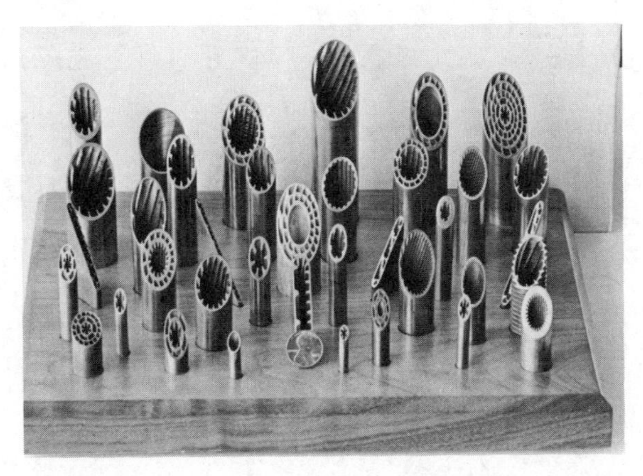

Figure 3-11 Internally finned tubes. (Used with permission, from Forged-Fin Division, Noranda Metal Industries, Inc., Newton, Conn.)

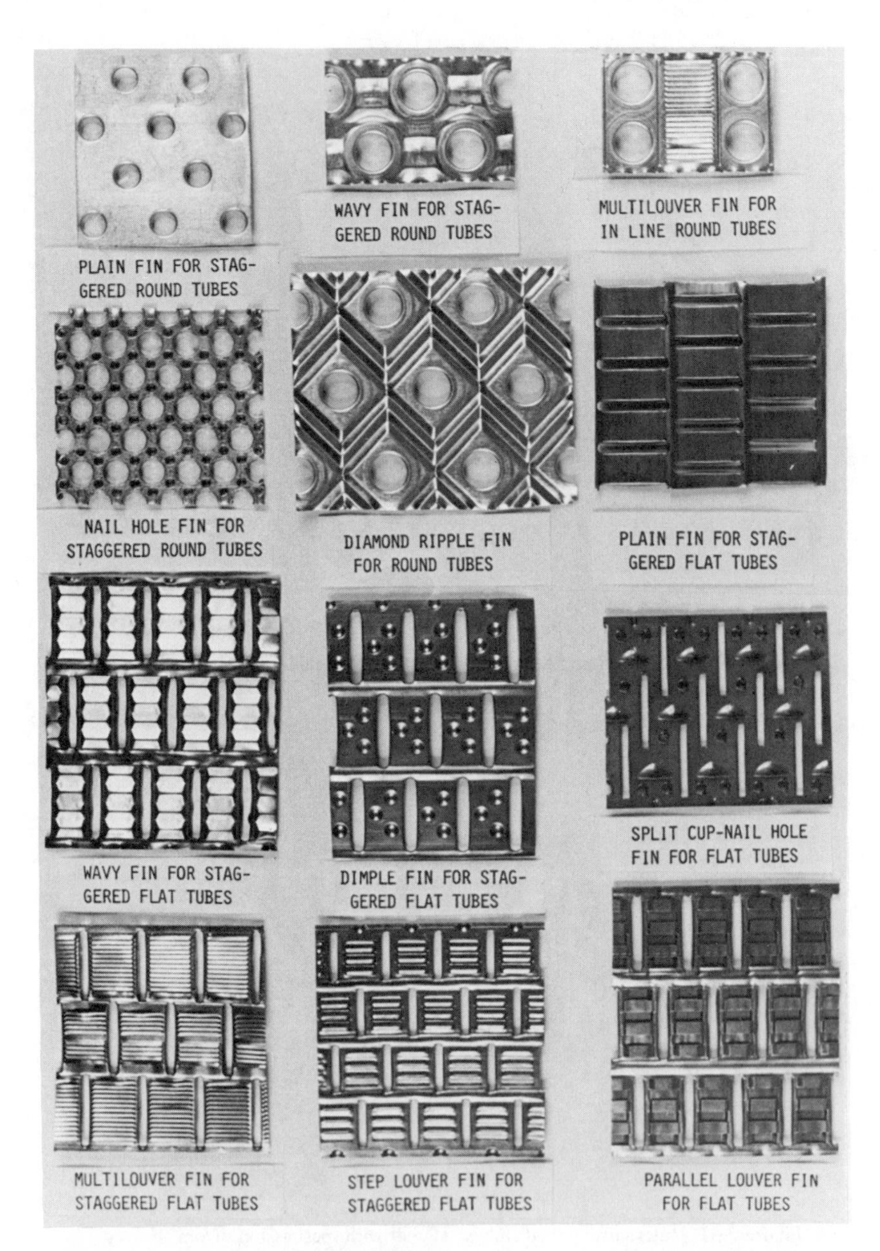

Figure 3-12 Continuous fin examples. (Used with permission, from Harrison Radiator Division, GMC, Lockport, New York.)

fins can take. Deviations from helical and circumferential fins are usually made with the idea that serrations, cuts, or waves on the fins will act to increase the turbulence of the fluid, thereby resulting in an increased convective heat transfer coefficient. Such increases in the convective heat transfer are not free, since the increased frictional losses result in larger pumping power. Longitudinal fins also come in several geometries. Figure 3-10 illustrates several of these types: (a) continuous plane, (b) cut and twisted, (c) perforated, and (d) internal and external longitudinal fins. Again the purpose is increased convective heat transfer and the penalty is additional pumping power. Internally, finned tubes likewise come in a variety of shapes. Figure 3-11 shows a wide range of geometries and sizes. Fouling is often a problem for finned surface as the fins, particularly serrated, slotted, or cut, offer an enhanced opportunity for scale development and growth.

Another category of fins is continuous from one tube, either circular or flat, to another. Typical geometries for continuous fins are illustrated in Fig. 3-12. Such fins can be plain, wavy, multilouvered, rippled, dimpled, and cup nail-holed. As in the case of internally or externally finned tubes, the purpose of the continuous fins is to increase the heat transfer from a high convective resistance surface; and, as in the previous cases, the fins are often devised and arranged such that the turbulence level and, thereby, the convective heat transfer coefficients are enhanced.

All the aforementioned fins are used to enhance the heat transfer from a high resistance surface. Webb (5) has pointed out that the practice of removing material from continuous fins in order to increase turbulence can be counterproductive since the increase in the convective heat transfer coefficient may not be sufficient to offset the loss of fin area.

Surfaces such as we have examined are used in devices such as tube fin heat exchangers [Fig. 3-13(a)], bar and plate heat exchangers [Fig. 3-13(b)], and formed plate fin heat exchangers [Fig. 3-13(c)].

Surface Geometry Data Base

Almost all heat transfer and pressure drop data relating to core surface geometries are presented in *graphical* form. The extensive data of Kays and London are typically presented with Reynolds number as the dependent variable and the Fanning friction factor f and the product St $\mathrm{Pr}^{2/3}$ as dependent variables. Where possible, the experimental data points as well as the suggested best interpretation line are presented. Figure 2-24 is typical of the mode of data presentation by Kays and London. Their data base includes St $\mathrm{Pr}^{2/3}$ and f data for 56 plate-fin surfaces and 21 tube-fin surfaces. Tabular values of all useful surface geometry characteristics are also presented. Shah lists 23 other sources of data for plane fins, wavy fins, strip offset fins, louvered fins, perforated fins, and pin fins. Webb presents an excellent contemporary (1980) assessment of various tube-fin geometries. The data base is rather

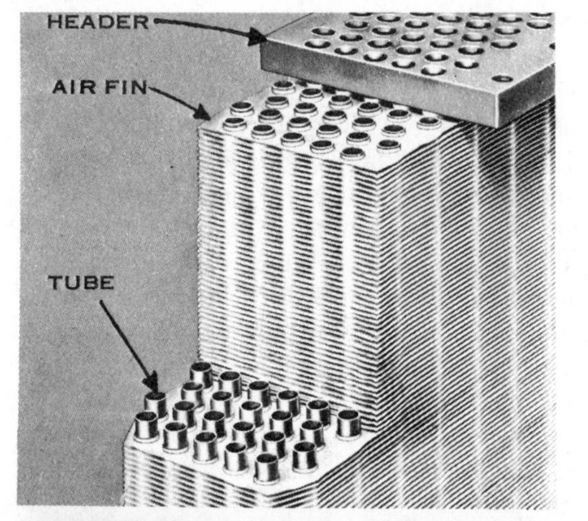

(a)

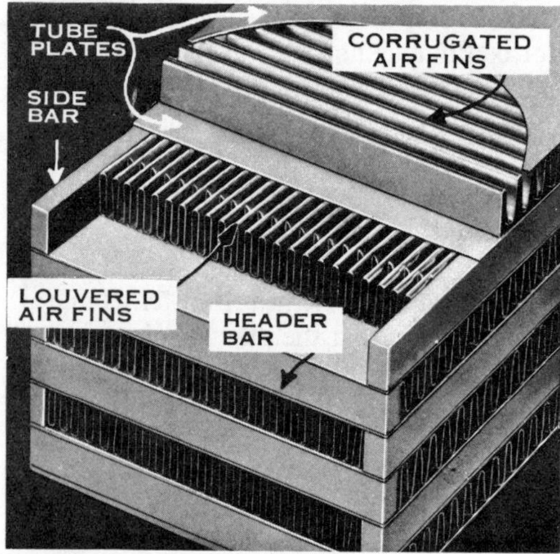

(b)

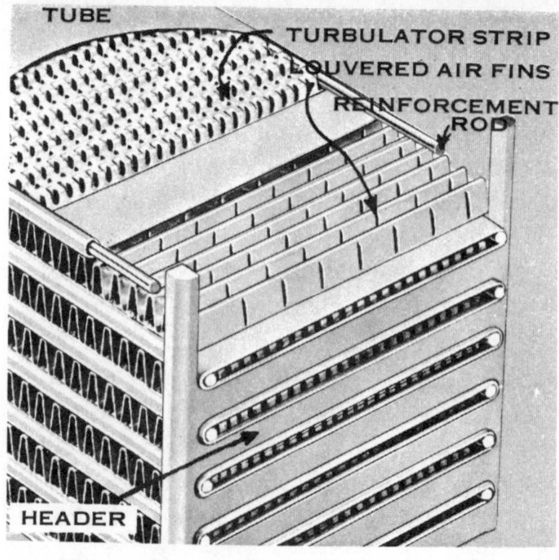

(c)

Figure 3-13 Examples of finned cross-flow heat exchangers: (a) flat tube and plain fin; (b) bar and plate; (c) formed plate fin. (Used with permission, from Harrison Radiator Division, GMC, Lockport, New York.)

extensive, and the references are generally available in the readily accessible open literature, such as *Transactions of the ASME*, *Journal of Heat Transfer*, *Heat Transfer Engineering*, and *International Journal of Heat and Mass Transfer*.

We shall examine further several typical surface geometries as presented by Kays and London. They present information about a given heat transfer surface in three portions: (1) tabular geometrical characteristics for a given family of surfaces, (2) schematics illustrating the surface geometry with salient dimensions indicated, and (3) graphical presentations of the St $\mathrm{Pr}^{2/3}$ and f data for a wide range of Reynolds numbers. Even with the increasingly accurate and sophisticated techniques available from computational fluid dynamics, there is no hope of accurately computing St $\mathrm{Pr}^{2/3}$ and f from the surface geometry description alone, and for the forseeable future we shall have to rely on experimental data for design purposes.

Since Kays and London's data are classic and the presentation concise, we shall examine portions of it in detail. Samples of both tabular information and data are given in Figs. 3-14 and 3-15 for several categories of heat exchanger surface geometries: (1) tubular surfaces, FT-2 and FTD-2, (2) plate fins, 9.03 and 12.00T, (3) louvered fins, $\frac{3}{16}$-11.1, (4) strip fins, $\frac{1}{8}$-13.95, (5) wavy fins, 17.8-$\frac{3}{8}$W, (6) pin fins, AP-2, (7) circular fins on circular tubes, CF-7.0-$\frac{5}{8}$J, (8) continuous fins on circular tubes, 8.0-$\frac{3}{8}$T, and (9) continuous fins on flat tubes, 9.29-0.737–SR.

Figure 3-14 contains an enormous amount of helpful information, which we shall find useful in design and analysis problems. Usual information such as the surface designation, plate spacing (if applicable), and the fins per inch is given along with the hydraulic diameter and fin thickness. The last two columns present either β (for plate-fin surfaces) or α (for finned tubes) and the fin area to total area. Beta, β, represents the heat transfer area per unit volume between plates; alpha, α, represents the heat transfer area per unit total volume. Both β and α can be used to obtain the heat transfer area for a given volume, and the presence of β or α for a surface relieves a great deal of tedious calculations. The units on α and β are $\mathrm{ft}^2/\mathrm{ft}^3$. The fin area divided by the total area, A_f/A, is used directly in finding η_t from Eq. (2.110).

The heat transfer data, St $\mathrm{Pr}^{2/3}$, and the friction factor data, f, as presented in Fig. 3-15 are given as functions of Reynolds number. In Fig. 3-15, $\mathrm{N_R}$ and $\mathrm{N_{Pr}}$ are the Reynolds number and Prandtl number, respectively, and the term (h/GC_p) is the Stanton number, St. The heat transfer data were reduced to the form St $\mathrm{Pr}^{2/3}$ by assuming a constant convective heat transfer coefficient over the entire surface. Use of these data must be made in the same way, that is, $\bar{h}_c$ computed, then η_f for the fins, and finally η_t for the entire surface. Most of these data were reduced with all thermophysical properties evaluated at T_m. For application to situations where T_s and T_m lead to a more than moderate $(T_s - T_m)$, the correction for property variations, presented at the beginning of this chapter, should be made.

Surface Geometry, Flow Normal to Banks of Bare Tubes

Surface designation	Pattern	Type of tests	Tube diameter		Transverse spacing		Longitudinal spacing		Hydraulic diameter $4r_h$		Free-flow/frontal area σ	Heat transfer area/total volume α, ft²/ft³
			ft	in.	ft	in.	ft	in.	ft	in.		
FT-2	Staggered	Steady state	Flattened		See surface diagrams				0.01433	0.1722	0.386	108
FTD-2	Staggered	Steady state	Flattened		See surface diagrams				0.0160	0.192	0.423	108

FINS

Plain fins

Surface designation	Plate spacing b		Fins/in.	Hydraulic diameter $4r_h$		Fin thickness δ, in.	Flow length of uninterrupted fin, in.	Heat transfer area/volume between plates, β, ft²/ft³	Fin area/total area
	ft	in.		ft	in.				
6.2	0.0337	0.405	6.2	0.0182	0.218	0.010	1.20	204	0.728
9.03	0.0686	0.823	9.03	0.01522	0.1828	0.008	1.19	244	0.888
12.00T	0.0208	0.250	12.00	0.009412	0.113	0.006	2.5	392.7	0.773

Louvered fins

Surface designation	Plate spacing b		Fins/in.	Hydraulic diameter $4r_h$		Fin thickness δ, in.	Louver spacing in.	Louver gap in.	Heat transfer area/volume between plates, β, ft²/ft³	Fin area/total area
	ft	in.		ft	in.					
$\frac{3}{8}$–6.06	0.0208	0.250	6.06	0.01460	0.1753	0.006	0.375	0.055	256	0.640
$\frac{3}{8}$(a)–6.06	0.0208	0.250	6.06	0.01460	0.1753	0.006	0.375	0.130	256	0.640
$\frac{1}{2}$–6.06	0.0208	0.250	6.06	0.01460	0.1753	0.006	0.500	0.055	256	0.640
$\frac{1}{2}$(a)–6.06	0.0208	0.250	6.06	0.01460	0.1753	0.006	0.500	0.130	256	0.640
$\frac{3}{8}$–8.7	0.0208	0.250	8.7	0.01196	0.1437	0.006	0.375	0.055	307	0.705
$\frac{3}{8}$(a)–8.7	0.0208	0.250	8.7	0.01196	0.1437	0.006	0.375	0.080	307	0.705
$\frac{3}{16}$–11.1	0.0208	0.250	11.1	0.01012	0.1214	0.006	0.1875	0.055	367	0.756

Strip fins

Surface designation	Plate spacing b		Fins/in.	Hydraulic diameter $4r_h$		Fin thickness δ, in.	Flow length of uninterrupted fin, in.	Heat transfer area/volume between plates, β, ft²/ft³	Fin area/total area
	ft	in.		ft	in.				
$\frac{1}{4}$(s)–11.1	0.0208	0.250	11.1	0.01012	0.1214	0.006	0.25	367	0.756
$\frac{3}{32}$–12.2	0.0404	0.485	12.2	0.01120	0.1343	0.004	0.094	340	0.862
$\frac{1}{8}$–15.2	0.0346	0.414	15.2	0.00868	0.1042	0.006	0.125	417	0.873
$\frac{1}{8}$–13.95	0.0313	0.375	13.95	0.00879	0.1055	0.010	0.125	381	0.840
$\frac{1}{2}$–11.94(D)	0.0198	0.237	11.94	0.007436	0.0892	0.006	0.500	461	0.796

Figure 3-14 Heat transfer surface characteristics. (Used with permission, from W. M. Kays and A. L. London, *Compact Heat Exchangers*, 2nd ed., McGraw-Hill Book Company, 1964.)

Wavy fins

Surface designation	Plate spacing b		Fins/in.	Hydraulic diameter $4r_h$		Fin thickness δ, in.	Wavelength, in.	Double wave amplitude, in.	Heat transfer area/volume between plates, β, ft²/ft³	Fin area/total area
	ft	in.		ft	in.					
11.48–3/8W	0.0345	0.413	11.44	0.01060	0.1272	0.006	0.375	0.0775	351	0.847
11.5–3/8W	0.0313	0.375	11.5	0.00993	0.119	0.010	0.375	0.078	347	0.822
17.8–3/8W	0.0345	0.413	17.8	0.00696	0.0836	0.006	0.375	0.0775	514	0.892

Pin fins

Surface designation	Plate spacing b		Pin pattern	Hydraulic diameter $4r_h$		Pin diameter in.	Transverse pin spacing, in.	Longitudinal pin spacing, in.	Heat transfer area/volume between plates, β, ft²/ft³	Fin area/total area
	ft	in.		ft	in.					
AP-1	0.020	0.240	In-line	0.01444	0.1734	0.040	0.125	0.125	188	0.512
AP-2	0.0332	0.398	In-line	0.01172	0.1408	0.040	0.12	0.096	204	0.686
PF-3	0.0625	0.750	In-line	0.00536	0.0644	0.031	0.0602	0.0602	339	0.834

Circular tubes, circular fins

Surface designation	Tube arrangement	Tube diameter, in.	Fin outside diameter, in.	Transverse tube spacing, in.	Longitudinal tube spacing, in.	Fins/in.	Hydraulic diameter $4r_h$		Fin thickness δ, in.	Free-flow/frontal area σ	Heat transfer area/total volume, α, ft²/ft³	Fin area/total area
							ft	in.				
CF-7.34	Staggered	0.38	0.92	0.975	0.800	7.34	0.0154	0.187	0.018	0.538	140	0.892
CF-8.72	Staggered	0.38	0.92	0.975	0.800	8.72	0.01288	0.1547	0.018	0.524	163	0.910
CF-8.72(c)	Staggered	0.42	0.861	0.975	0.800	8.72	0.01452	0.1742	0.019	0.494	136	0.876
CF-11.46	Staggered	0.38	0.92	0.975	0.800	11.46	0.00976	0.1173	0.016	0.510	209	0.931
CF-7.0-$\frac{5}{8}$J	Staggered	0.645	1.121	1.232	1.35	7.0	0.0219	0.263	0.010	0.449	82	0.830

Circular tubes, continuous fins

Surface designation	Tube arrangement	Tube diameter, in.	Fin type	Fins/in.	Hydraulic diameter $4r_h$		Fin thickness δ, in.	Free-flow/frontal area σ	Heat transfer area/total volume, α, ft²/ft³	Fin area/total area
					ft	in.				
8.0-$\frac{3}{8}$T	Staggered	0.402	Plain	8.0	0.01192	0.1430	0.013	0.534	179	0.913
7.75-$\frac{5}{8}$T	Staggered	0.676	Plain	7.75	0.0114	0.137	0.016	0.481	169	0.950

Flat tubes, continuous fins

Surface designation	Tube arrangement	Fin type	Tube length (parallel to flow), in.	Tube width (normal to flow), in.	Fins/in.	Hydraulic diameter $4r_h$		Fin thickness δ, in.	Free-flow/frontal area σ	Heat transfer area/total volume, α, ft²/ft³	Fin area/total area
						ft	in.				
9.68-0.87	In-line	Plain	0.870	0.120	9.68	0.01180	0.1416	0.004	0.697	229	0.795
9.1-0.7378	Staggered	Plain	0.737	0.100	9.1	0.01380	0.1656	0.004	0.788	224	0.813
9.68-0.87R	In-line	Ruffled	0.870	0.120	9.68	0.01180	0.1416	0.004	0.697	229	0.795
9.29-0.7378R	Staggered	Ruffled	0.737	0.100	9.29	0.01352	0.1622	0.004	0.788	228	0.814
11.32-0.7378R	Staggered	Ruffled	0.737	0.100	11.32	0.01152	0.1382	0.004	0.780	270	0.845

Figure 3-14 (continued)

Flow normal to flattened tubes, surface FT-2.

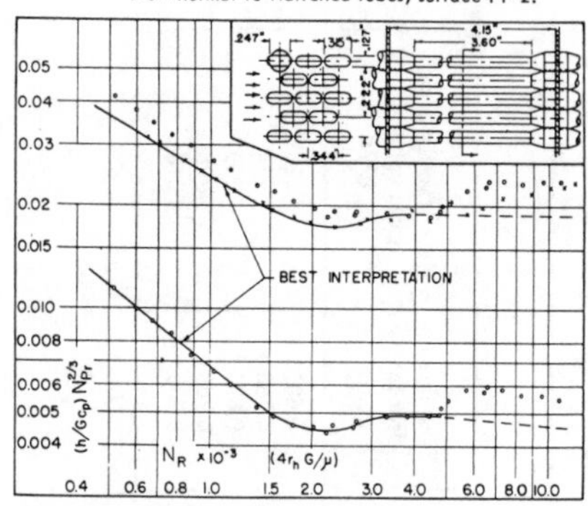

Tube OD before flattening = 0.247 in.
Distance between centers parallel to flow = 0.344 in.
Distance between centers perpendicular to flow = 0.222 in.
Tube dimension parallel to flow = 0.315 in.
Tube dimension perpendicular to flow = 0.127 in.
Distance between spacing plates = 4.15 in.
Length of flat along tube = 3.60 in.
Flow passage hydraulic diameter, $4r_h = 0.01433$ ft
Total transfer area/total volume, $\alpha = 108\ \text{ft}^2/\text{ft}^3$
Free-flow area/frontal area, $\sigma = 0.386$

(a)

Flow normal to dimpled flattened tubes, surface FTD-2.

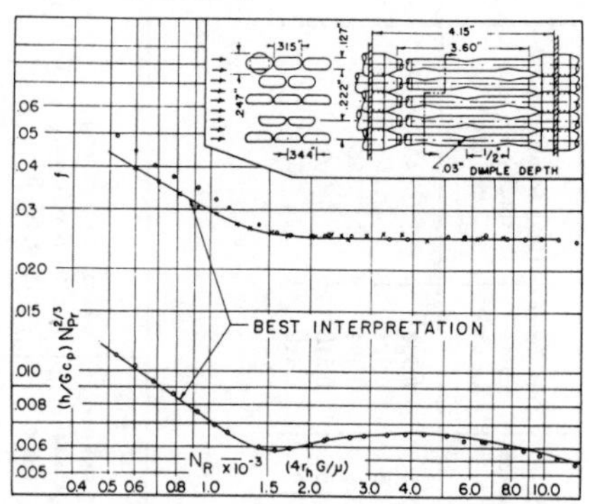

Tube OD before flattening = 0.247 in.
Distance between centers parallel to flow = 0.344 in.
Distance between centers perpendicular to flow = 0.222 in.
Tube dimension parallel to flow = 0.315 in.
Tube dimension perpendicular to flow = 0.127 in.
Distance between spacing plates = 4.15 in.
Length of flat along tube = 3.60 in.
Dimple depth = 0.03 in.
Minimum distance between dimples along tube = 0.5 in.
Flow passage hydraulic diameter, $4r_h = 0.016$ ft
Total transfer area/total volume, $\alpha = 108\ \text{ft}^2/\text{ft}^3$
Free-flow area/frontal area, $\sigma = 0.423$

(b)

Plain plate-fin surface 9.03.

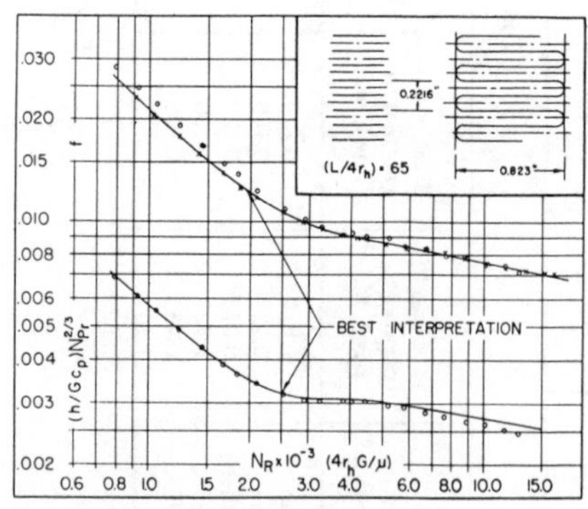

Fin pitch = 9.03 per in.
Plate spacing, $b = 0.823$ in.
Flow passage hydraulic diameter, $4r_h = 0.01522$ ft
Fin metal thickness = 0.008 in., aluminum
Total transfer area/volume between plates, $\beta = 244\ \text{ft}^2/\text{ft}^3$
Fin area/total area = 0.888

(c)

Plain plate-fin surface 12.00 T.

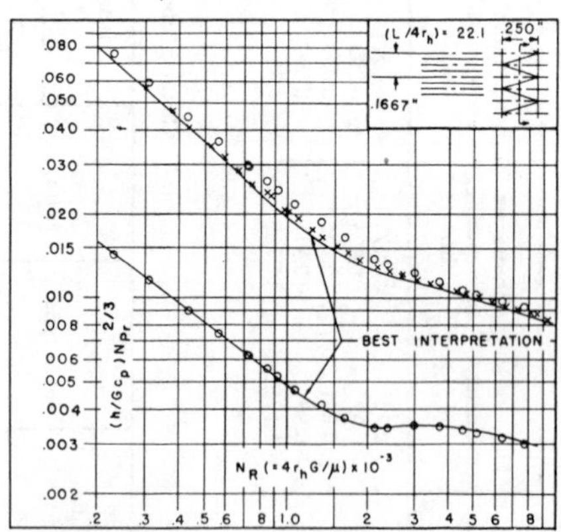

Fin pitch = 12.00 per in.
Plate spacing, $b = 0.250$ in.
Fin length flow direction = 2.50 in. strips (two = 5.00 in.)
Flow passage hydraulic diameter, $4r_H = 0.009412$ ft
Fin metal thickness = 0.006 in., nickel
Total heat transfer area/volume between plates, $\beta = 392.7\ \text{ft}^2/\text{ft}^3$
Fin area/total area = 0.773

(d)

Figure 3-15 Selected heat transfer surface data. (Used with permission, from W. M. Kays and A. L. London, *Compact Heat Exchangers*, 2nd ed., McGraw-Hill Book Company, 1964.)

Louvered plate-fin surface 3/16-11.1.

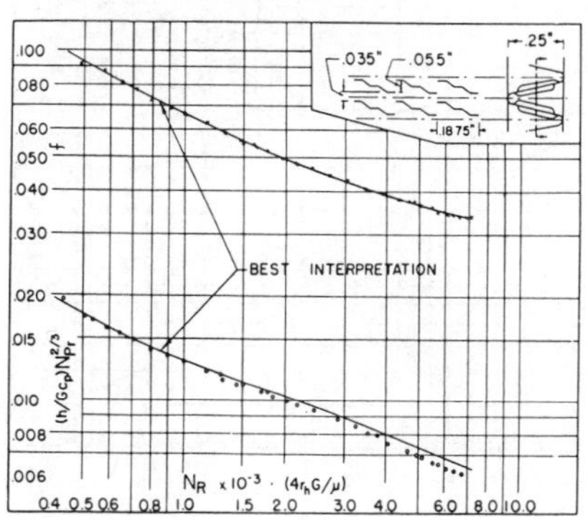

Fin pitch = 11.1 per in.

Plate spacing, $b = 0.250$ in.

Louver spacing = 0.1875 in.

Fin gap = 0.035 in.

Louver gap = 0.055 in.

Flow passage hydraulic diameter, $4r_h = 0.01012$ ft

Fin metal thickness = 0.006 in., aluminum

Total heat transfer area/volume between plates, $\beta = 367\ \text{ft}^2/\text{ft}^3$

Fin area/total area = 0.756

(e)

Strip-fin plate-fin surface 1/8-13.95.

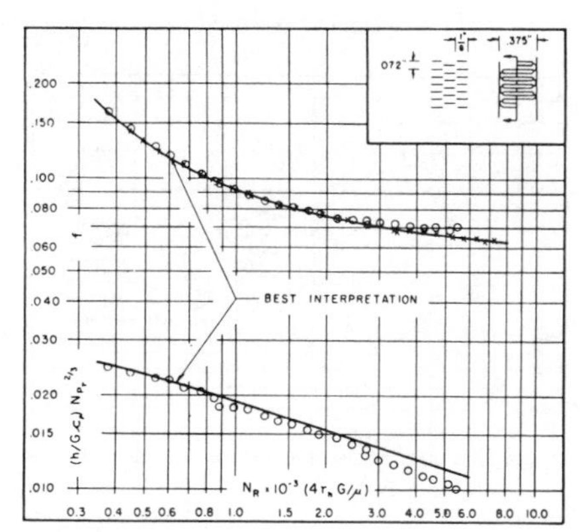

Fin pitch = 13.95 per in.

Plate spacing, $b = 0.375$ in.

Fin length = 0.125 in.

Flow passage hydraulic diameter, $4r_h = 0.00879$ ft

Fin metal thickness = 0.010 in., aluminum

Total heat transfer area/volume between plates, $\beta = 381\ \text{ft}^2/\text{ft}^3$

Fin area/total area = 0.840

Note: The fin surface area on the leading and trailing edges of the fins have not been included in area computations.

(f)

Wavy-fin plate-fin surface 17.8-3/8W.

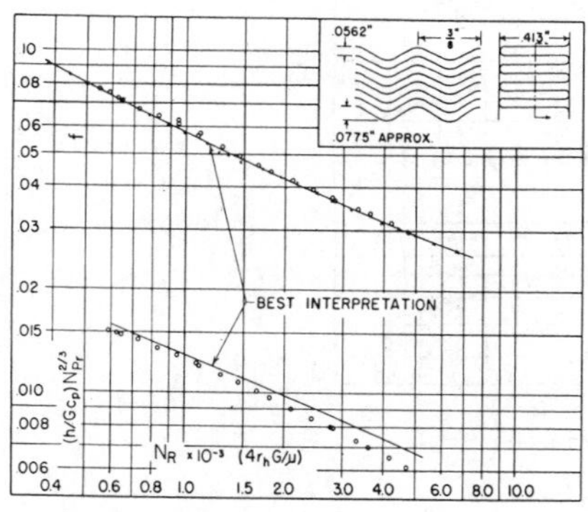

Fin pitch = 17.8 per in.

Plate spacing, $b = 0.413$ in.

Flow passage hydraulic diameter, $4r_h = 0.00696$ ft

Fin metal thickness = 0.006 in., aluminum

Total heat transfer area/volume between plates, $\beta = 514\ \text{ft}^2/\text{ft}^3$

Fin area/total area = 0.892

Note: Hydraulic diameter based on free-flow area normal to mean flow direction.

(g)

Pin-fin plate-fin surface AP-2.

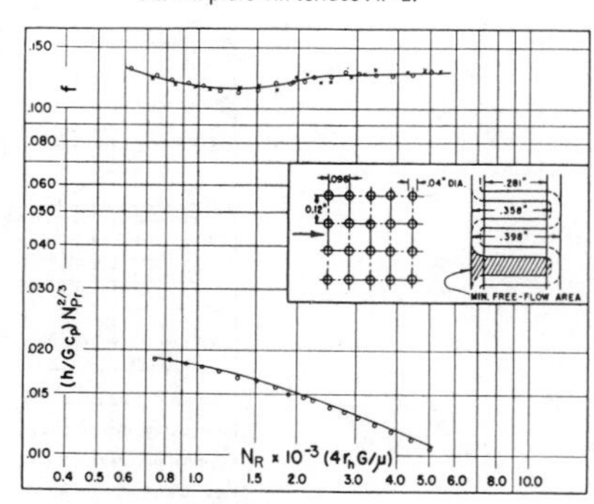

Pin diameter = 0.04 in., copper

Pin pitch parallel to flow = 0.096 in.

Pin pitch perpendicular to flow = 0.12 in.

Plate spacing, $b = 0.398$ in.

Flow passage hydraulic diameter, $4r_h = 0.01172$ ft

Total heat transfer area/volume between plates, $\beta = 204\ \text{ft}^2/\text{ft}^3$

Fin area/total area = 0.686

Note: Plates to which the pins are attached are corrugated with the pins soldered into the corrugations.

(h)

Figure 3-15 (continued)

Finned circular tubes, surface CF-7.0-5/8 J.

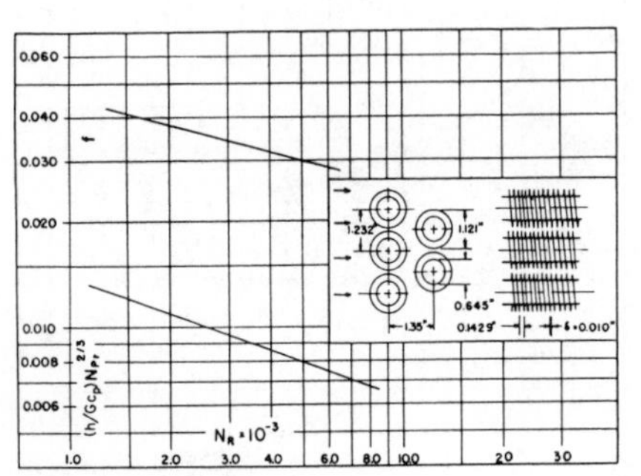

Tube outside diameter = 0.645 in.

Fin pitch = 7.0 per in.

Flow passage hydraulic diameter, $4r_h$ = 0.0219 ft

Fin thickness = 0.010 in.

Free-flow area/frontal area, σ = 0.449

Heat transfer area/total volume, α = 82 ft²/ft³

Fin area/total area = 0.830

Note: Minimum free-flow area is in spaces transverse to flow.

(i)

Finned circular tubes, surface 8.0-3/8T.

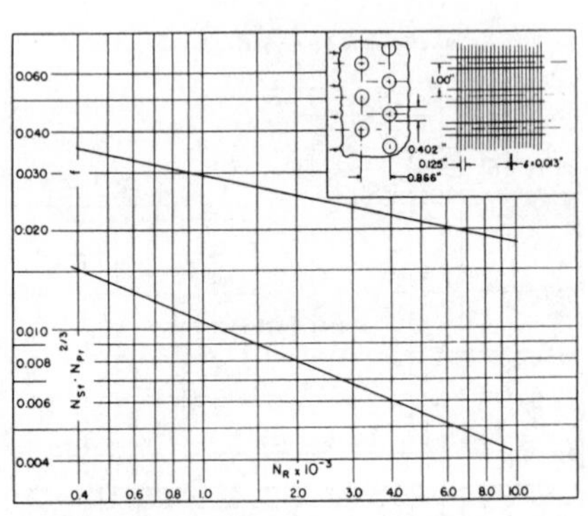

Tube outside diameter = 0.402 in.

Fin pitch = 8.0 per in.

Flow passage hydraulic diameter, $4r_h$ = 0.01192 ft

Fin thickness = 0.013 in.

Free-flow area/frontal area, σ = 0.534

Heat transfer area/total volume, α = 179 ft²/ft³

Fin area/total area = 0.913

Note: Minimum free-flow area in spaces transverse to flow.

(j)

Finned flat tubes, surface 9.29-0.737-SR.

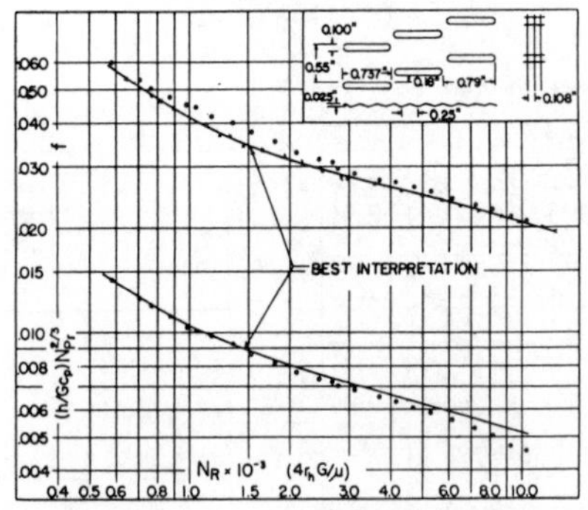

Fin pitch = 9.29 per in.

Flow passage hydraulic diameter, $4r_h$ = 0.01352 ft

Fin metal thickness = 0.004 in., copper

Free-flow area/frontal area, σ = 0.788

Total heat transfer area/total volume, α = 228 ft²/ft³

Fin area/total area = 0.814

(k)

Figure 3-15 (continued)

An example is appropriate at this point.

Example 3-2:

A cross-flow heat exchanger uses surfaces 9.29-0.737–SR. If air at a mean bulk temperature of 150°F and at a pressure of 1 atm flows at 30 ft/s into the exchanger and if water at 30 psia and 90°F flows through the tubes with a mean velocity of 4 ft/s, find the product *UA* per unit volume of the heat exchanger. The tubes are aluminum and the fins copper.

Solution. The product *UA* is given by

$$UA = \frac{1}{\dfrac{1}{A_a \bar{h}_a \eta_{t_a}} + R_{\text{cond}} + \dfrac{1}{A_w \bar{h}_w}}$$

R_{cond} is normally neglected since the other resistances are much larger. An air side analysis and a water side analysis will be made. Information about the surface is given in Table 3-15 and Fig. 3-16.

Air Side

A typical flattened tube and the fin about it appear as in Fig. 3-16. A mean surface temperature of 100°F will be assumed. This must be verified later. At 150°F, the important properties are

$$C_p = 0.240 \text{ Btu/lbm °F}$$

$$\rho = 0.0896 \text{ lbm/ft}^3$$

$$\text{Pr} = 0.72$$

$$\mu = 1.362 \times 10^{-5} \text{ lbm/ft s}$$

The mass velocity G within the core is

$$G = \rho V = \rho \frac{V_\infty}{\sigma} = (0.0896)\frac{30}{0.788} = 3.4112 \text{ lbm/s ft}^2$$

$$\sigma = \frac{A_c}{A_{fr}}$$

and

$$\text{Re}_{D_h} = \frac{G 4 r_h}{\mu} = \frac{(3.4112)(0.01352)}{1.3625 \times 10^{-5}} = 3385$$

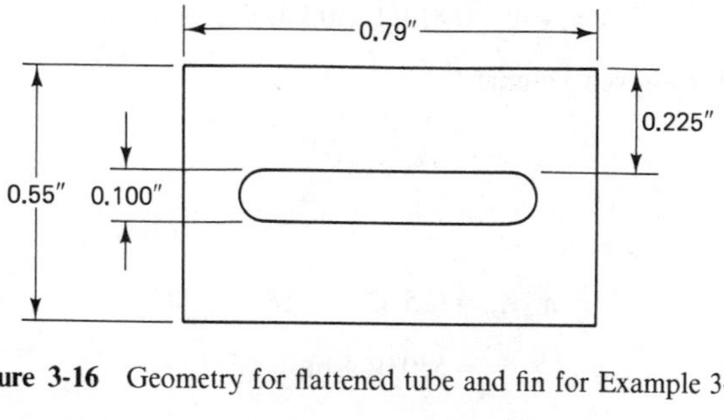

Figure 3-16 Geometry for flattened tube and fin for Example 3-2.

From the information given in Fig. 3-15 for surface 9.29-0.737-SR

$$\frac{\bar{h}}{GC_p}\Pr^{2/3} = 0.0069$$

$$f = 0.027$$

From Table 3-1, no correction need be made for property variations since $m = n = 0$ for flow normal to circular tubes. Then

$$\bar{h} = \frac{GC_p}{\Pr^{2/3}}0.0069 = \frac{3.4112}{(0.72)^{2/3}}(0.24)(0.0069)(3600)$$

$$= 25.32 \text{ Btu/h ft}^2 \text{ °F}$$

The total surface effectiveness is

$$\eta_t = 1 - \frac{A_f}{A}(1 - \eta_f)$$

and the fin parameter for Fig. 2-20 is

$$L_c^{3/2}\left(\frac{\bar{h}}{kA_p}\right)^{1/2}$$

with $L_c = L$ and $A_p = tL$. (The $t/2$ is not added since at the symmetry line between tubes no additional edge surface is exposed.)

Then

$$L = L_c = 0.225 \text{ in.} = 0.01875 \text{ ft}$$

$$A_p = \frac{(0.225)(0.004)}{144} = 0.0000063 \text{ ft}^2$$

and

$$(0.01875)^{3/2}\left(25.32\frac{1}{221}\frac{1}{6.3\times 10^{-6}}\right)^{1/2} = 0.346$$

From Fig. 2-20,

$$\eta_f = 0.90$$

and

$$\eta_t = 1 - 0.814(1 - 0.90) = 0.9186$$

From Fig. 3-14 for a given volume V,

$$A = 228V\frac{\text{ft}^2}{\text{ft}^3}$$

and for a given volume V

$$\bar{h}_a\eta_{ta}A_a = (25.32)(0.9186)(228V)$$

$$= 5303V \text{ Btu/h °F ft}^2$$

Water Side

From Kays and London for the flattened tube, we get (tube thickness = 0.01 in.)

$$\text{free-flow area of one tube} = 0.0560 \text{ ft}^2$$
$$\text{free-flow/frontal area}, \ \sigma_w = 0.129$$
$$\text{transfer area/total volume}, \ \alpha_w = 42.1 \text{ ft}^2/\text{ft}^3$$
$$\text{hydraulic radius}, \ r_h = 0.00306 \text{ ft}$$

For water at 90°F,

$$C = 0.997 \text{ Btu/lbm °F}$$
$$\rho = 62.1 \text{ lbm/ft}^3$$
$$\mu = 0.514 \times 10^{-3} \text{ lbm/ft s}$$
$$\text{Pr} = 5.13$$

The Reynolds number is

$$\text{Re}_{D_h} = \frac{4r_h \rho V}{\mu} = \frac{(4)(0.00306)(62.1)(4)}{0.514 \times 10^{-3}} = 5927$$

and from Eq. (2-70)

$$\overline{\text{Nu}}_{D_h} = 0.023 \text{ Re}_{D_h}^{0.8} \text{ Pr}^{0.4}$$
$$= 0.023(5927)^{0.8}(5.13)^{0.4} = 46.14$$

A temperature correction could be made, but no significant change would result.

$$\bar{h}_w = \overline{\text{Nu}}_{D_h} \frac{k}{4r_h}$$
$$= 46.14 \frac{0.359}{4(0.00306)} = 1353 \text{ Btu/h ft}^2 \text{ °F}$$

For a given volume V, the heat transfer area is

$$A_w = 42.1V \text{ ft}^2/\text{ft}^3$$

and

$$\bar{h}_w A_w = 1353(42.1V) = 56961V \text{ Btu/h ft}^2 \text{ °F}$$

Then

$$UA = \frac{1}{\dfrac{1}{5303V} + \dfrac{1}{56961V}}$$
$$\frac{UA}{V} = 4850 \text{ Btu/h °F} \frac{1}{\text{ft}^3}$$

Thus, for a given volume, UA can be calculated. The estimated mean surface temperature for the preceding is 113°F instead of the assumed 100°F. Another pass could be made using this new surface temperature, but the change in UA/V would be small.

Design Strategy

Much of the previous discussion of design strategy for shell-and-tube heat exchangers is appropriate for cross-flow heat exchangers. Device specifications are generally expressed in terms of rating and allowable pressure drops. Strategies for simultaneously satisfying the rating and pressure drop constraints are as for the shell-and-tube exchangers. However, the cross-flow procedure has at least one more level of iteration since specification of the free-flow area $A_{c,1}$ and length (L_1) for one side does not uniquely determine the free-flow area $A_{c,2}$ and length (L_2) for the other side. This is illustrated by consideration of Fig. 3-17.

For the situation portrayed in Fig. 3-17, the determination of the *product* of $L_2 \cdot L_3$ and the length L_1 is sufficient to satisfy both the rating and pressure drop requirement for fluid 1, but the *individual* values of L_2 and L_3 are not known. The length L_2 as well as the frontal area $L_1 \cdot L_3$ must then be computed to satisfy the rating and pressure drop for fluid 2. Hence the need for at least an additional level of iteration.

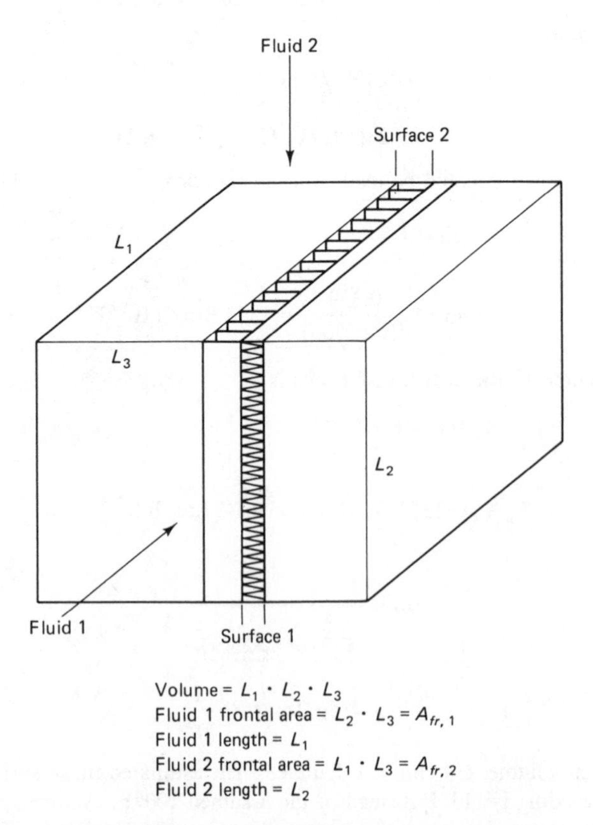

Figure 3-17 Cross-flow heat exchanger geometry.

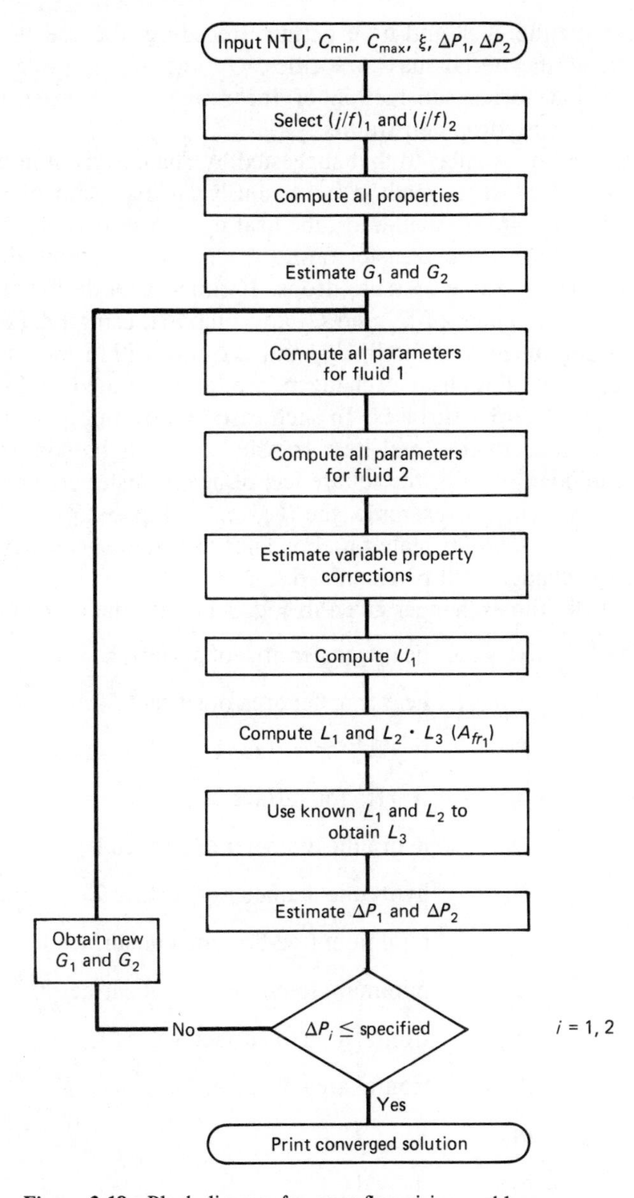

Figure 3-18 Block diagram for cross-flow sizing problem.

As with the shell-and-tube design procedure, the use of Eq. (3-5) for prediction of the initial mass velocities, G_1 and G_2, is preferred, as is the procedure that forces satisfaction of the rating and iterates to satisfy the allowable pressure drop constraints.

One method, similar to that suggested by Shah, is given in block diagram form in Fig. 3-18. This method is obviously an extension of the procedure delineated earlier in the shell-and-tube heat exchanger discussion. Fundamentally, it satisfies the heat transfer rating requirement for both fluids and then checks for the allowable pressure drops. If either or both pressure constraints are not met, new values of G_1 and G_2 are computed using Eq. (2-115), and the process is continued until both ΔP_1 and ΔP_2 are within the allowable limits.

Many cross-flow heat exchangers are, as illustrated in Fig. 3-17, composed of two different surfaces. In such cases, an overall α relating the *entire* exchange volume to the total heat transfer area can be computed. A much more useful quantity is β, the square feet of heat transfer area per unit volume of a given surface. For example, the β given in Fig. 3-15(f) is for that single surface, $\frac{1}{8}$-13.95 strip-fin plate-fin. For later use, some geometry relations for cross-flow exchanges will be examined.

Consider the exchanger given in Fig. 3-17 with the following definitions:

A_1	heat transfer area of surface 1
A_2	heat transfer area of surface 2
β_1	ft^2/ft^3 for surface 1
β_2	ft^2/ft^3 for surface 2
D_{h_1}	hydraulic diameter of surface 1
D_{h_2}	hydraulic diameter of surface 2
$A_{c,1}$	minimum free-flow area of surface 1
$A_{c,2}$	minimum free-flow area of surface 2
$A_{fr,1}$	frontal area of surface 1
$A_{fr,2}$	frontal area of surface 2
σ_1	$A_{c,1}/A_{fr,1}$
σ_2	$A_{c,2}/A_{fr,2}$
L_1	depth parallel to fluid 1
L_2	depth parallel to fluid 2
L_3	length of remaining side
b_1	width of one reach of surface 1

b_2	width of one reach of surface 2
a	wall thickness (parting plate)
N	number of passages
V	total volume of exchanger
V_1	volume of surface 1
V_2	volume of surface 2

Then

$$V_1 = L_1 L_2 b_1 N \tag{3-12a}$$
$$V_2 = L_2 L_1 b_2 N \tag{3-12b}$$
$$A_1 = \beta_1 V_1 \tag{3-12c}$$
$$A_2 = \beta_2 V_2 \tag{3-12d}$$

and the total exchanger volume can be written

$$V = (b_1 N + b_2 N + 2aN) L_1 L_2 \tag{3-12e}$$

With the preceding,

$$\sigma_1 = \frac{A_{c,1} L_1}{A_{fr,1} L_1} = \frac{A_1 D_{h_1}/4}{V} = \frac{\beta_1 V_1 D_{h_1}/4}{V} \tag{3-13a}$$

$$= \frac{L_1 L_2 b_1 \beta_1 N D_{h_1}/4}{(b_1 N + b_2 N + 2aN) L_1 L_2} \tag{3-13b}$$

$$= \frac{b_1 \beta_1 D_{h_1}/4}{b_1 + b_2 + 2a} \tag{3-13c}$$

and in similar fashion

$$\sigma_2 = \frac{b_2 \beta_2 D_{h_2}/4}{b_1 + b_2 + 2a} \tag{3-14}$$

Thus

$$\alpha_1 = \frac{b_1 \beta_1}{b_1 + b_2 + 2a} \tag{3-15}$$

$$\alpha_2 = \frac{b_2 \beta_2}{b_1 + b_2 + 2a} \tag{3-16}$$

Equations (3-12) through (3-16) completely define the cross-flow heat exchanger geometry in terms of parameters typically given for heat transfer surfaces in the literature and as presented by Kays and London (1) and shown in Fig. 3-15.

To assimilate the cross-flow heat exchanger design approach, consider the following example:

Example 3-3

A cross-flow heat exchanger is to be designed that will cool 2 lbm/s of air from 200° to 100°F. The cooling flow is 2 lbm/s of air with an inlet temperature of 50°F. The hot fluid is to be cooled in surface 17.8-$\frac{3}{8}$W, a wavy-fin plate surface; and the cold fluid is to flow through surface 9.03, a plain-fin surface. Data and information for both surfaces were presented earlier. The inlet pressure for both sides is 16 psia, with the prescribed drops of 8 lbf/ft^2 for the hot fluid side and 35 lbf/ft^2 for the cold fluid side. The parting plate that separates the surfaces is 0.0015 ft thick.

Solution. This is the classical cross-flow heat exchanger design problem. The block diagram of Fig. 3-18 is thus applicable. For these conditions

$$\dot{m}_1 = \dot{m}_2 = 2.00 \text{ lbm/s}$$

and

$$q = 48 \text{ Btu/s}$$
$$T_{c_{\text{out}}} = 150°F$$

Then

$$C_c = C_h = C_{\text{min}} = C_{\text{max}} = 0.48 \text{ Btu/s °F}$$

and

$$\xi = \frac{q}{C_{\text{min}}\left(T_{h_{\text{in}}} - T_{c_{\text{in}}}\right)} = \frac{48 \text{ Btu/s}}{0.48 \text{ Btu/s °F } (200 - 50)\text{°F}}$$

$$\xi = 0.6667$$

Surfaces 17.8-$\frac{3}{8}$W and 9.03 are unmixed, so that Eq. (2-32) or Fig. 2-8(e) can be used to obtain

$$\text{NTU} = 2.679$$

The mean temperatures are

$$T_{h_m} = 150\text{°F}$$
$$T_{c_m} = 100\text{°F}$$

and the following chart can be developed

	μ(lbm/ft s)	Pr	C_p(Btu/lbm °F)
Air at 150°F	1.370×10^{-5}	0.72	0.24
Air at 100°F	1.285×10^{-5}	0.72	0.24

$$\rho_{h_{\text{in}}} = 0.06543 \text{ lbm/ft}^3$$
$$\rho_{h_m} = 0.07114 \text{ lbm/ft}^3$$
$$\rho_{h_{\text{out}}} = 0.07685 \text{ lbm/ft}^3$$
$$\rho_{c_{\text{in}}} = 0.08468 \text{ lbm/ft}^3$$
$$\rho_{c_m} = 0.07720 \text{ lbm/ft}^3$$
$$\rho_{c_{\text{out}}} = 0.06972 \text{ lbm/ft}^3$$

Information for the surfaces is as follows:

1. Surface 17.8-$\frac{3}{8}$W:
   $b_1 = 0.03442$ ft
   $\beta_1 = 514.0$ ft^2/ft^3
   $D_{h_1} = 0.00696$ ft
   fin thickness $= \delta_1 = 0.0005$ ft
   fin length $= 0.413$ in./2 $= 0.01721$ ft
2. Surface 9.03:
   $b_2 = 0.06858$ ft
   $\beta_2 = 244.0$ ft^2/ft^3
   $D_{h_2} = 0.01522$ ft
   fin thickness $= \delta_2 = 0.000667$ ft
   fin length $= 0.823$ in./2 $= 0.03429$ ft

From which Eqs. (3-13b), (3-14), (3-15) and (3-16) yield

$$\alpha_1 = \frac{b_1\beta_1}{b_1 + b_2 + 2a} = \frac{(0.03442)(514)}{0.03442 + 0.06858 + 2(0.0015)} \text{ ft}^2/\text{ft}^3$$

$$= 166.89 \text{ ft}^2/\text{ft}^3$$

$$\alpha_2 = 157.87 \text{ ft}^2/\text{ft}^3$$

$$\sigma_1 = \frac{\alpha_1 D_{h_1}}{4} = \frac{(166.89)(0.01522)}{4} = 0.2904$$

and

$$\sigma_2 = \frac{\alpha_2 D_{h_2}}{4} = 0.600$$

With the preceding geometrical and thermophysical property information, the iteration sequence can begin. Air and finned surfaces on both the hot and cold fluid sides suggest that the resistances should be about equal, and that for the initial guess on G_1 and G_2

$$\text{NTU}_h = \text{NTU}_c = 2\text{NTU} = 5.358$$

With no knowledge of the Reynolds number at this point

$$\left(\frac{j}{f}\right)_{17.8\text{-}\frac{3}{8}W} \sim 0.224(\text{Re}_D \sim 1000)$$

$$\left(\frac{j}{f}\right)_{9.03} \sim 0.260(\text{Re}_D \sim 1000)$$

The computations then proceed as the diagram in Fig. 3-18 suggests. For each surface, Re_D is calculated and then $(\bar{h}/GC_p)\text{Pr}^{2/3}$ and f are obtained from the appropriate data presentation. The convective coefficient is recovered; η_f and η_t are calculated; and, finally, U_1 is computed. With U_1, $C_{\min}$, and NTU known, A_1 (and A_2) can be calculated. Then

$$L_1 = \frac{D_{h_1}A_1}{4A_{c,1}} \qquad \text{where } A_{c,1} = \frac{\dot{m}_1}{G_1}$$

and

$$L_2 = \frac{D_{h_2} A_2}{4A_{c,2}} \qquad \text{where } A_{c,2} = \frac{\dot{m}_2}{G_2}$$

Finally,

$$L_3 = \frac{A_{fr,1}}{L_2} = \frac{A_{c,1}}{\sigma_1 L_2}$$

or

$$L_3 = \frac{A_{c,2}}{\sigma_2 L_1}$$

Once L_1, L_2, and L_3 are known, the pressure drops ΔP_1 and ΔP_2 are computed using Eq. (2-115) and the K_e and K_c values from Fig. 2-25(d). If these pressure drops are not within the specified limits, new values of G_1 and G_2 are estimated using

$$G_i = \sqrt{\frac{\Delta P_i 2\rho_{i_{\text{in}}} g_c}{\left(K_c + 1 - \sigma^2\right) + 2\left(\frac{\rho_{i_{\text{in}}}}{\rho_{i_{\text{out}}}} - 1\right) + f\frac{A_i}{A_{c,i}}\frac{\rho_{i_{\text{in}}}}{\rho_{i_m}} - \left(1 - \sigma^2 - K_e\right)\frac{\rho_{i_{\text{in}}}}{\rho_{i_{\text{out}}}}}} \qquad i = 1,2 \tag{3.17}$$

and the iteration continued.

Table 3-6 presents some results from the iteration sequence. Although nine iterations were required for convergence, the results of the second iteration were essentially the converged solution, and the results of iterations 5 through 8 were essentially those of iteration 9. The limit of the accuracy of the iterative procedure is the accuracy of the experimental input, that is, the $(\bar{h}/GC_p)\text{Pr}^{2/3}$ and f from the heat transfer surface information. As Table 3-6 verifies, the iterative process is well behaved and rapid. The results of the design process are presented in Fig. 3-19.

The number of passages N is

$$N = \frac{1.291}{0.03442 + 0.06858 + 2(0.0015)} = 12.17$$

Thus 13 alternate rows of surface 17.8-$\frac{3}{8}$W and surface 9.03 will satisfy the specifications of rating and pressure drops.

TABLE 3-6 Approach to Convergence for Cross-Flow Design Problem

Iteration	G_1 (lbm/ft^2 s)	G_2 (lbm/ft^2 s)	Re_{D_1}	Re_{D_2}	η_{f_1}	η_{f_2}	L_1 (ft)	L_2 (ft)	L_3 (ft)	ΔP_1 (PSF)	ΔP_2 (PSF)
1	1.2348	2.8987	627	3433	0.933	0.916	0.870	4.224	1.320	12.28	18.04
2	0.9967	4.0374	506	4782	0.942	0.888	0.605	5.072	1.362	6.13	41.49
3	1.1382	3.7084	578	4392	0.938	0.896	0.706	4.759	1.271	8.908	32.989
4	1.0787	3.8197	548	4524	0.941	0.893	0.671	4.912	1.300	7.619	36.042
⋮											
9	1.0981	3.7782	558	4475	0.940	0.894	0.683	4.856	1.291	8.04	34.89

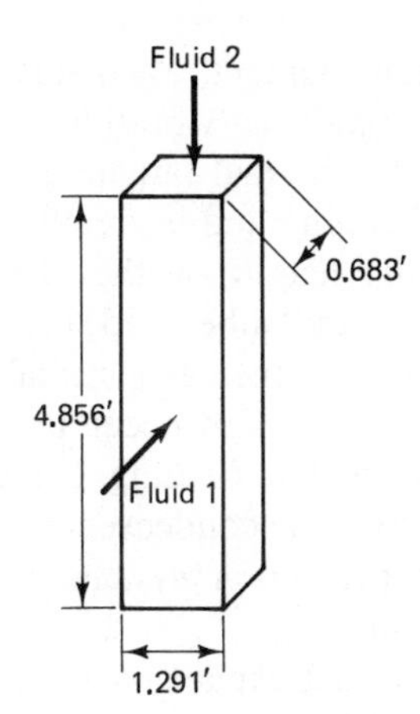

Figure 3-19 External core dimensions for Example 3-3.

3-5 MECHANICAL PROBLEMS

Heat exchangers, even in hostile environments, usually require little maintenance and have a long service life. Problem areas and failures that do occur are usually [Schwartz (6)] identified by type as (1) mechanical, (2) corrosion, (3) mechanical plus chemical-induced corrosion, or (4) scale, mud, and algae fouling.

The survey paper by Schwartz (6) lists seven forms of mechanical failure (1) metal erosion, (2) waterhammer, (3) vibration, (4) thermal fatigue, (5) freeze-up, (6) thermal expansion, and (7) loss of cooling water. Metal erosion, on either the shell or tube side of the heat exchanger, is caused by excessive fluid velocities and results in the metal wearing away. Existing corrosion problems are usually exacerbated, since the wearing continuously exposes fresh metal to further corrosion attack. Most metal erosion problems occur inside tubes, either in the entrance areas or the U bend of U-tube-type heat exchangers.

The tube entrance area metal erosion is caused by high velocity, turbulence, and vortices that are formed as the fluid is divided into smaller streams at the tube sheet. Schwartz suggests that fluid velocity inside tubes be limited to 7.5 ft/s for copper and 11 ft/s for steel, stainless steel, or copper–nickel.

Metal erosion on the outside of tubes is generally caused by the impingement of wet high-velocity gases (such as steam) and can be controlled by impingement baffles or oversizing inlets.

Waterhammer, examined in Chapter 6, can result in mechanical failure in heat exchangers, as well as in other piping system components. Schwartz states that waterhammer pressure surges up to 20,000 psi have been measured. Such pressure surges are of sufficient magnitude to rupture or collapse tubing.

Vibration, either from rotating equipment or flow induced, can cause failures in the form of fatigue stress cracking or tubing erosion at the baffle–tube contact. Shell side velocities less than 4 ft/s are recommended to limit flow-induced vibrations.

Tube thermal fatigue, especially in U-bend areas, is caused by repeated thermal cycling and is greatly aggravated by increasing temperature difference across the length of the U-bend tube. The temperature difference causes flexing of the tube, and the resulting additive stress eventually results in cracking. In some cases a total tube break can result.

Freeze-up can occur in any heat exchanger in which the temperatures drop below the freezing points of either fluid, but it is more common in evaporators or condensers. Freeze-up is generally caused by failure of a thermal protection system, improper winter operation, or sudden loss of refrigerant.

Thermal expansion failures can happen in any type of heat exchanger when fluid that is being heated is shut off without provisions to absorb thermal expansion. The continued heating causes thermal expansion, which can, if severe enough, exceed the strength of tube sheets and other heat exchanger components. Cast-iron parts usually fracture, whereas steel components become permanently distorted.

The last of the seven modes of mechanical failure, loss of cooling water, can result in melting or warping of tubing if adequate cooling fluid is not supplied.

As with mechanical failure, Schwartz recognizes seven types of chemically induced corrosion failures: (1) general corrosion, (2) pitting corrosion, (3) stress corrosion, (4) dezincification, (5) galvanic corrosion, (6) crevice corrosion, and (7) condenstate grooving. General corrosion is characterized by relatively uniform corrosion over a given component. Its uniformity may make general corrosion very difficult to spot until some component becomes so thin that rupture occurs. Proper chemical treatment and material selection can minimize general corrosion.

Pitting corrosion, which is usually encountered in ferrous and nonferrous metals, results from the electrochemical potential set up by differences in oxygen concentration within and without the pit. In a "concentration cell," the oxygen-starved pit acts as an anode while the metal surface is the cathode. The danger of pitting corrosion is that any *one* pit can cause failure. Scratches, surface defects, breaks in protective scale layers, and grain boundary conditions increase the susceptibility for pitting corrosion.

Stress corrosion attacks grain boundaries in stressed areas. Failures take the form of fine cracks that follow stress lines or grain boundaries. Stainless steel is particularly susceptible to stress corrosion when chlorine is present. Ammonia seems to cause or intensify this form of corrosion on copper or copper alloys. Copper–nickel alloys have good resistances to stress corrosion in the presence of ammonia.

Dezincification occurs when copper–zinc alloys (containing less than 85 percent copper) are exposed to water with a high oxygen and carbon dioxide content. This corrosion creates a porous surface in which the zinc is removed chemically from the alloy. Dezincification can be prevented by using brass with a low zinc content.

Galvanic corrosion is caused when dissimilar metals are joined in the presence of an electrolyte. The less noble metal corrodes faster. Proper material selection can decrease galvanic corrosion.

Crevice corrosion, as its name implies, originates in and around hidden and secluded areas such as baffles and tubes. This corrosion, which produces local pits, is in sharp contrast to vibration, which produces metal that is sharply cut. Crevice corrosion develops local cells that have metal loss.

Condensate grooving is caused by condensate draining in riverlets from the outside of tubes. An irregular groove or channel is cut into the tube by a corrosion cell in the wetted area. Condensate grooving can be reduced by ensuring uniform wetting of the outside of tubes.

Combinations of mechanical and chemically induced corrosion failure are common. Often, the combination produces a quicker failure than either the mechanical or corrosion mode. Schwartz suggests two common types of combination mechanical and corrosion failures: (1) erosion–corrosion and (2) corrosion–fatigue. As we might have suspected, any erosion of protective film will cause enhanced corrosion, thus the erosion–corrosion combination. The corrosion–fatigue combination results from cyclic stresses rupturing protected areas, leaving them permeable and open to accelerated corrosion.

The last mechanical problem area is scale, mud, and algae fouling. Fouling results in the deposition of a film on heat transfer surfaces. Scale results when dissolved minerals are precipitated out of heat transfer fluids. The rate of precipitation is *reduced* with increasing fluid velocity; but remember that metal erosion is *increased* with increasing fluid velocity. A trade-off obviously must be made to decide the optimum fluid velocity that will give the least scale–erosion combination.

Suspended solids offer the same choice: velocities that are high enough to keep particles in suspension result in accelerated erosion. Again, a compromise must be effected.

Algae can cause serious problems if they get inside a heat exchanger. In many instances, the environment *inside* the exchanger promotes the rapid growth of such algae. Flow is restricted and heat transfer impeded by algae growth on heat transfer surfaces. Chemicals and high fluid velocities discourage algae and other marine growth and attachment.

This section has presented a short overview of mechanical problems in heat exchangers. It was abstracted from Schwartz (6) and is only an introduction. Further information is available in the current heat exchanger literature and, for corrosion and corrosion-coupled problems, the materials literature. Various heat exchanger component suppliers also provide information relating to these problems. The recommended good practices section of the *TEMA Standards* is excellent for shell-and-tube heat exchangers.

3-6 CONCLUDING REMARKS

The information presented in this chapter is sufficient for detailed design considerations of many different types of heat exchangers. Specialized heat exchanger devices are not covered, and the design of such devices will, in general, require additional supporting information from the literature. The heat exchanger literature base is enormous, continues to expand, and must be continually perused if the interested individual is to stay abreast of state-of-the-art developments.

Longitudinal conduction effects have not been addressed in this discussion. Articles by Chiou (7) and Mondt (8) are appropriate for an introductory study into longitudinal heat conduction effects.

REFERENCES

1. Kays, W. M., and London, A. L., *Compact Heat Exchangers*, 2nd ed. New York: McGraw-Hill, 1964.
2. Shah, R. K., "Compact Heat Exchangers" and "Compact Heat Exchanger Design Procedures," in Kakac, Bergles, and Mayinger, eds., *Thermal-Hydraulic Fundamentals and Design*, New York: Hemisphere, 1981.
3. Rubin, F. L., "Introduction to the Interrelationships between Codes, Standards, and Customer Specifications for Process Heat Transfer Equipment," in "The Interrelationships between Codes, Standards, and Customer Specifications for Process Heat Transfer Equipment," presented at the 1979 Winter Annual Meeting of the ASME, December 2–7, 1979, New York, pp. 1–2.
4. *Standards of the Tubular Exchanger Manufacturers Association*, 6th ed. New York: TEMA, 1978.
5. Webb, R. L., "Air-Side Heat Transfer in Finned Tube Heat Exchangers," *Heat Transfer Engineering*, Vol. 1, No. 3, Jan.–March 1980, pp. 33–49.
6. Schwartz, M. P., "Four Types of Heat Exchanger Failures," *Plant Engineering*, File 3050, December 1982.
7. Chiou, J. P., "The Advancement of Compact Heat Exchanger Theory Considering the Effects of Longitudinal Heat Conduction and Flow Nonuniformity," in *Compact Heat Exchangers—History, Technological Advancement and Mechanical Design Problems*, ASME HTD-Vol. 10, November 1980.
8. Mondt, J. R., "Correlating the Effects of Longitudinal Heat Conduction on Exchanger Performance," in *Compact Heat Exchangers—History, Technological Advancement and Mechanical Design Problems*, ASME HTD-Vol. 10, November 1980.

REVIEW QUESTIONS

1. For what type of use are TEMA class C heat exchangers specified?

2. Sketch and label the important dimensions for a TEMA size 26-163 type AFN.
3. What does the word "CODE" mean in the TEMA standards?
4. Is surface FTD-2 (Fig. 3-15b) compact?
5. Why are fins used in compact cross-flow heat exchangers?
6. A shell-and-tube heat exchanger is designed for a given (specified) pressure drop along each tube. What happens to (1) the number of tubes required, (2) the friction factor, and (3) the heat transfer coefficient as the tube length is increased?
7. Explain why fouling degrades UA.
8. Sketch a TEMA size 20–90 type BGU exchanger.
9. Is a surface with $\alpha = 125$ ft^2/ft^3 compact?
10. For surface 8.0-$\frac{3}{8}$T, if $\eta_f = 0.60$, what is η_t?
11. What is a tube sheet?

PROBLEMS

1. If the tube inside diameter is 0.300 in., find α for the tube side flow for surface 8.0-$\frac{3}{8}$T.
2. A surface 8.0-$\frac{3}{8}$T is used as a heat exchanger core. A cubic section of 1 ft^3 is used. If $\bar{h}_{\text{air}} = 10$ Btu/h ft^2 °F and $\bar{h}_w = 400$ Btu/h ft^2 °F, evaluate *UA*. The fin material is aluminum; tube thickness is 0.01 in.
3. For pin-fin plate-fin surface AP-2, specify the following:
   (a) f at $G4r_h/\mu = 2000$.
   (b) Sr Pr$^{2/3}$ at $G4r_h/\mu = 2000$.
   (c) η_t for $\eta_f = 0.60$.
   (d) Heat transfer area for a 400-ft^3 block core of this surface.
4. A compact heat exchanger using surface 8.0-$\frac{3}{8}$T is to be designed. The entrance and exit temperatures, as well as the flow rates of each fluid, are given. List the steps and describe how you would design a heat exchanger (if air flows over the tubes and water through the tubes) such that a specified *water* pressure drop must be maintained.
5. Write a computer program to solve the following problem: In a shell-and-tube heat exchanger consisting of 1-in. schedule 40 steel pipe, 10,000 standard cubic feet per minute of air is to be heated from 1 atm and 70° to 205°F. On the outside of the tubes, steam is condensing at 220°F. A fan is available that will deliver cold air to the entrance header at 70°F at a static pressure of 1.5 in. of water gauge. The hot air is to be discharged to the atmosphere. The cross-sectional areas of the entrance and exit headers are twice the total internal cross-sectional area of the tubes.
   (a) Verify that the resistance on the steam side and the tube wall conductive resistance are negligible.
   (b) Calculate the number and length of each tube required.
   (c) If the fan–motor combination has an efficiency of 0.39, what size motor is required?
6. Water is to be heated from 55° to 105°F at the rate of 275,000 gal/h in a single-pass shell-and-tube heat exchanger consisting of 1-in. schedule 40 steel pipe. The surface

coefficient on the steam side is estimated to be 2000 Btu/h ft^2 °F. A pump is available that can deliver the desired quantity of water provided the pressure drop through the pipes does not exceed 1 psi. Water is supplied at 100 psia. The pump is to be used to make up the pressure drop. Saturated steam is supplied at 30 psia to the shell.

(a) Calculate the number of tubes in parallel and the length of each tube required to fulfill the requirement.
(b) Present the information on the TEMA form.
(c) How much steam is condensed on the shell side?

7. A direct-transfer intercooler for a gas turbine is to be designed using either surface 8.0-$\frac{3}{8}$T or surface 9.29-0.737-SR. The requirements are:

Air side	Water side
$\dot{m} = 150{,}000$ lbm/h	$\dot{m} = 300{,}000$ lbm/h
$T_{in} = 725$°R	$T_{in} = 65$°F
$T_{exit} = 500$°R	
$P_{in} = 40.0$ psia	

The heat exchanger can be the necessary width, but the air inlet side must be 1.5 ft × 4 ft. Water flows through the tubes. Select the better surface based on economy of operation by calculating the following items:

(a) The heat exchanger width required for both surfaces.
(b) The pressure drop (air side) for both designs.
(c) If electricity is 15¢/kWh and the air pump is 53 percent efficient, what are the operating costs for each pump for 2000 h/year operation?

4

PRIME MOVERS

4-1 INTRODUCTION

An integral part of many energy systems is a prime mover, a device that can overcome head losses and/or a device that can increase the pressure of a fluid. We usually call prime movers pumps or fans. Pumps come in an amazing variety of sizes and types and present many diverse characteristics. However, a pump does not operate in isolation, and the complementary characteristics of the system into which a pump is placed are of equal concern and importance as the pump characteristics. In fact, it is the combination of pump and system characteristics that determines at what level the pump and, hence, the system will operate.

Although we shall be primarily concerned with centrifugal pumps in this chapter, the characteristics of other types will be examined. Extensive use will be made of manufacturers' performance data and specifications. Pump placement in systems will be examined, as will pump operation in systems. Fans, a special category of prime movers, and their operating characteristics will be explored.

As in the previous chapters, very little new information or original concepts are presented here, but many sources have been sifted for pertinent material. *Pump Application Engineering* by Hicks and Edwards (1), the *Pump Handbook* by Karassik et al. (2), *Fundamentals and Application of Centrifugal Pumps* by Benaroya (3), and *Fans* by Osborne (4) are frequently used, as are short survey and applications articles from *Plant Engineering* and *Power*. For many industrial uses the *Standards of the Hydraulic Institute* (5) provides a common set of nomenclature and standards. The problem of accepted standards for centrifugal pumps is immense if for no other reason than the large number of manufacturers of such devices. The *Thomas Register* (6) lists over 180 pages of manufacturers for these items! The *Thomas Register* is a valuable source of information on manufacturers' products and addresses. Extensive listings are provided for all the hardware items considered in this chapter.

4-2 PUMP CHARACTERISTICS

Pump nomenclature and classification are dominated by the concepts of class and type. Classification according to *class* is based on the hydraulic characteristics of a device, while classification according to *type* is based on the specific physical attribute of the application. Table 4-1 attempts to delineate between the better known classes and types.

The classes, *centrifugal*, *rotary*, and *reciprocating*, refer only to the mechanics associated with moving the fluid; no designation of service is implied. Each class is further subdivided into a number of different types. A third descriptor, detail, is sometimes included but the Hydraulic Institute recommends that only class and type be used. The *Standards of the Hydraulic Institute* (5) classify centrifugal pumps by physical attributes such as (1)

TABLE 4-1 Pump Classes and Types

Class	Type
Centrifugal	Volute
	Diffuser
	Regenerative turbine
	Vertical turbine
	Mixed flow
	Axial flow (propeller)
Rotary	Gear
	Vane
	Cam and piston
	Screw
	Lobe
	Shuttle block
Reciprocating	Direct acting
	Power
	Diaphragm
	Rotary piston

Used with permission, from T. G. Hicks and T. W. Edwards, *Pump Application Engineering*, McGraw-Hill Book Company, 1971.

number of stages (single or multistage), (2) casing type (volute, circular, or diffuser), (3) shaft position (horizontal or vertical), (4) suction (single or double), and (5) construction material (bronze fitted, all bronze, specific composition bronze, all iron, stainless steel fitted, and all stainless steel).

Different classes of pumps have different characteristics; the variations that exist between classes determine which class (and type) is appropriate for a given use. Table 4-2 indicates the salient features of various characteristics for each class. We should note that the discharge of both centrifugal and rotary pumps is steady, whereas the discharge of reciprocating pumps is pulsating. In terms of the versatility of liquids handled, the rotary and reciprocating are not nearly so versatile as the centrifugal. The discharge pressure–capacity relationship of each class is dependent on the mechanical principle used to add energy (enthalpy) to the fluid and will be investigated later.

TABLE 4-2 Pump-Type Attributes

	Centrifugal		Rotary	Reciprocating		
	Volute and Diffuser	Axial Flow	Screw and Gear	Direct-acting Steam	Double-acting Power	Triplex
Discharge flow	Steady	Steady	Steady	Pulsating	Pulsating	Pulsating
Usual max suction lift, ft	15	15	22	22	22	22
Liquids handled	Clean, clear; dirty, abrasive; liquids with high solids content		Viscous, nonabrasive	Clean and clear		
Discharge pressure range	Low to high		Medium	Low to highest produced		
Usual capacity range	Small to largest available		Small to medium	Relatively small		
How increased head affects:						
Capacity	Decrease		None	Decrease	None	None
Power input	Depends on specific speed		Increase	Increase	Increase	Increase
How decreased head affects:						
Capacity	Increase		None	Small increase	None	None
Power input	Depends on specific speed		Decrease	Decrease	Decrease	Decrease

Used with permission, from T. G. Hicks and T. W. Edwards, *Pump Application Engineering*, McGraw-Hill Book Company, 1971

We shall be primarily concerned with centrifugal pumps in this chapter because their inherent simplicity, low initial cost, quiet flow, and low maintenance costs make them the most widely used class for general industrial operation. Nonetheless we shall examine in some detail the rotary and reciprocating pumps. This is desirable both from a complementary sense for comparison with centrifugal pumps and because of the significant use made of the classes.

Both rotary pumps and reciprocating pumps are usually classed as positive-displacement pumps since the pumping action is not continuous, as in a centrifugal pump, but a result of action on an identifiable finite volume of fluid. We usually think of a centrifugal pump as "throwing" the fluid outward, whereas the rotary pump traps a given volume of fluid and pushes it around a closed casing. The output is basically steady, since the volumes of the trapped fluid are small and since the cyclic rate of discharge of the trapped volumes is large.

Many different tactics have been proposed and used in rotary pumps. The most common rotary pump types are (1) cam and piston (plunger), (2)

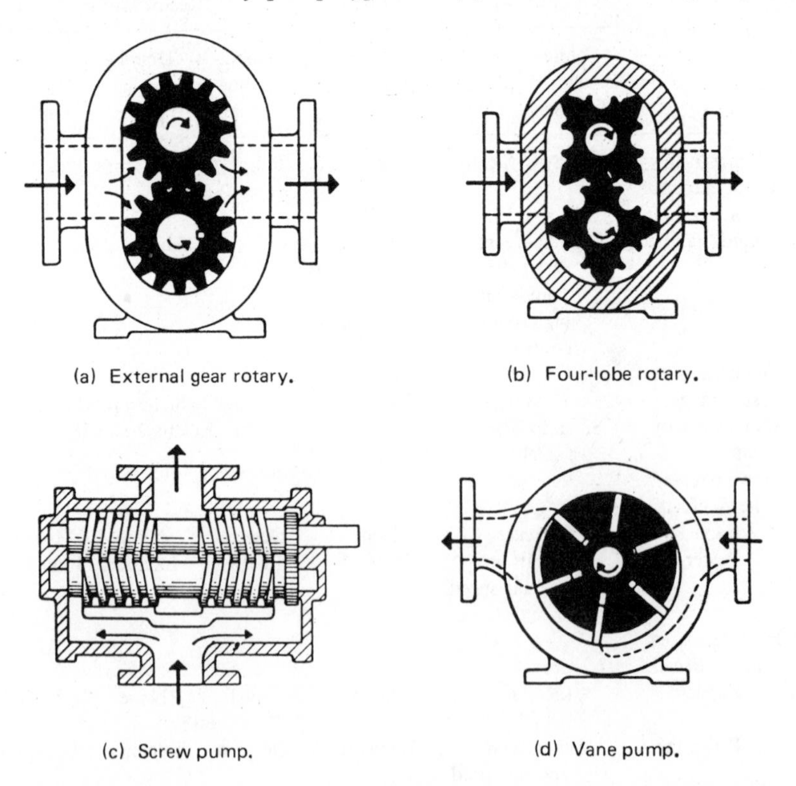

(a) External gear rotary.

(b) Four-lobe rotary.

(c) Screw pump.

(d) Vane pump.

Figure 4-1 Rotary pump types. (Used with permission, from T. G. Hicks and T. W. Edwards, *Pump Application Engineering*, McGraw-Hill Book Company, 1971.)

external gear, (3) internal gear, (4) lobular, (5) screw, and (6) vane. Figure 4-1 shows schematics of several types of rotary pumps. The *external-gear* rotary pump, schematically shown in Fig. 4-1(a), develops pumping action by entrapping liquid between the rotating teeth, moving the fluid around the interior edge of the casing, and then squeezing out the fluid as the teeth mesh. The steadiness of the discharge results from the relatively large number of teeth involved, each tooth discharging a small amount of liquid. The *lobular pump*, shown in Fig. 4-1(b), is similar to the external-gear pump except that the smaller number of lobes delivers larger quantities of fluid than the gears and, as a result, the discharge is not quite so constant as from the gear type.

The *screw pump*, shown in Fig. 4-1(c), traps fluid within the threads and eventually forces the fluid out as discharge. *Vane pumps*, schematically illustrated in Fig. 4-1(d), trap the liquid when centrifugal force holds the vanes against the casing core. The fluid trapped between two adjacent vanes is forced around and discharged. While other variations are possible with rotary pumps, the strategy of trap, transport, and discharge is followed.

Rotary pumps deliver almost constant capacity against a variable discharge pressure. Displacement of a rotary pump varies directly as the speed to a first-order approximation. The dominant source of loss in capacity of a rotary pump is through the clearances between the casing and rotating ele-

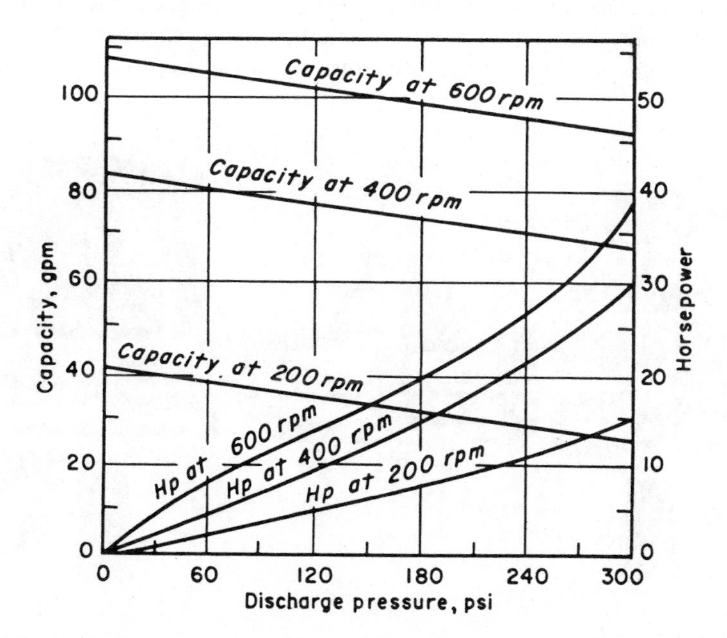

Figure 4-2 Rotary pump performance characteristics. (Used with permission, from T. G. Hicks and T. W. Edwards, *Pump Application Engineering*, McGraw-Hill Book Company, 1971.)

ments. Figure 4-2 presents typical performance curves for an external-gear rotary pump. Both the capacity and the horsepower (hp) are shown as a function of discharge pressure. The parameter rpm is used to illustrate speed effects. At any given speed (rpm), the capacity is a weak function of discharge pressure, slowly decreasing as the discharge pressure increases. For a given discharge pressure, the capacity increases with speed. Horsepower required increases with both discharge pressure and speed. Rotary pumps are widely used for transfer, recirculation, and metering of many different liquids.

The other class of positive-displacement pumps is reciprocating. As the name implies, a piston moving the stroke distance is responsible for the pumping action. The devices are positive-displacement since an identifiable fluid volume is taken in and discharged by the action of the piston (and various valves). The two dominant types of this class are direct acting and power. *Direct-acting reciprocating* pumps are steam driven and are built as simplex (one steam and one liquid piston) or duplex (two steam and two liquid pistons). The pumps are called direct acting since they are steam powered and require no other source of energy input. A cutaway sketch of a duplex piston is shown in Fig. 4-3. The steam end is at the left, and the fluid end at the right. The discharge is pulsating.

Power pumps are the other type of reciprocating pumps we shall examine. Power pumps are so named because external energy is required to run them. At a constant speed, power pumps deliver nearly constant capacity over a wide range of heads. Power pumps will develop a high pressure before they stall and are commonly fitted with a discharge relief valve to protect the pump and

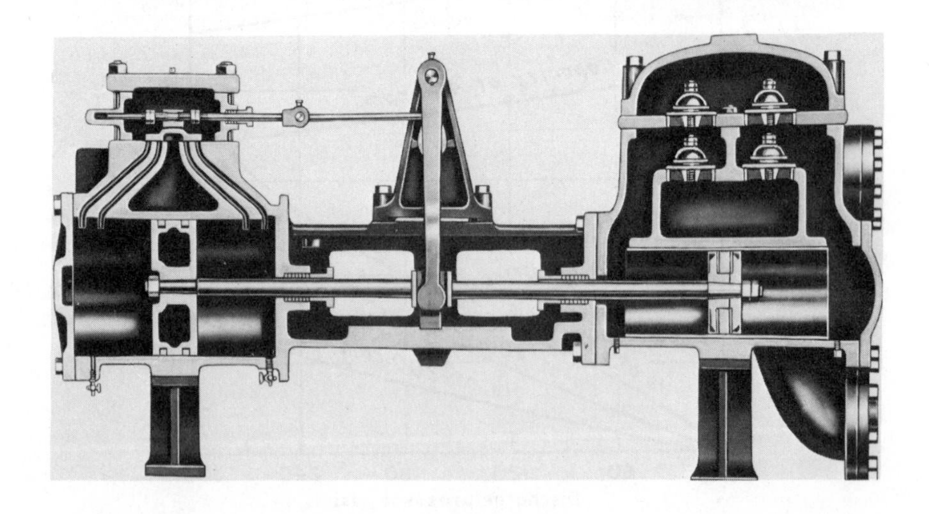

Figure 4-3 Direct-acting duplex pump. (Used with permission, from Warren Pump Division, Houdaille Industries, Inc., Warren, Mass.)

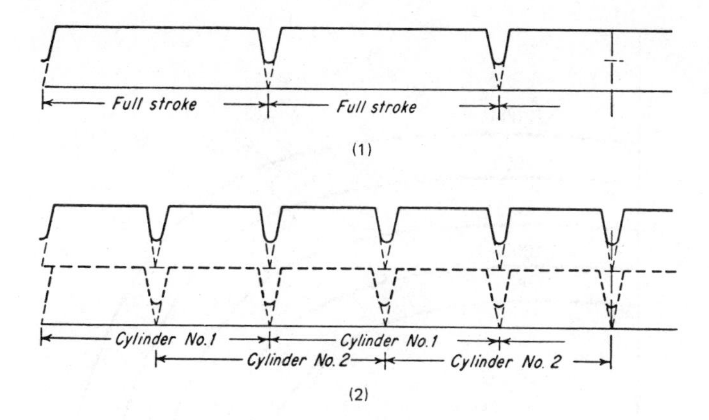

(a) Simplex (1) and duplex (2) direct acting pumps.

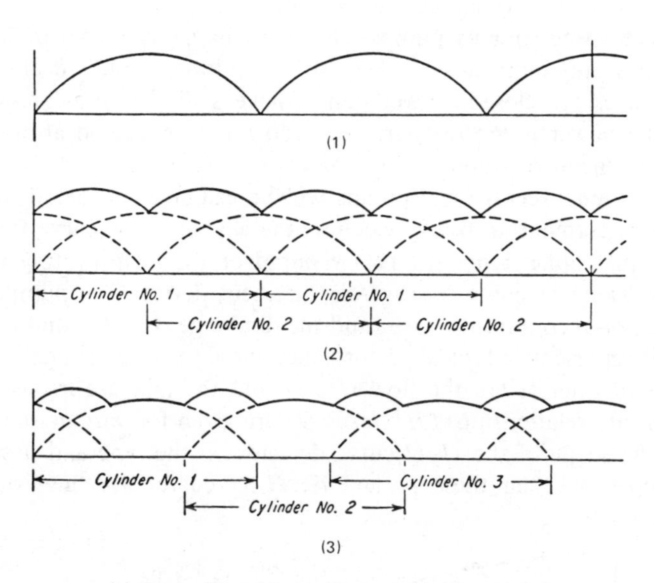

(b) Simplex (1), and duplex (2), and triplex (3) power pumps.

Figure 4-4 Reciprocating pump discharge characteristics. (Used with permission, from T. G. Hicks and T. W. Edwards, *Pump Application Engineering*, McGraw-Hill Book Company, 1971.)

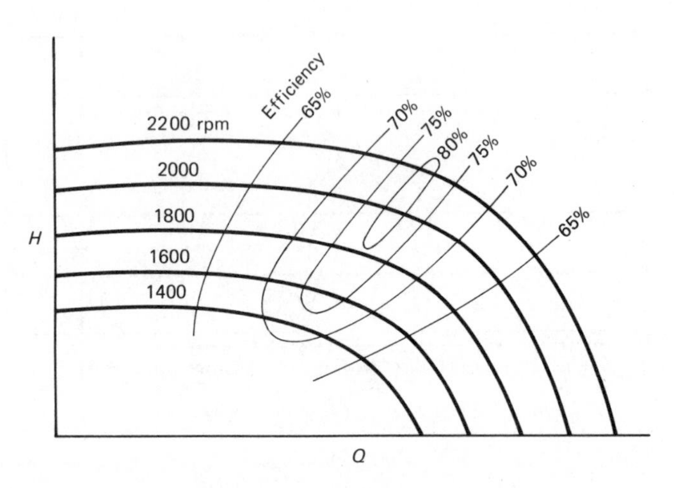

Figure 4-5 Variable-speed head–capacity curves.

piping within the system. Another type of power pump is the *diaphragm* pump, which is used to handle low capacities.

Direct-acting simplex pumps have a discharge as shown in Fig. 4-4(a). The flow is steady until the end of the stroke and then rapidly decreases as the exhaust valves are closed. Power pumps have a discharge as shown in Fig. 4-4(b). The power pump discharge flow does not change so abruptly as the discharge of the direct-acting pump.

Even though centrifugal pumps will be examined in detail in the next section, it is germane to briefly examine the characteristic curve for a centrifugal pump. Unlike either the rotary pump or the reciprocating pump, the *centrifugal pump* is not a positive-displacement device. The pumping action takes place on a continuous basis, and the discharge is steady and continuous. Figure 4-5 represents a typical performance curve for a centrifugal pump. The quantities of interest are the flow rate Q and the increase in head H. The head–capacity relationships (H versus Q) are given for various speeds (rpm). The specific shape of the H–Q curve depends on the type and design of the pump. Superposed on these parametric H–Q curves are lines of constant

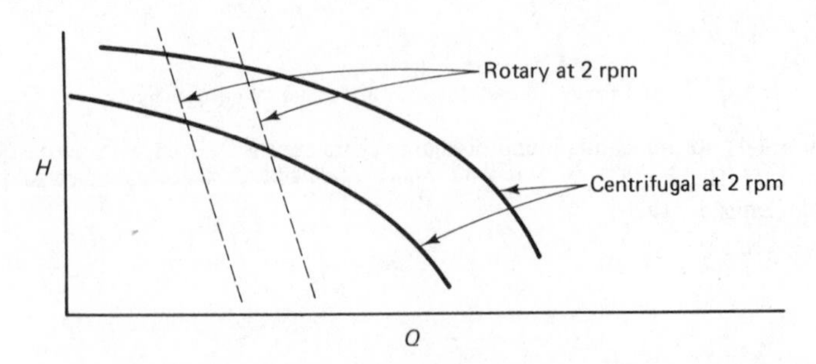

Figure 4-6 Centrifugal and rotary pump head–capacity relationships.

efficiency (isoefficiency curves). These are sometimes called *efficiency islands* since the isoefficiency lines would form closed curves if a sufficiently wide range of speeds and discharge rates were considered.

The head–capacity curve for a centrifugal pump generally has a negative slope (Fig. 4-5, for example), with the slope becoming more negative as the capacity increases. The rotary pump, on the other hand, delivers almost constant capacity against a variable discharge pressure. The difference between the centrifugal and rotary pump head–capacity curve is illustrated in Fig. 4-6. The small change in capacity for a wide range of head for the rotary pump is evident. Performance curves for all classes of pumps are available from the manufacturers.

4-3 CENTRIFUGAL PUMPS

Centrifugal pumps constitute one of the three classes of prime movers we examined in the previous section of this chapter. In the remaining portion of this chapter, we shall use the expression centrifugal pump in a slightly more restrictive sense; hereinafter centrifugal pump will refer to a turbomachine in which the dominant direction of flow during the energy transfer process is *radial*. Figure 4-7 shows by means of a cutaway schematic the velocity pathlines for a typical diffuser-type centrifugal pump. The rotating element of the device is called the impeller and is where the energy *transfer* process occurs. The diffuser is stationary and is responsible for the *transformation* of velocity head, $V^2/2g_c$, into static pressure. The pumped fluid is normally introduced in the eye through an inlet normal to the plane of the paper. The path of the fluid is then directed radially outward.

The energy transfer *mechanism* in the rotor results from the change in angular momentum of the fluid. The integral moment of momentum conserva-

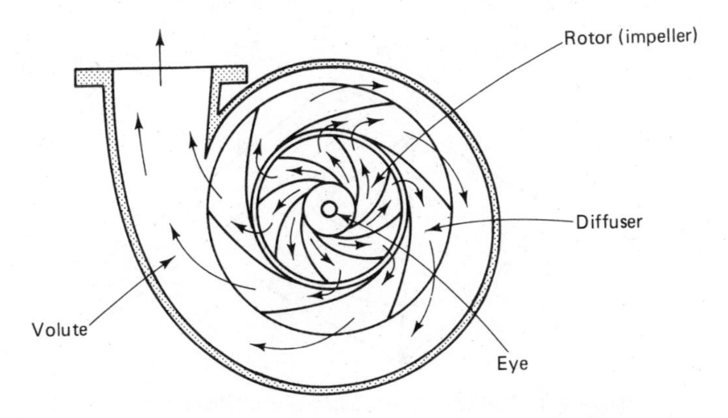

Figure 4-7 Centrifugal pump cutaway schematic.

tion equation provides the basis for all turbomachinery analysis. Recalling from Chapter 1 the integral form

$$\sum \overline{M} = \frac{\partial}{\partial t}\int_{cv}(\bar{r}\times\overline{V})\frac{\rho}{g_c}\,d\text{Vol} + \int_{cs}(\bar{r}\times\overline{V})\frac{\rho\overline{V}}{g_c}\cdot d\overline{A} \tag{4-1}$$

We wish to apply Eq. (4-1) about the axis of rotation of the control volume illustrated in Fig. 4-8, where the resultant torque on the shaft is τ and the symbol V_u denotes the component of the vector $\overline{V}$ in the direction of U, the tangential wheel speed (at a given $U = r\omega$). Applying Eq. (4-1) about the axis of rotation of the control volume under the assumptions of a steady-state frictionless flow results in

$$\tau = \int_{cv} rV_u\frac{\rho\overline{V}}{g_c}\cdot d\overline{A} \tag{4-2}$$

Equation (4-2) shows explicitly that the time rate of change of angular momentum is equal to the shaft torque. If we make the further assumptions of *uniform* flow at the inlet and exit and of an effective *mean* radius, then

$$\tau = \frac{\dot{m}}{g_c}\left(r_2V_{u_2} - r_1V_{u_1}\right) \tag{4-3}$$

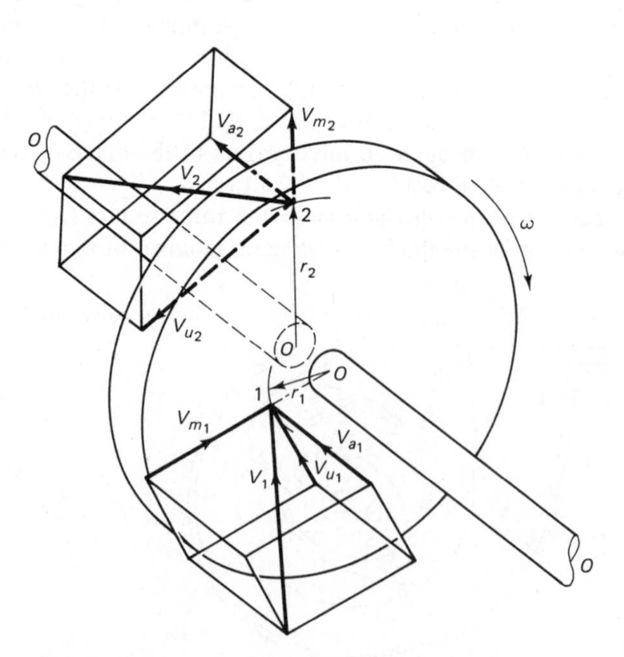

Figure 4-8 Turbomachinery analysis control volume.

and the power becomes

$$P = \tau\omega = \frac{\dot{m}\left(r_2\omega V_{u_2} - r_1\omega V_{u_1}\right)}{g_c}$$
$$= \frac{\dot{m}\left(U_2V_{u_2} - U_1V_{u_1}\right)}{g_c} \tag{4-4}$$

where $U = r\omega$. The increase in head is

$$H = \frac{P}{\dot{m}} = \frac{U_2V_{u_2} - U_1V_{u_1}}{g_c} \tag{4-5}$$

If $U_2V_{u_2}$ is greater than $U_1V_{u_1}$, then the device functions as a compressor; if $U_2V_{u_2}$ is less than $U_1V_{u_1}$, then energy is extracted from the flow and the device functions as a turbine. Equation (4-5) is known as the Euler pump (or turbine) equation and is the most fundamental expression in turbomachinery. Although the assumptions made in the derivation of Eq. (4-5) are severe and are restrictive, they are "nearly" satisfied in a number of situations, and the results are adequate for many preliminary purposes.

The usual way of applying Eqs. (4-3) to (4-5) is via the velocity triangle. Three different velocities can be identified in a rotating machine: the *absolute* velocity of the fluid, denoted by V, the *tangential wheel* speed, denoted by U, and the *relative velocity* of the fluid, denoted by V_r. Vectorially, we express the relationship among these velocities as

$$\overline{V} = \overline{U} + \overline{V}_r$$

and represent them schematically as in Fig. 4-9. Placed in velocity triangle form, V_u is readily identified.

Let us consider, as depicted in Fig. 4-10(a), a simple centrifugal pump impeller from a pump that has radial flow. The fluid enters the rotor (impeller) with an absolute velocity that is completely radial. Such an inlet condition is called zero *prewhirl* since the fluid entering the rotor has no angular momentum associated with it and V_{u_1} is identically zero. The fluid leaves the rotor with a velocity V_{r_2} with respect to the rotor exit plane. The absolute velocity of

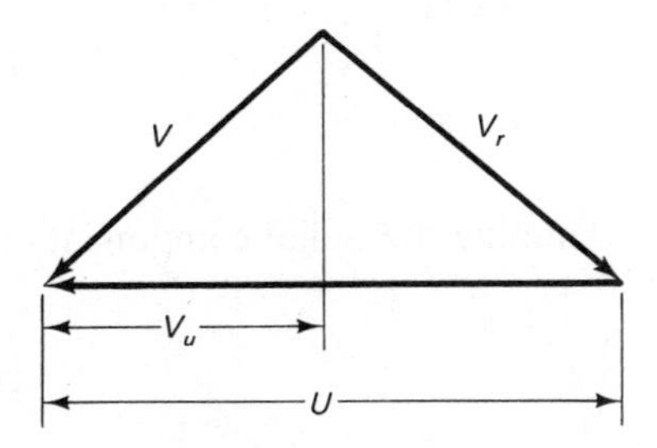

Figure 4-9 Velocity triangle nomenclature.

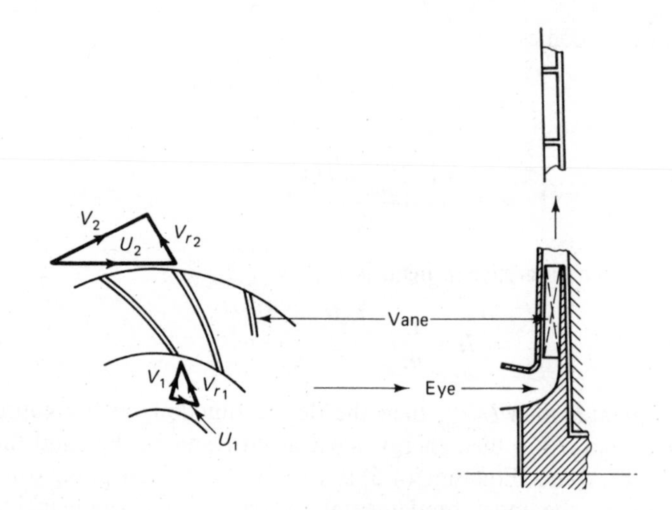

(a) Impeller schematic.

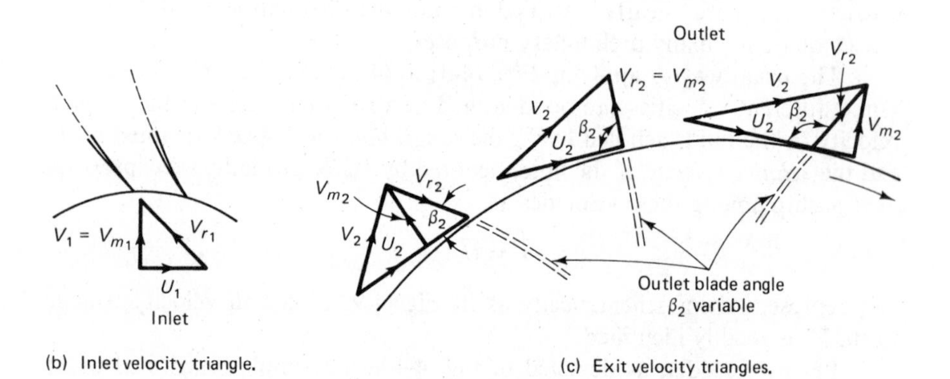

(b) Inlet velocity triangle.

(c) Exit velocity triangles.

Figure 4-10 Centrifugal compressor schematic and velocity triangle.

the exiting fluid is then $\overline{U} + \overline{V}_r$, as indicated by the velocity triangle drawn for the rotor exit.

The increase in head is

$$H = \frac{U_2 V_{u_2}}{g_c} - \frac{U_1 V_{u_1}}{g_c} \quad (V_{u_1} \to 0) \tag{4-6}$$

Denoting the radial component of the exit velocity as V_m, we can write

$$\cot \beta_2 = \frac{V_{r_{u2}}}{V_m}$$

and from the exit velocity triangle [see Fig. 4-10(c)]

$$V_{u_2} = U_2 - V_{r_{u_2}} = U_2 - V_m \cot \beta_2 \tag{4-7}$$

Then Eq. (4-6) becomes

$$H = \frac{U_2^2 - U_2 V_m \cot \beta_2}{g_c} \tag{4-8}$$

For an impeller of width w, the mass flow rate is

$$Q = \pi D_2 w V_m \tag{4-9}$$

The increase in head can then be expressed as a function of the volumetric flow rate Q:

$$H = \frac{U_2^2}{g_c} - \frac{U_2}{\pi D w g_c} Q \cot \beta_2 \tag{4-10}$$

Defining

$$K_1 = U^2/g_c \quad \text{and} \quad K_2 = \frac{U_2}{\pi D w g_c} \cot \beta_2$$

the expression becomes

$$H = K_1 - K_2 Q \tag{4-11}$$

The sign on the term K_2 then establishes the compressor characteristics (i.e., the behavior of H as a function of Q).

Three separate cases can be considered: (1) *radial exit blades* for which $\beta_2 = 90°$, (2) *backward curved blades* for which $\beta_2 < 90°$ and K_2 is positive,

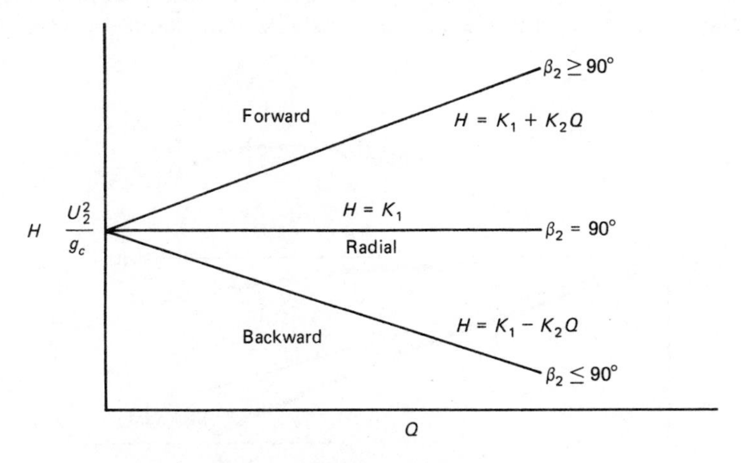

Figure 4-11 Head versus capacity for variable exit blade angle.

and (3) *forward curved blades* for which $\beta_2 > 90°$ and K_2 is negative. These three H–Q characteristics are illustrated in Fig. 4-11.

The expression U_2^2/g_c is called the *shut-off head* and represents the head developed by a compressor if the flow is cut off or the exit lines plugged. For blading with radial exit, the H–Q curve is ideally constant; for forward curved blading, the ideal head increases with flow rate; and for backward curved blades, the ideal head decreases with flow rate. These derived relationships, as illustrated in Fig. 4-11, all represent the H–Q behavior with zero losses and as such must show some deviations from reality.

If we were to test an actual centrifugal pump with the geometry described, the results might appear as in Fig. 4-12. In Fig. 4-12 the solid lines represent the "ideal" linear relationships, the short dashed line represents the actual relationship, and the heavy solid line represents the range of stable flow. Point c represents for each of the three cases the maximum head produced and is a very important point insofar as system operation is concerned.

Consider a system that is operating to the right of point c. If the volumetric flow rate is momentarily reduced, the compressor head is increased, since the decrease in flow rate can occur only by moving to the left along the characteristic curve. This increased head tends to overcome the reduction in flow rate and drives the flow rate back to the original. Thus, the attribute of *stability* is exhibited when head increases with decreasing flow rate. Had the system been operating to the left of point c, exactly the opposite would have happened; a decrease in flow rate would result in a decrease in head, which would further reduce the flow rate. This mode of operation is *unstable* and should be avoided. Unstable operation is called *surge* or *compressor stall* and is characterized by rapid fluctuations in flow rate and pressure or, in extreme instances, reversal of flow. The portion of the characteristic curve that exhibits decreased head for decreasing flow rate is a region of unstable operation. Point c is the *surge point* and demarcates unstable and stable operation for a

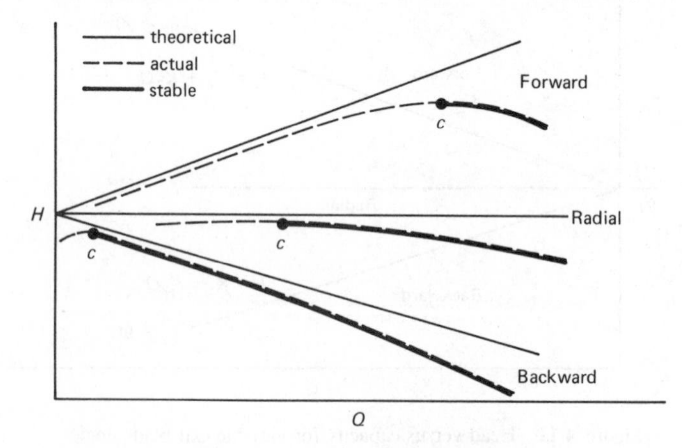

Figure 4-12 Actual H–Q relationships.

constant rpm characteristic. The loci of surge points define the *surge line* for variable rpm.

Figure 4-12 also points out the relative extent of stable operation for forward, radial, and backward curved blades. Forward curved blades have the smallest range, followed by radial blades. The backward curved blades offer the largest stable operating range. Systems that must operate over a wide range of flow rates usually use pumps with backward curved blades because of stability considerations.

We have developed a lot of understanding about centrifugal pumps by just exploring a few basic concepts. However, this is about as far as we can progress with simple analysis. From this point on we shall examine characteristics of actual pumps.

4-4 MANUFACTURERS' PUMP CHARACTERISTICS

Not surprisingly, the best and most complete set of technical data on a given pump is from the manufacturer. Most companies making and selling pumps design, manufacture, and market several different lines of pumps, each line containing families of geometrically similar pumps, each pump having a unique range of flow rates and heads. Catalogs providing geometrical data and performance characteristics for all lines and all pumps become rather lengthy. The 1981 *Goulds Pump Manual* (7) from Goulds Pumps, Inc., one of the largest manufacturers of process and general-purpose centrifugal pumps, is 480 pages long and includes sufficient (about 100 pages) ancillary information relating to pumps and piping systems to serve as a virtually complete source of design data for flow networks.

The information relating strictly to the Gould pump lines is contained in 383 pages. Obviously, with such a plethora of useful (and needed) information, a logical and orderly presentation of pump data is necessary. The Goulds manual divides their line into 15 different pump types, each type named for either a specific use (e.g., chemical/process) or a dominant physical characteristic (submersible pumps). Each of these types is further broken down into the various families available. Figure 4-13, taken from the *Goulds Pump Manual* Index, shows three families (3196, 3199, and 3604) are available in the chemical/process type. The attributes of each family are briefly described and the appropriate sections and pages indicated for further detail.

Page 9 is indicated as the beginning of information on chemical/process pump family 3196. The flow rate–head (H–Q) relationship for all pumps of a single family is usually shown on a *composite rating* chart. Figure 4-14 is the composite rating chart for the Goulds 3196 chemical/process pump family when operating at 3500 rpm. The nominal operating ranges for each pump are presented, the maximum efficiency point indicated, and the Goulds pump size

PUMP TYPE	APPLICATIONS	MODEL	RATINGS	SECTION	PAGE
1 CHEMICAL/ PROCESS	Corrosive/abrasive liquids, slurries and solids, high temperature, general purpose pumping, process and transfer.	3196	□ Q to 4250 GPM (965 m^3/h) □ H to 720 feet (220 m) □ T to 500°F (360°C) □ P to 375 PSIG (2586 kPa)	1A	9
		3199	□ Q to 220 GPM (50 m^3/h) □ H to 150 feet (46 m) □ T to 350°F (177°C) □ P to 150 PSIG (1034 kPa)	1B	27
		3604	□ Q to 16 GPM (3.6 m^3/h) □ H to 28 feet (8.5 m) □ T to 220°F (105°C) □ P to 75 PSIG (517 kPa)	1C	35
2 SELF-PRIMING PROCESS	Corrosive/abrasive liquids, slurries and suspensions, high temperature, industrial sump, mine dewatering, tank car unloading, bilge water removal, filter systems, chemical transfer.	3796	□ Q to 1500 GPM (340 m^3/h) □ H to 375 feet (114 m) □ T to 500°F (260°C) □ Suction Lifts to 25 feet (7.6 m)	2A	41
3 NON-METALLIC CHEMICAL/PROCESS	Severe corrosive/abrasive services. Moderate to large percentages of industrial solids and stringy materials.	NM4100–4120	□ Q to 2200 GPM (500 m^3/h) □ H to 290 feet (88 m) □ T to 250°F (120°C) □ P to 125 PSIG (862 kPa)	3A	53
		3107/3198	□ Q to 800 GPM (182 m^3/h) □ H to 440 feet (134 m) □ T to 300°F (150°C) □ P to 225 PSIG (1550 kPa)	3B	71

Figure 4-13 Goulds pump index. (Used with permission, Goulds Pumps, Inc., Seneca Falls, N.Y.)

shown ($1\frac{1}{2} \times 3$-13 A20, for example). The composite charts are useful for locating a particular pump if the flow rate–head requirements are known. Generally, the information available from composite charts is not sufficient for pump selection or system analysis purposes, and complete performance characteristics are provided for each pump within the family.

Figure 4-15 presents several performance curves for specific Goulds Model 3196 pumps. Information is typically presented for operation at 3500 and 1750 rpm. As can be seen from an examination of the curves shown in Fig. 4-15, these presentations are extremely "busy" since a lot of useful data is packed into a small area. Let's examine the useful data for one particular pump, the MT 2×3-10 A60.

The specific pump number 2×3-10 contains three pieces of information: (1) the first number is the volute discharge radius, 2 in.; (2) the second number is the suction (inlet) radius, 3 in.; and (3) the last number is the maximum impeller (rotor) diameter, 10 in. The trailing A60 is the ANSI standard pump designation within which the 2×3-10 is contained. Rotor diameters of 6, 7, 8, 9, and 10 in. can be used within the basic 2×3-10 pump frame. The effect on the H–Q curve of changing rotor diameter is shown for both 3500 and 1750

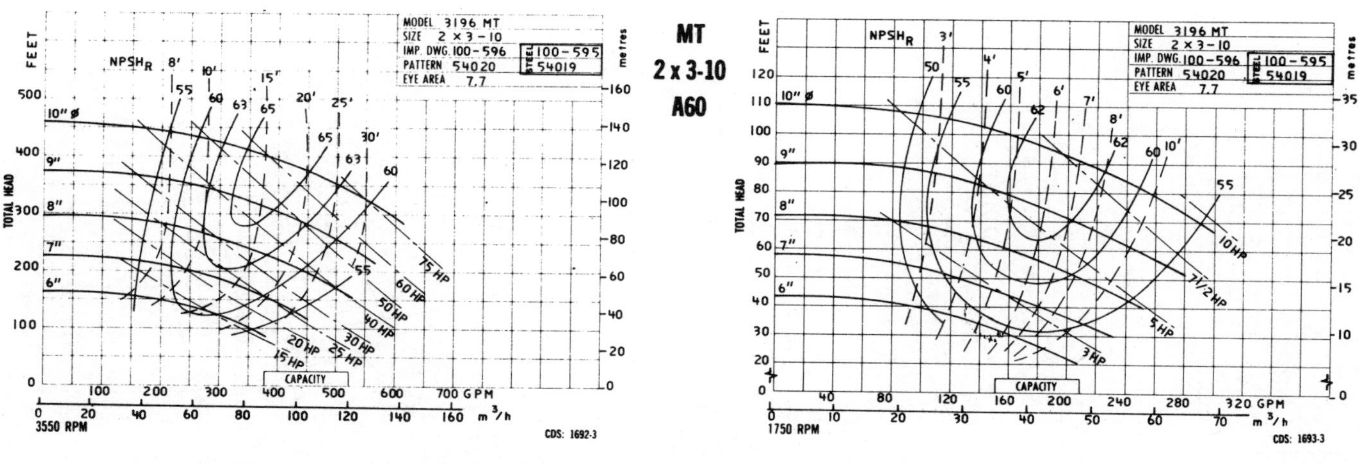

Figure 4-15 Selected Goulds 3196 pump performance charts. (Used with permission, Goulds Pumps, Inc., Seneca Falls, N.Y.)

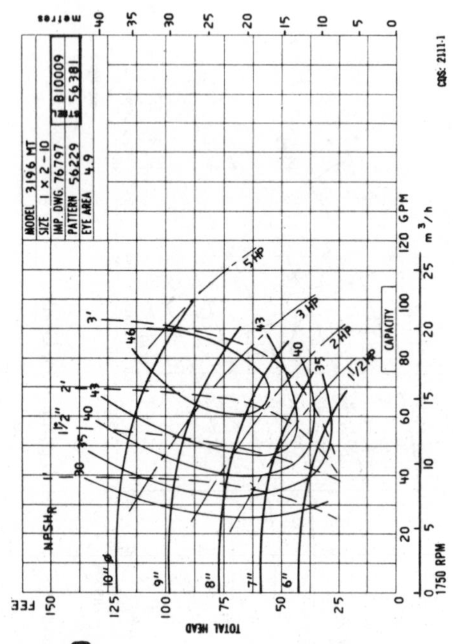
MODEL 3196 MT
SIZE 1 x 2-10
IMP. DWG. 76797
PATTERN 56229
EYE AREA 4.9
B10009
STEEL 56381
NPSH$_R$
1750 RPM
TOTAL HEAD
FEET
metres
CAPACITY
GPM
m^3/h
CDS: 2111-1

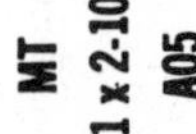
MT
1 x 2-10
A05

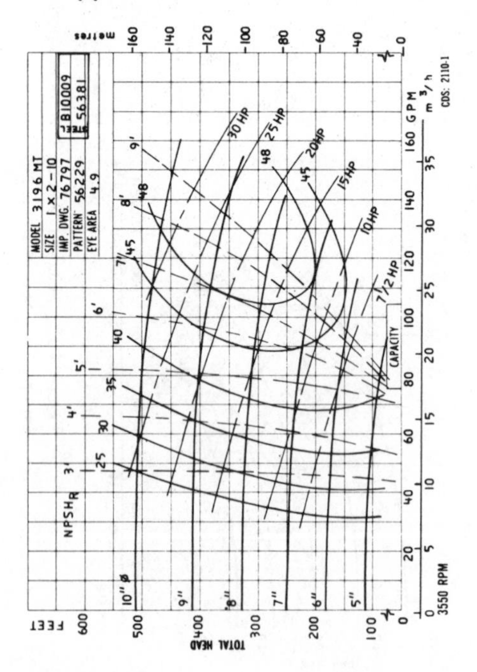
MODEL 3196 MT
SIZE 1 x 2-10
IMP. DWG. 76797
PATTERN 56229
EYE AREA 4.9
B10009
STEEL 56381
NPSH$_R$
3550 RPM
TOTAL HEAD
FEET
metres
CAPACITY
GPM
m^3/h
CDS: 2110-1

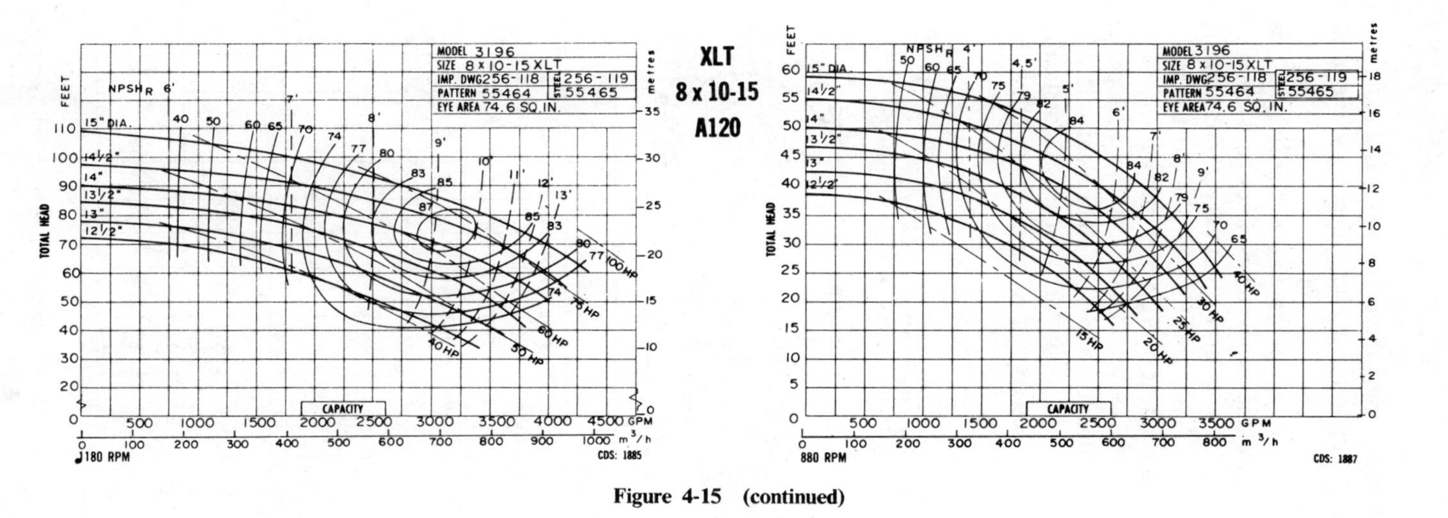

Figure 4-15 (continued)

rpm operation. Changing rotor diameters has a dramatic effect on the H–Q performance curve. As the rotor diameter decreases, the head decreases for a given flow rate.

The parabolalike curves superposed over the H–Q curves and designated 55, 60, 63, and 65 are the isoefficiency lines and denote the *mechanical efficiency* at the particular operating point (H–Q) for a given diameter rotor. Except for very viscous fluids, the H–Q characteristics and the isoefficiency lines are relatively independent of Reynolds number.

The remaining two curves shown are the horsepower (hp) and net positive suction head required ($NPSH_R$). The hp curve is valid only for water at normal temperature while the $NPSH_R$ curve is theoretically independent of the liquid handled. The horsepower required curves allow the required power

All dimensions in inches and (mm). Not to be used for construction.
Model 3196 MT illustrated. Dimensions apply to 3196 ST, 3196 MT and 3196 XLT.
Dimensions apply to both 150 and 300 pound flanges. Flanges are drilled to ANSI dimensions.

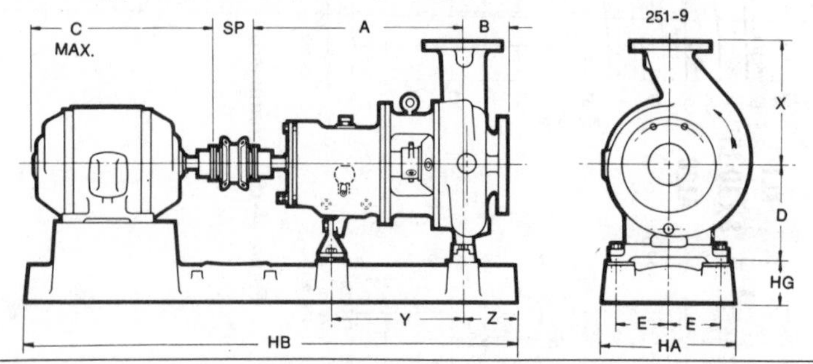

DIMENSIONS DETERMINED BY PUMP

Model	Pump Size	ANSI Designation	Discharge Size	Suction Size	X	A	B	D	Y	Z	E	SP	Shaft Diameter at Cplg.	Key-Way	Pump Weight lbs. (kg.)
3196 ST	1 x 1½-6	AA	1 (25.4)	1½ (38.1)											84 (38)
	1½ x 3-6	AB	1½ (38.1)	3 (76.2)											92 (42)
	2 x 3-6		2 (50.8)	3 (76.2)	6½ (165)	13½ (343)	4 (102)	5¼ (133)	7¼ (184)	4½ (114)	3 (76.2)	3¾ (95.3)	⅞ (22.2)	3/16 x 3/32 (4.8 x 2.3)	95 (43)
	1 x 1½-8	AA	1 (25.4)	1½ (38.1)											100 (45)
	1½ x 3-8	AB	1½ (38.1)	3 (76.2)											108 (49)
3196 MT**	2 x 3-6	A10	2 (50.8)	3 (76.2)	8¼ (210)										180 (82)
	2 x 3-8	A60	2 (50.8)	3 (76.2)	9½ (241)										200 (91)
	3 x 4-8	A70	3 (76.2)	4 (102)	11 (279)										220 (100)
	3 x 4-8G	A70	3 (76.2)	4 (102)	11 (279)	19½ (495)	4 (102)	8¼ (210)							[illegible]
	1 x 2-10	A05	1 (25.4)	2 (50.8)	8½ (216)										200 (91)
	1½ x 3-10	A50	1½ (38.1)	3 (76.2)	8½ (216)										220 (100)
	2 x 3-10	A60	2 (50.8)	3 (76.2)	9½ (241)				12½ (318)	4½ (114)	4⅞ (124)	3¾ (95.3)	1⅛ (28.6)	¼ x ⅛ (12.7 x 3.2)	230 (104)
	3 x 4-10	A70	3 (76.2)	4 (102)	11 (279)										265 (120)
	4 x 6-10	A80	4 (102)	6 (152)	13½ (343)										305 (138)
	1½ x 3-13	A20	1½ (38.1)	3 (76.2)	10½ (267)										245 (111)
	2 x 3-13	A30	2 (50.8)	3 (76.2)	11½ (292)	19½ (495)	4 (102)	10 (254)							275 (125)
	3 x 4-13	A40	3 (76.2)	4 (102)	12½ (318)										330 (150)
	4 x 6-13	A80	4 (102)	6 (152)	13½ (343)										405 (184)
3196 XLT	6 x 8-13	A90	6 (152)	8 (203)	16 (406)										560 (254)
	8 x 10-13	A100	8 (203)	10 (254)	18 (457)										670 (304)
	6 x 8-15	A110	6 (152)	8 (203)	18 (457)	27⅞ (708)	6 (152)	14½ (368)	18¾ (476)	6½ (165)	8 (203)	5¼ (133)	2⅜ (60.3)	⅝ x 5/16 (15.9 x 7.9)	610 (277)
	8 x 10-15	A120	8 (203)	10 (254)	19 (483)										740 (336)
	8 x 10-15G	A120	8 (203)	10 (254)	19 (483)										710 (322)

**On LTC frame, shaft diameter at coupling is 1⅞", (47.6) keyway is ½" x ¼".)(12.7) x (6.4) ps.

Figure 4-16 Series 3196 nominal dimensions. (Used with permission, Goulds Pumps, Inc., Seneca Falls, N.Y.)

for pumping water at a given flow rate and head to be determined. Given a fluid density, the efficiency and H–Q operating point are sufficient to determine the power required; but since water is often the pumped fluid, these hp curves are useful. The meaning of net positive suction head will be covered in the next section, so suffice it to say that the *net positive suction head required* must be adhered to if proper operation is to result. The $NPSH_R$ is determined by the manufacturer.

Nominal geometric characteristics for the 3196 family of pumps are also provided. These are given in Fig. 4-16. These dimensions are not to be used for construction drawing, but they do allow us some feel for the various sizes involved in the 3196 series.

This section has examined the logic and information available from a typical manufacturer's catalog. Various corrections and extrapolations may be made to these data for other conditions and fluids. Some of these are examined in the next section.

4-5 PERFORMANCE CHARACTERISTICS RELATIONS

The performance curves and characteristics discussed in the previous section were generally concerned with water as the pumped fluid and with operation at some specified rpm; however, in many instances water is not the fluid and the speed is different from that of the supplied performance characteristics. Short of physically running a test with the appropriate fluid at the specified rpm, some means of obtaining this information from that provided by the manufacturer is needed. Two approaches that allow manufacturers' data to be extrapolated are generally available: (1) *dimensional analysis and similitude*, and (2) *correction factors for very viscous fluids*.

The basic idea of dimensional analysis is a familiar concept to most engineers and will not be elaborated on here. A turbomachine such as a pump, using an incompressible fluid, possesses the following important parameters:

Symbol	Parameter	Unit
H	Head	ft lbf/lbm
Q	Flow rate	ft^3/s
N	Speed	rpm(1/s)
M	Mechanical efficiency	none
D	Characteristic dimension	ft
ρ	Density	lbm/ft^3
μ	Viscosity	lbm/ft s
P	Power	ft lbf/s

A complete set of dimensionless parameters for the preceding is as follows (8):

$$\pi_1 = \frac{Q}{ND^3} \tag{4-12}$$

$$\pi_2 = \frac{H}{N^2D^2} \tag{4-13}$$

$$\pi_3 = \frac{P}{\rho N^3D^5} \tag{4-14}$$

$$\pi_4 = \frac{\mu}{\rho ND^2} \tag{4-15}$$

$$\pi_5 = \eta \tag{4-16}$$

The group π_5 is really redundant since π_1, π_2, and π_3 determine π_5. *Similarity* is often used to estimate performance curves for slightly different conditions from those provided by the manufacturer. The similarity argument for a given pump is based on the hypothesis that if all pertinent dimensionless groups at two operating points have the same values then the efficiency is the same. This can be inverted to say that at the same efficiency "similar" conditions will occur when each of the dimensionless parameters has the same values; that is, $\pi_{1_1} = \pi_{1_2}$, $\pi_{2_1} = \pi_{2_{2_1}} \ldots \pi_{n_1} = \pi_{n_2}$. This idea is approximately true even if the impeller in diameter (characterized by D) is altered slightly.

Straight application of the similarity concept can be dangerous if experimental data are not consulted. Shepherd (8) points out that just looking at π_1 we would expect flow rate to be proportioned to D^3 or at π_3 we would expect power to be proportional to D^5. These are *not* true. In π_1 and π_2 the D^3 and D^5 are really AD and AD^3, where A is a cross-sectional pump inlet (or exit) area and would not change if only the impeller diameter changed. With this realization, the correct relationships for impeller diameter change in the *same* casing are

$$\frac{Q_2}{Q_1} = \frac{D_2}{D_1} \tag{4-17}$$

$$\frac{H_2}{H_1} = \left(\frac{D_2}{D_1}\right)^2 \tag{4-18}$$

$$\frac{P_2}{P_1} = \left(\frac{D_2}{D_1}\right)^3 \tag{4-19}$$

If only the *speed* changes, then

$$\frac{Q_2}{Q_1} = \frac{N_2}{N_1} \tag{4-20}$$

$$\frac{H_2}{H_1} = \left(\frac{N_2}{N_1}\right)^2 \tag{4-21}$$

$$\frac{P_2}{P_1} = \left(\frac{N_2}{N_1}\right)^3 \tag{4-22}$$

Example 4-1 (From the *Goulds Pump Manual*)

The performance data for a Goulds 2 × 3-13 pump at 1750 rpm are shown below in Fig. 4-17. Estimate the pump performance at 2000 rpm for a 13-in. impeller.

Solution. Equations (4-20) through (4-22) will be used to determine the new performance, with $N_1 = 1750$ rpm and $N_2 = 2000$ rpm. The first step is to read the capacity, head, and horsepower at several points on the 13-in.-diameter curve in Fig. 4-17. For example, one point may be near the best efficiency point where the capacity is 300 gpm,

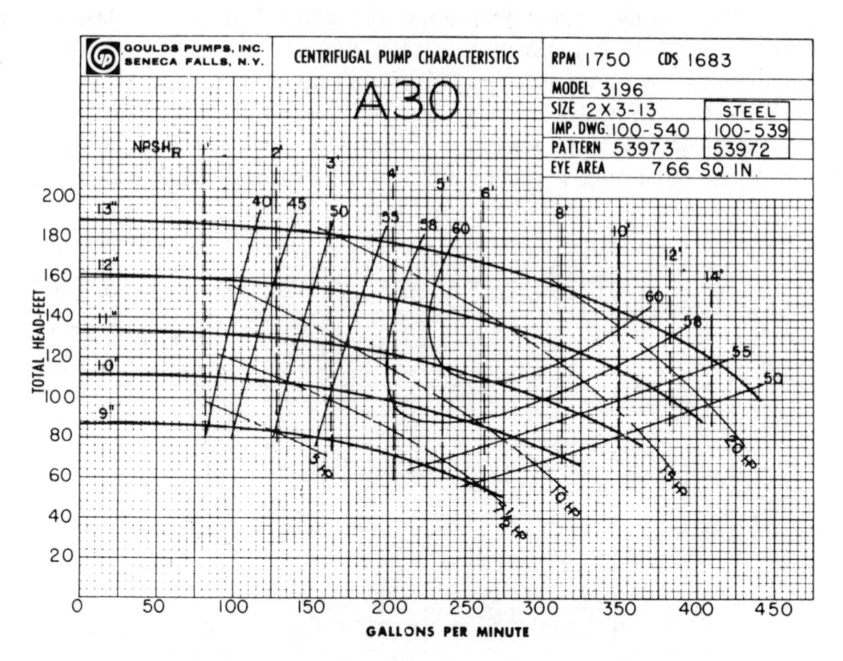

Figure 4-17 Performance data for a Goulds model 3196 2 × 3-13 pump at 1750 rpm. (Used with permission, Goulds Pumps, Inc., Seneca Falls, N.Y.)

the head is 160 ft, and the power is approximately 20 hp. Then

$$\frac{300}{Q_2} = \frac{1750}{2000} \qquad Q_2 = 343 \text{ gpm}$$

$$\frac{160}{H_2} = \left(\frac{1750}{2000}\right)^2 \qquad H_2 = 209 \text{ ft lbf/lbm}$$

$$\frac{20}{P_2} = \left(\frac{1750}{2000}\right)^3 \qquad P_2 = 30 \text{ hp}$$

This will then be the best efficiency point on the new 2000 rpm curve. By performing the same calculations for several other points on the 1750 rpm curve, a new curve (shown in Fig. 4-18) can be drawn that shows the pump's performance at 2000 rpm.

One other dimensionless group is of special interest. If D is eliminated in π_1 and π_2 [Eqs. (4-12) and (4-13)], then the *specific speed*, N_s, results and is

$$N_s = \frac{N\sqrt{Q}}{H^{3/4}} \qquad (\eta_{\text{max}}) \tag{4-23}$$

where by tradition the following units are used: N in rpm, Q in gallons per minute, and H in ft lbf/lbm. When evaluated at the point of *maximum efficiency*, the specific speed determines the general shape or class of the impeller. Figure 4-19 graphically illustrates this and also points out that the maximum efficiency occurs for $N_s \sim 2500$ for a rather wide range of flow rates.

All the preceding discussion on similarity generally holds true for fluids of relatively low viscosity. Then π_4 is relatively small (or Re_D is relatively large) and the remaining π groups are nearly invariant with respect to π_4. If

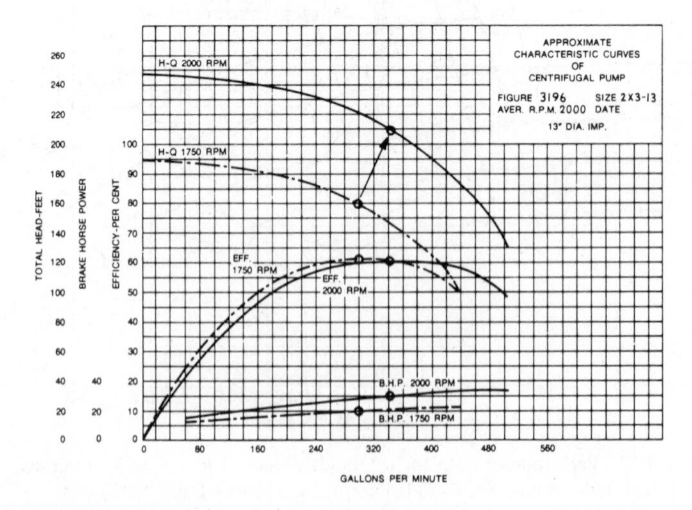

Figure 4-18 Solution for Example 4-1. (Used with permission, Goulds Pumps, Inc., Seneca Falls, N.Y.)

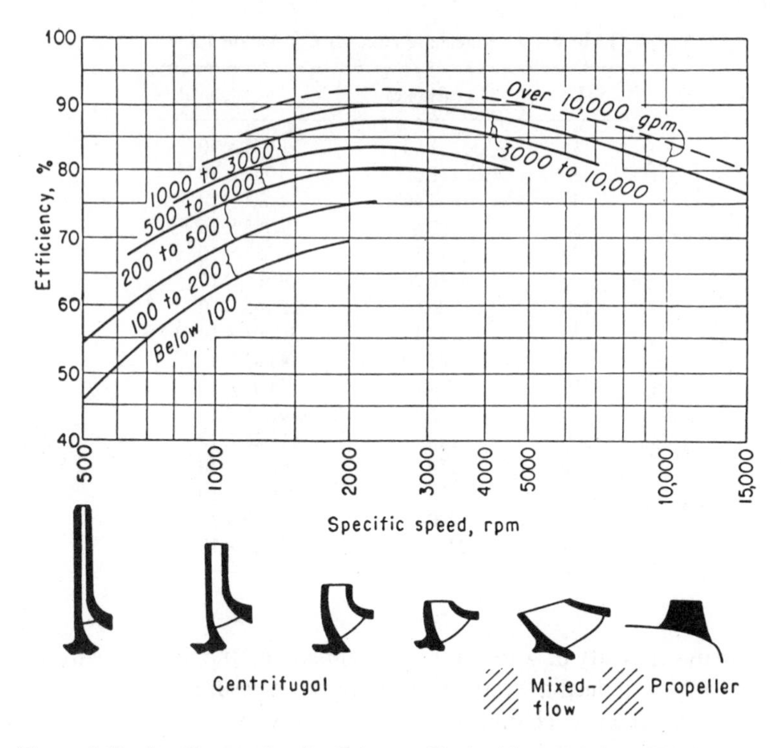

Figure 4-19 Specific speed and efficiency. (Used with permission, Worthington Division, McGraw-Edison Co.)

the viscosity is large, this is not true and similarity cannot be used to extend the manufacturer's data to highly viscous fluids. As defined by the *Standards of the Hydraulic Institute* (5), correction factors must be used.

Fluids that are very viscous can markedly affect the performance of centrifugal pumps. We are accustomed to stating viscosity in terms of lbm/ft s or centistokes or poise, as they are appropriate for fluids such as water that have small values of viscosity. However, for fluids that are very viscous (some crude oils, for example), viscosity units are used that reflect these higher viscosities. The most common of these units is the SSU, *Saybolt seconds universal.* In the SSU viscometer, the time required for a certain volume of fluid to flow (under stated condition of head) through a short tube of small bore is measured. This time is the viscosity in SSU and is directly relatable to more common units of viscosity. The SSU viscometer has a vertical tube 0.483 ± 0.004 in. long and 0.0695 ± 0.0006 in. in diameter. Quite often the viscosity of very viscous liquids is stated in SSU, and the *Standards of the Hydraulic Institute* express their viscous correction factors for centrifugal pumps in terms of SSU. Table 4-3 presents a tabular comparison of SSU values and the kinematic viscosity in centistokes and in ft^2/s, the kinematic unit of viscosity with which we are most familiar. Considering that at 100°F

TABLE 4-3 Kinematic Viscosity Conversion Table

SSU	Centistokes	ft^2/s
31	1.00	1.075×10^{-5}
35	2.56	2.752×10^{-5}
40	4.30	4.623×10^{-5}
50	7.40	7.955×10^{-5}
70	13.10	14.083×10^{-5}
90	18.20	19.565×10^{-5}
100	20.60	22.145×10^{-5}
150	32.10	34.508×10^{-5}
200	43.20	46.44×10^{-5}
250	54.	58.05×10^{-5}
300	65.	69.88×10^{-5}
500	110.	118.25×10^{-5}
700	154.	165.55×10^{-5}
1,000	220.	236.50×10^{-5}
2,000	440.	473.0×10^{-5}
10,000	2200.	2365.0×10^{-5}

the kinematic viscosity of water is 0.6875 centistokes, the SSU, which can go as high as 30,000, is indeed a "very" viscous unit of measure.

The performance characteristics for a very viscous fluid for a pump whose characteristics are known for water can be estimated by

$$Q_{\text{vis}} = C_Q \times Q_w \tag{4-24}$$

$$H_{\text{vis}} = C_H \times H_w \tag{4-25}$$

$$\eta_{\text{vis}} = C_\eta \times \eta_w \tag{4-26}$$

$$P_{\text{vis}} = \frac{Q_{\text{vis}} \times H_{\text{vis}} \times \text{S.G.}}{3960 \times \eta_{\text{vis}}} \tag{4-27}$$

where the subscript "vis" indicates the parameter for a viscous fluid, the subscript w indicates the parameter for water, and where C_Q, C_H, and C_η are empirically obtained *correction factors* for the effects of the very viscous fluid upon capacity (Q), head (H), and efficiency (E). Values of these correction factors are given in Figs. 4-20 and 4-21. Information for the results presented in these figures is from the *Standards of the Hydraulic Institute* and resulted from a series of tests run with pumps of 2- to 8-in. size handling various oils. Do not forget that these correction factors are *approximations* and that exact pump performance for a viscous fluid can only be obtained by tests on the pump with the viscous fluid.

The correction factors are obtained by reading upward from the capacity to the intersection with the head; then a horizontal traverse is made either to the right or left to intersect the appropriate viscosity. A vertical line drawn

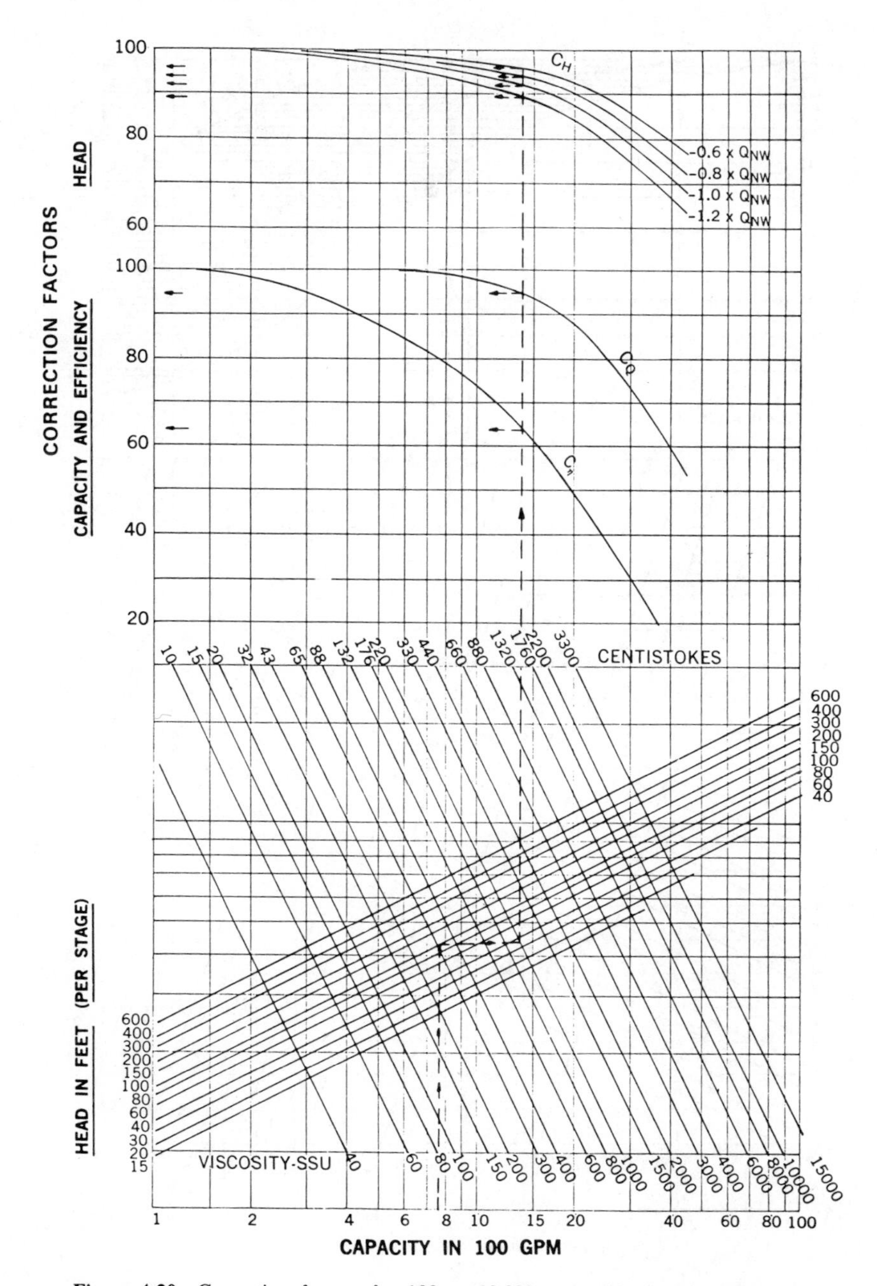

Figure 4-20 Correction factors for 100 to 10,000 gpm. (Used with permission, Hydraulic Institute.)

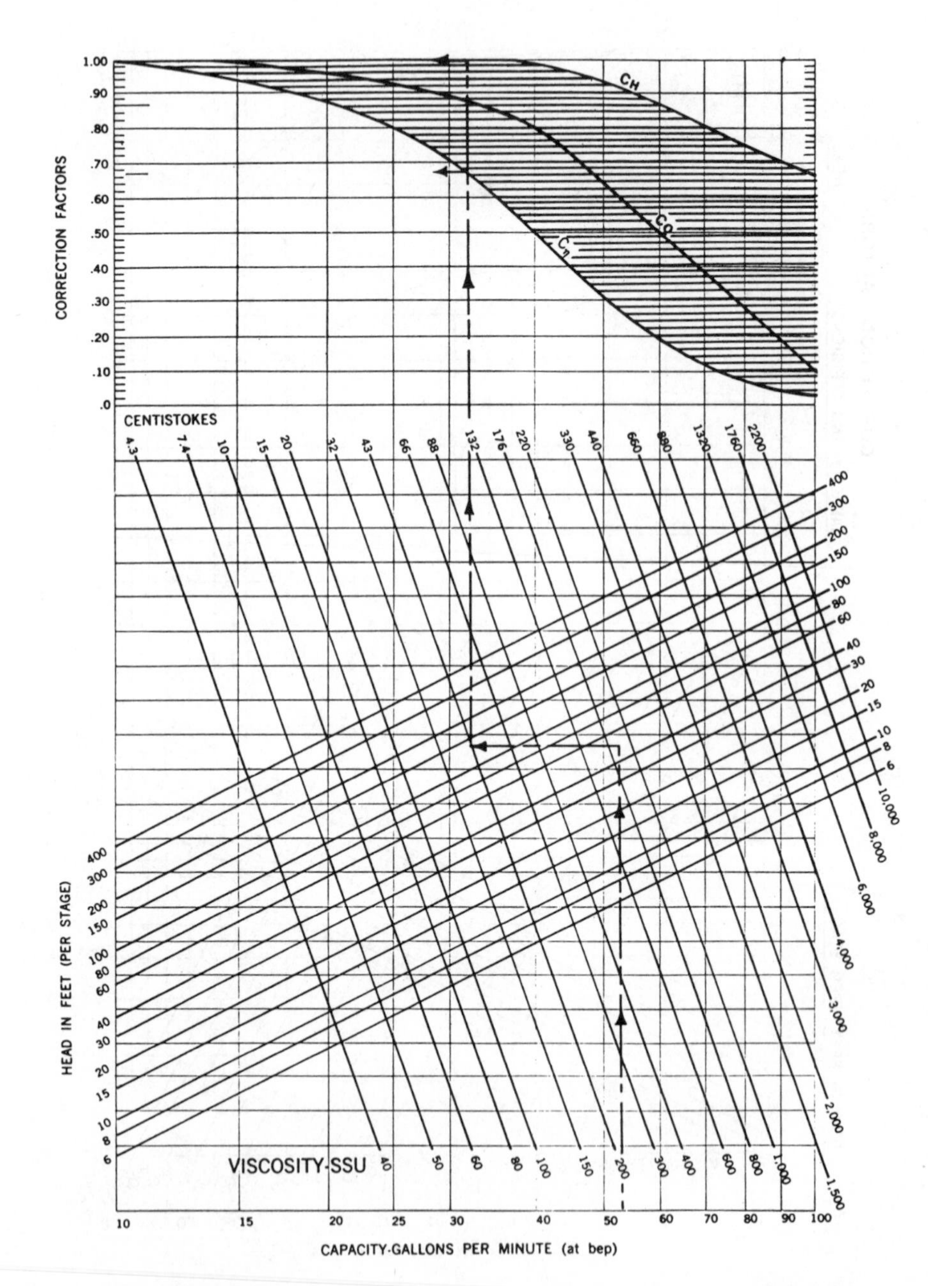

Figure 4-21 Correction factors for less than 100 gpm. (Used with permission, Hydraulic Institute.)

from this intersection then crosses the C_η, C_Q, and C_H curves from which the values of the correction factors are read. The methodology for translating a given pump performance curve for water to that for a viscous fluid is as follows:

1. Locate the point of maximum efficiency on the H–Q curve for water. If this point is Q_w, then obtain $0.6 \times Q_w$, $0.8 \times Q_w$, and $1.2 \times Q_w$.
2. Enter either Fig. 4-20 or 4-21 with the capacity at η_{max}, go upward to the head developed, then horizontally to the desired viscosity. From that point proceed upward to C_H, C_Q, and C_η.
3. Read C_η and C_Q, and then C_H for all four capacities.
4. Multiply each head (water) by the corresponding C_H, each efficiency (water) by C_η, and each capacity (water) by C_Q to obtain the corresponding corrected values.
5. Plot the corrected values and draw smooth curves through them. Take the head at shut-off to be the same as for shut-off with water.
6. Calculate the power required from Eq. (4-27). The results will be a good approximation of the pump's performance characteristics for the viscous fluid. S.G. is the specific gravity.

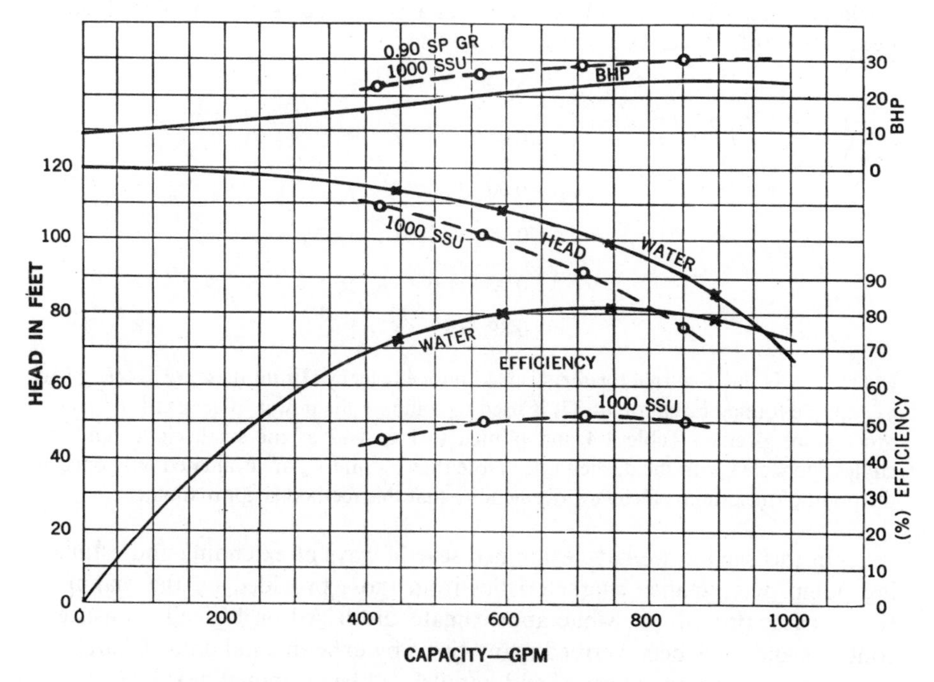

Figure 4-22 Solution for Example 4-2. (Used with permission, Hydraulic Institute.)

TABLE 4-4 Viscous Performance Estimate

	$0.6 \times Q_{Nw}$	$0.8 \times Q_{Nw}$	$1.0 \times Q_{Nw}$	$1.2 \times Q_{Nw}$
Water capacity (Q_w)	450	600	750	900
Water head in feet (H_w)	114	108	100	86
Water efficiency (E_w)	72.5	80	82	79.5
Viscosity of liquid	1000 SSU	1000 SSU	1000 SSU	1000 SSU
C_Q from chart	0.95	0.95	0.95	0.95
C_H from chart	0.96	0.94	0.92	0.89
C_η from chart	0.635	0.635	0.635	0.635
Viscous capacity, $Q_w \times C_Q$	427	570	712	855
Viscous head, $H_w \times C_H$	109.5	101.5	92	76.5
Viscous efficiency, $\eta_w \times C_\eta$	46.0	50.8	52.1	50.5
Specific gravity of liquid	0.90	0.90	0.90	0.90
bhp viscous	23.1	25.9	28.6	29.4

Used with permission, Hydraulic Institute.

Example 4-2 (from the *Standards of the Hydraulic Institute*)

Given the pump characteristics for water as the working fluid, determine the approximate characteristics if a 1000 SSU fluid were pumped. The characteristics for water as the pumped fluid are given in Fig. 4-22.

Solution. The maximum efficiency occurs at a Q_w of 750 gpm, an η of 0.82, and a head of 100 ft lbf/lbm. Using Fig. 4-20, we find

$$C_Q = 0.95$$

$$C_\eta = 0.635$$

$$C_H = 0.96 \quad (0.6 \times C_{Nw})$$

$$0.94 \quad (0.8 \times C_{Nw})$$

$$0.92 \quad (1.0 \times C_{Nw})$$

$$0.89 \quad (1.2 \times C_{Nw})$$

where C_{Nw} is the flow rate (water) at maximum efficiency. From these, Q_{vis}, H_{vis}, and η_{vis} can be found. Equation (4-27) is used to compute the power. The results of such a process are given in Table 4-4 and plotted in Fig. 4-22 as the small circles and the circles connected with the dashed line. Note that C_Q and C_η are evaluated only once at the point of maximum efficiency. Remember that this method is approximate.

In this section we have examined several ways of extending and estimating pump performance characteristics from those provided by the manufacturer. These procedures, while approximate, are based on logically consistent concepts and have been verified many times by experimental data. Characteristic curves for pumps in series and parallel will be examined next.

4-6 PUMPS IN SERIES AND PARALLEL

Many times for either added head or added flow rate, pumps are operated in series or parallel. Examples of *series* and *parallel* pump arrangements are given in Fig. 4-23. Usual practice is to provide check valves to prevent backflow and shut-off valves on the suction (inlet) and discharge (outlet) lines for complete isolation when needed. The minor losses associated with these valves and the necessary piping are usually neglected, although the losses could be accounted for using the techniques described in Chapter 1.

Analysis or selection of pumps in series or parallel is greatly facilitated by using an *effective performance curve* for the entire series and/or parallel pump arrangement. This can be achieved by providing a single head–capacity relationship between positions A and B that would be the equivalent to that from the pump network. We are already familiar with the concept that would allow this. These equivalent "pumps" are indicated in Fig. 4-23. Pumps in series, like lines and devices in series, have the same flow rate going through each pump and have a total increase in head that is the sum of the increase in heads of the individual pumps. For example, in Fig. 4-23(a), at the flow rate Q

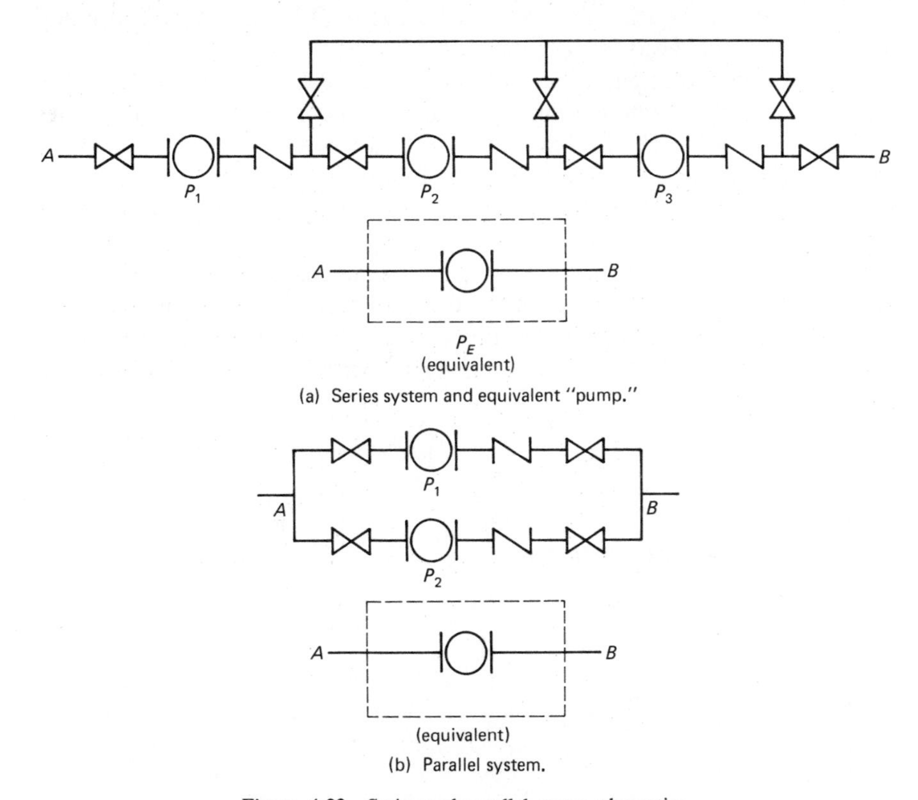

(a) Series system and equivalent "pump."

(b) Parallel system.

Figure 4-23 Series and parallel pump schematics.

the total increase in head is $H_1 + H_2 + H_3$ for pumps 1, 2, and 3. Pumps in parallel legs, like lines and devices in parallel, must have pressure changes that are the same, but the total flow rate is the sum of the individual flow rate in each leg. Figure 4-23(b), for example, would have the same head increase for pumps 1 and 2, but the total flow rate would be $Q_1 + Q_2$. Hence, for pumps in series, head increases add for a given flow rate, while for pumps in parallel, flow rates are additive at a given head. These simple concepts are all that are needed to reduce a series–parallel combination to a single equivalent H–Q curve. Examples are given in the next paragraph.

We can identify four distinct cases for series and parallel arrangements of two pumps: (1) two pumps with identical performance characteristics in series, (2) two pumps with different performance characteristics in series, (3) two pumps with identical performance characteristics in parallel, and (4) two pumps with different performance characteristics in parallel. We shall examine each of these cases in turn. Consider, as presented in Fig. 4-24, the equivalent performance characteristics curve for two identical pumps in series. The single pump H–Q curve is shown. For series operation, the construction is made by adding heads at equal capacities. The point at which a given pump is operating can be found by using the prescribed capacity Q for the equivalent pump to locate the actual head for each individual pump.

If two dissimilar pumps are placed in series, the equivalent performance characteristics construction is similar to that of case 1. Figure 4-25 illustrates this procedure with the H–Q curves for the two dissimilar pumps shown, as well as the resulting equivalent performance curves. As in case 1, the operating point of the individual pump is found by using the prescribed capacity Q to locate the actual head for each pump.

The case of similar pumps in parallel is examined in Fig. 4-26. Here the flow rates (or capacities) are added at a constant head. The operating point for an individual pump is found by locating the pump flow rate for the given head. The case for dissimilar pumps is somewhat more involved, as is seen in the next paragraph.

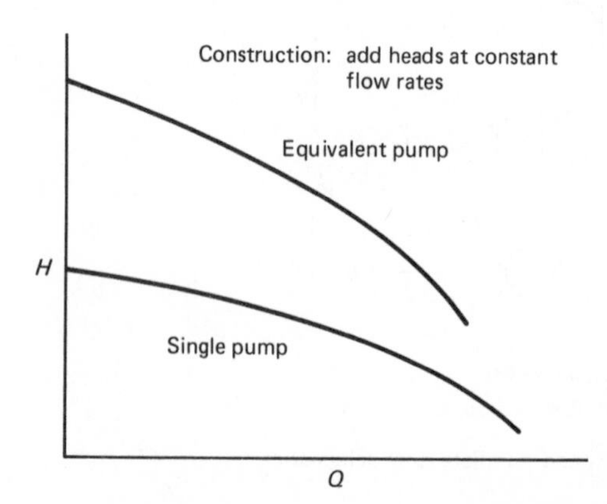

Figure 4-24 Similar pumps in series.

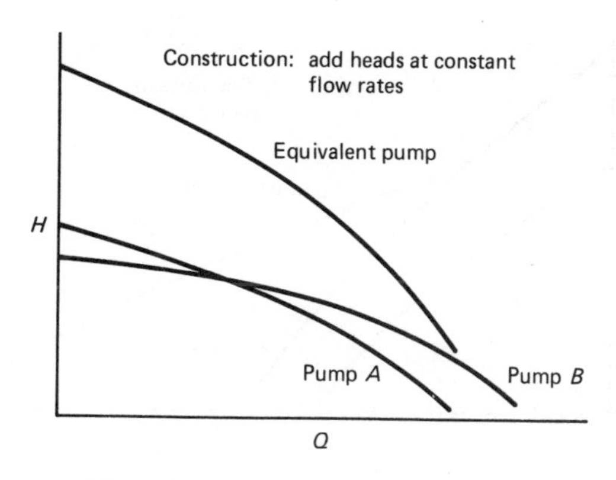

Figure 4-25 Dissimilar pumps in series.

Parallel operation of dissimilar pumps is illustrated in Fig. 4-27. Again, as in case 3, capacities were added at constant head, but for dissimilar pump characteristics, the pump with the lower shut-off head could not be brought into operation until the head of the larger was decreased below this lower shut-off head. Had this not been done, the larger head of the more powerful pump would have actually "bucked" the output of the lower shut-off head pump. The operating points for dissimilar pumps in parallel are located for each pump by finding the individual pump flow rate for the given head. The concepts developed in the preceding paragraphs are sufficient for generation of equivalent H–Q curves for virtually any parallel–series arrangement of similar or dissimilar pumps. The question is how do we use the performance characteristics. The answer is in the next section.

Example 4-3

Construct an equivalent head–capacity curve for two Goulds pumps in parallel. Both pumps are $1\frac{1}{2} \times 3 - 7$ Model 3198 and are run at 3500 rpm. One has a 7-in. rotor and

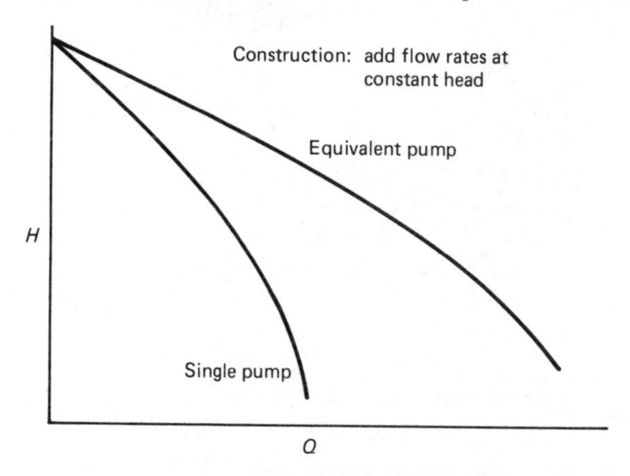

Figure 4-26 Similar pumps in parallel.

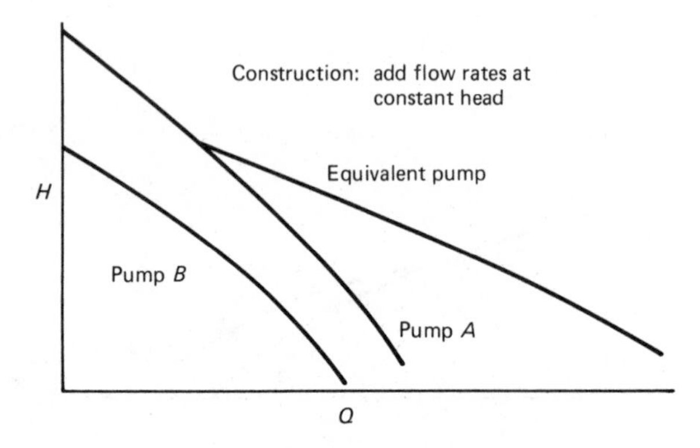

Figure 4-27 Dissimilar pumps in parallel.

the other a $6\frac{1}{2}$ in. rotor. The performance curve is given in Fig. 4-28. If the desired flow rate is 250 gpm, find the operating point of each pump.

Solution. This problem is solved in Fig. 4-29. First the appropriate characteristics for each pump are drawn. The pump with the $6\frac{1}{2}$ in. rotor has the lowest shut-off head; hence, fluid cannot be pumped by this pump until the head increase by the larger pump falls below this value. The dashed line in the figure shows this. Thus the flow rate of the parallel pumps is just the flow rate of the larger pump until the smaller shut-off head is reached; after that, the pumps' flow rates add at a given head.

To find the operating point of each pump, the total flow rate is first located on the equivalent curve. This is labeled as point T in Fig. 4-29. Since both pumps are operating at a head of 175 ft lbf/lbm, a horizontal line drawn from T to the left will intersect both pump characteristics. For a system flow rate of 250 gpm, each pump in the parallel arrangement operates at 175 ft lbf/lbm, with the 7-in. impeller pump producing 160 gpm and the $6\frac{1}{2}$-in. impeller pump producing 90 gpm.

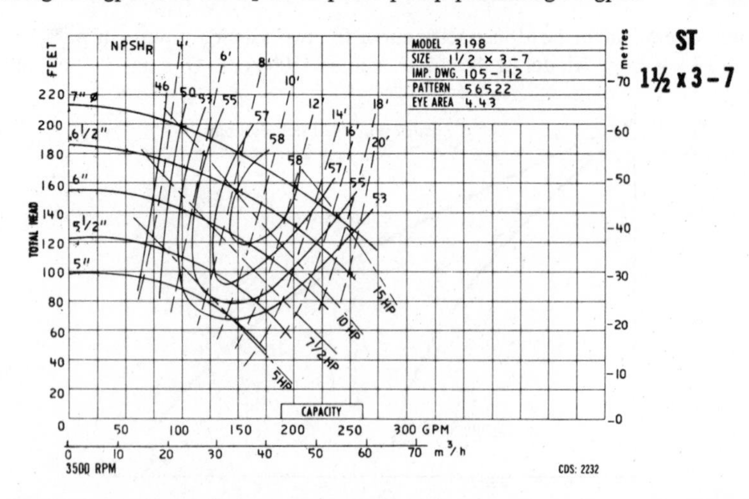

Figure 4-28 Performance data for Goulds model 3198 $1\frac{1}{2} \times 3$-7 pump. (Used with permission, Goulds Pumps, Inc., Seneca Falls, N.Y.)

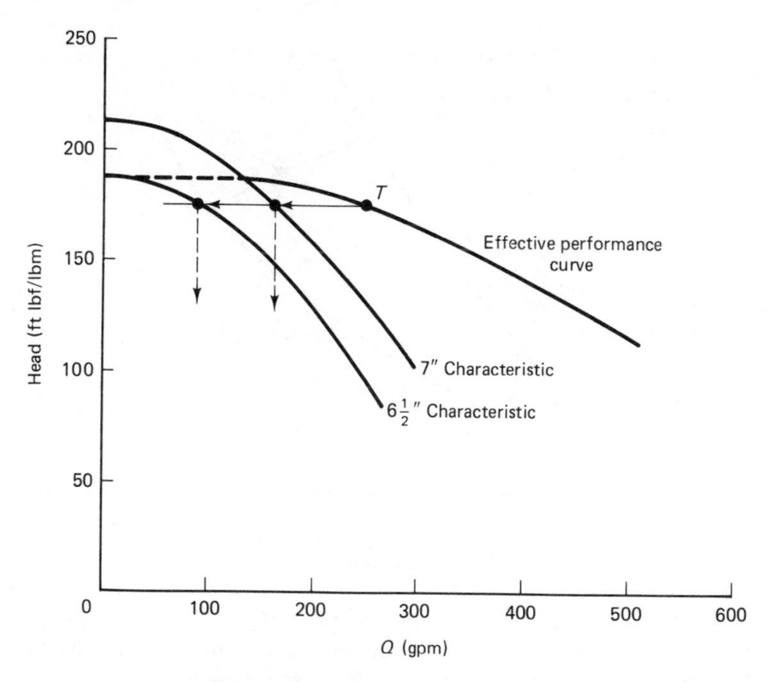

Figure 4-29 Solution for Example 4-3.

4-7 PUMP–SYSTEM OPERATION

We have examined in some detail the performance characteristics of centrifugal pumps, but we have said little about either system characteristics or how the pump–system characteristics can be used in quantitative assessments. In many fluid systems we are interested in quantities such as flow rate or capacity and pressure drop or head. The energy equation, Eq. (1-18), can be used to parametrically determine the head required versus flow rate for a specific system. Examples of graphical representations of such parametric studies are shown in Fig. 4-30. These curves are appropriately called *system characteristics* since they describe the H–Q curves of given systems. Curve A represents a system in which the initial and final elevations are the same so that the required pumping head is directly proportional to Q to some positive power, 1.852 for water in a pipeline (the Hazen–Williams equation), for example. Systems with characteristics such as represented by curve A require no pump energy input when the flow rate is reduced to zero. A pipe network connecting two reservoirs with the same surface elevations would yield a system characteristic of the form of curve A. A pipe network connecting two reservoirs with different surface elevations would yield a system characteristic of the form of curve B if the fluid were to be pumped from the low to the high reservoir.

System characteristic curves have two primary uses: (1) pump selection or specification for a desired system flow rate, and (2) determination of the

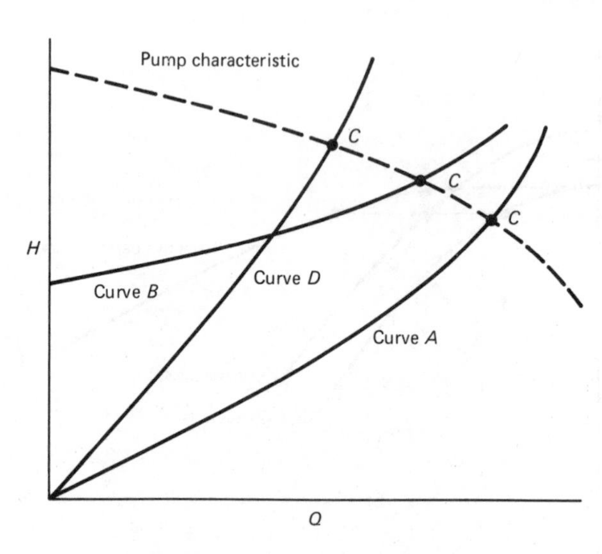

Figure 4-30 System characteristic examples.

system–pump operating point for a given pump and system combination. The former can be accomplished in conjunction with a pump manual composite chart by selecting the appropriate pump for a specified flow rate.

The system–pump operating point is the intersection of the system characteristic curve and the pump characteristic curve, since at this head–capacity point the system and the pump characteristics are simultaneously satisfied. These are indicated as points C in Fig. 4-30. As the system resistance increases, the system characteristic moves upward and to the left (curve D), which results in a decrease in the system flow rate at a new pump operating point. This procedure is also instructive in pointing out head–capacity regions where peculiar pump or system characteristics could lead to unstable and/or "hunting" behavior. Such a situation is illustrated in Fig. 4-31. Points A, B and C are multiple operating points but not necessarily stable operating points, since operation may drift from A to B to C and back again. Such regions and situations are to be avoided.

Example 4-4

A pump is used to pump water through a 3-in.-ID, 2000-ft-long schedule 40 pipe from one reservoir to another situated 20 ft higher. Minor losses have a combined loss coefficient of 1000. The pump characteristics (1 × 2-10, 9-in. rotor) are given in Fig. 4-32. Find the pump–system operating point if the fluid is water at 70°F.

Solution. A schematic of the system is given in Fig. 4-33. The energy equation written from station 1 to station 2 is

$$\frac{P_1}{\gamma}+\frac{V_1^2}{2g}+z_1=\frac{P_2}{\gamma}+\frac{V_2^2}{2g}+z_2+C\frac{V^2}{2g}+f\frac{L}{D}\frac{V^2}{2g}-H\frac{g_c}{g}$$

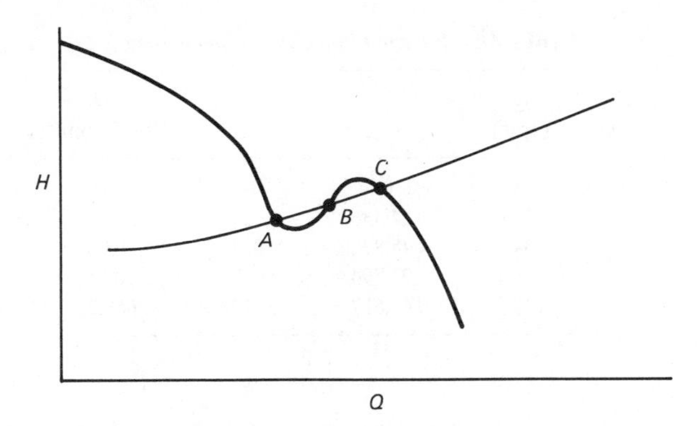

Figure 4-31 Multi-valued operating point illustration.

with the usual assumption ($P_1 = P_2$ and $V_1 = V_2 \sim 0$), the preceding equation becomes

$$\frac{g_c}{g} H = (z_2 - z_1) + \left(C + f\frac{L}{D}\right)\frac{V^2}{2g}$$

The expression $(g_c/g)H$ is the "head" in feet required by the system to pass the volumetric flow rate corresponding to the average velocity V. The results of evaluating this expression for several values of V are given by Table 4-5 and illustrated graphically in Fig. 4-34.

The pump characteristics are transcribed onto Fig. 4-34 from the manufacturer's data (given in Fig. 4-32), and the operating point occurs at the intersection of the system and pump characteristics. For this example, the operating point is 99 gpm and 387 ft lbf/lbm. The pump efficiency is 46 percent.

All the examples presented so far in this chapter have been worked essentially graphically. This was done because the graphical mode of presenta-

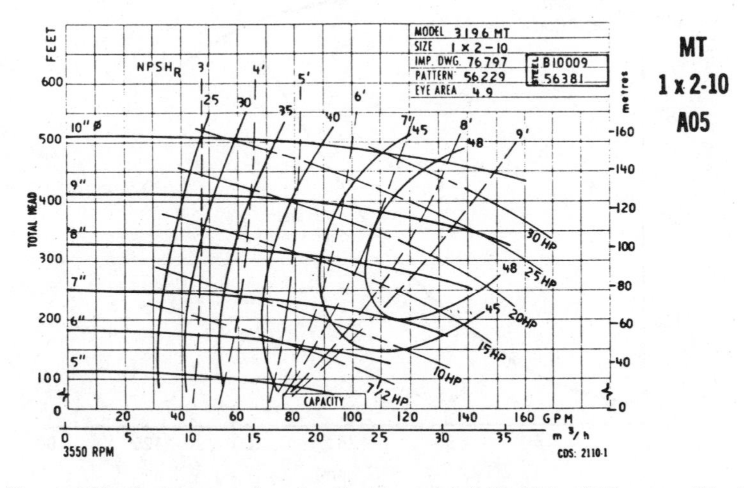

Figure 4-32 Performance data for Goulds model 3196MT 1 × 2-10 pump. (Used with permission, Goulds Pumps, Inc., Seneca Falls, N.Y.)

TABLE 4-5 System Characteristic Generation

V (ft/s)	Q (gpm)	Re_D	f	H (ft lbf/lbm)
0	0	0	0	20.0
1	22.0	24,436	0.027	38.9
2	44.0	48,872	0.023	93.6
4	88.1	97,794	0.0205	309.5
6	132.2	146,617	0.0198	668.2

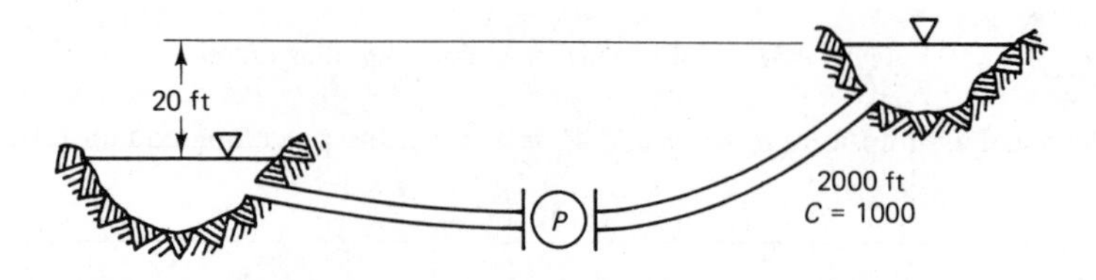

Figure 4-33 System schematic for Example 4-4.

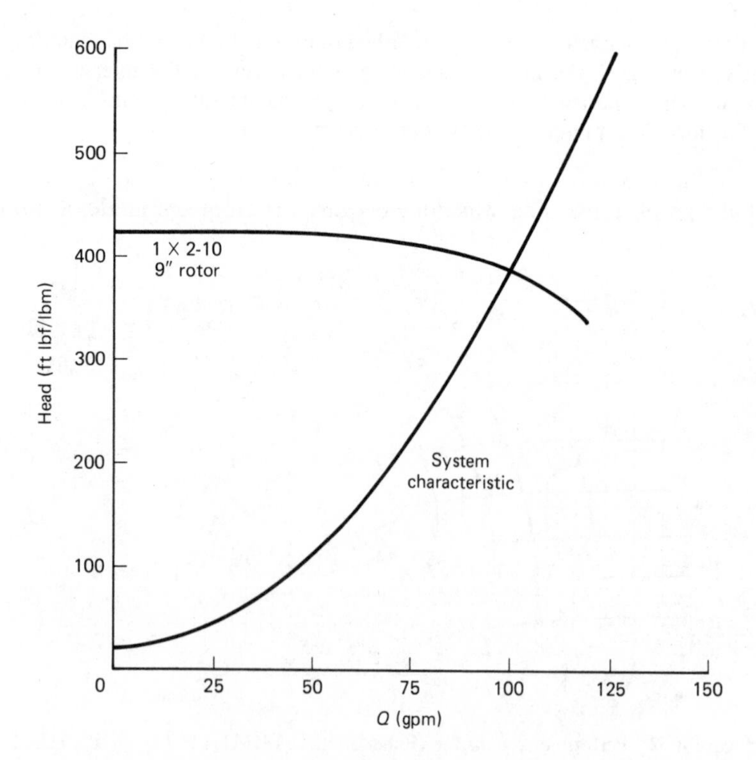

Figure 4-34 Solution for Example 4-4.

tion very clearly illustrates the concepts covered in this chapter. Once understanding is achieved, the processes can easily be adapted to the digital computer. Let's explore this adaptation briefly. The characteristic curve of a pump can be represented as a low-order polynomial such as

$$H = A_1 + A_2 Q + A_3 Q^2 \tag{4-28}$$

Curve fitting will be examined in the next chapter.

If the flow rate Q is needed as a function of the head, Eq. (4-28) can be "inverted" to yield

$$Q = B_1 + B_2 H + B_3 H^2 \tag{4-29}$$

The various situations encountered need to be examined and modeled using the preceding expressions.

The equivalent performance curve for series pumps is constructed by adding the pump heads at a given flow rate; hence

$$H = H_1 + H_2 + \cdots \tag{4-30}$$

or

$$\begin{aligned} H = (A_{11} + A_{12} + \cdots) + (A_{21} + A_{22} + \cdots)Q \\ + (A_{31} + A_{32} + \cdots)Q^2 \end{aligned} \tag{4-31}$$

where in A_{ij} the subscript i refers to the coefficient [A_1 or A_2 or A_3 in Eq. (4-28)] and the subscript j refers to the pump (1 or 2 or ...). Equation (4-28) for each pump and Eq. (4-31) for the series are sufficient for virtually any design or analysis problem statement.

The equivalent performance curve for parallel pumps is constructed by adding flow rates at a given head; hence,

$$Q = Q_1 + Q_2 + \cdots \tag{4-32}$$

or

$$\begin{aligned} Q = (B_{11} + B_{12} + \cdots) + (B_{21} + B_{22} + \cdots)H \\ + (B_{31} + B_{32} + \cdots)H^2 \end{aligned} \tag{4-33}$$

where the subscripts in B_{ij} are as in A_{ij}, which is described in the previous paragraph. As in the series case, Eq. (4-29) or (4-33) is sufficient for virtually any analysis or design problem.

Just as curve fits such as Eq. (4-28) or (4-29) can be generated for pumps, they can also be generated for systems. The simultaneous solution of analytical expressions for the pump and system characteristics yields the pump–system operating point. Numerical procedures that use interpolated tabular data for the pump characteristics and computed head–capacity for the system characteristics can be devised to locate operating points via iterative schemes such as Newton–Raphson. In such procedures, explicit curve fitting is avoided, but programming complexity is increased.

4-8 PUMP PLACEMENT TO AVOID CAVITATION

We have made rather full use of the manufacturer-supplied pump performance curves except for a single line mentioned earlier, the $NPSH_R$, the net positive suction head required. The $NPSH_R$ does not enter into the analysis of a given system, but it is *vital* to the design of a system if proper pump operation is to be assured. The NPSH parameter is used as an indicator of the susceptibility of a pump to cavitation. In particular, the NPSH is the difference between the static pressure at the pump suction (or inlet) nozzle and the vapor pressure of the fluid and is usually written as $P_s - P_v$.

When the static pressure of the fluid is reduced to the vapor pressure, pockets of vapor will form, and cavitation is said to take place. Cavitation is detrimental to both the pump's operating efficiency and its mechanical condition since the vapor pockets displace the liquid, disrupting the flow, and since the growth and collapse of these pockets can cause mechanical damage to the pump's components. Thus pump placement in the design of a system must be such that cavitation is avoided.

Cavitation can *always* be avoided by locating the pump where the NPSH available ($NPSH_A$) is greater than the NPSH required ($NPSH_R$), as specified by the manufacturer on pump performance curves. The $NPSH_R$ is thus the *minimum* value of $P_s - P_v$ (static pressure − vapor pressure), which can be tolerated at the pump inlet (suction nozzle) if cavitation is to be avoided. The $NPSH_R$ on the performance curves is that required if water at 60°F is used by the pump. The $NPSH_R$ is usually conservative; the *Standards of the Hydraulic Institute* (5) address corrections for other liquids. As the fluid enters the pump casing and is "picked up" by the impeller, the static pressure decreases. The positive value of $NPSH_R$ at the suction inlet is required if the static pressure within the turbomachine is to be maintained higher than the vapor pressure. Hence, $NPSH_A$ greater than ($>$) $NPSH_R$ will always ensure that $P_s - P_v$ is positive and that cavitation conditions will be avoided. The vapor pressure for water is given in Fig. 4-35.

The inlet suction pressure, P_s (or the static pressure at the inlet), is usually computed by starting at a position where the pressure is known and calculating the pressure changes in the downstream direction until the pump inlet is reached. Friction reduces the pressure, as do increases in elevation; decreases in elevation increase the static pressure. For example, the pressurized tank in Fig. 4-36 is drained by a line H feet below the liquid level and has a pressure loss due to friction of P_f. The suction pressure is then

$$P_s = P_t + \rho H_z \frac{g_c}{g} - P_f \tag{4-34}$$

and

$$NPSH_A = P_t + \rho H_z \frac{g_c}{g} - P_f - P_v \tag{4-35}$$

This also points out that the *lower* the pump placement the *higher* the value of

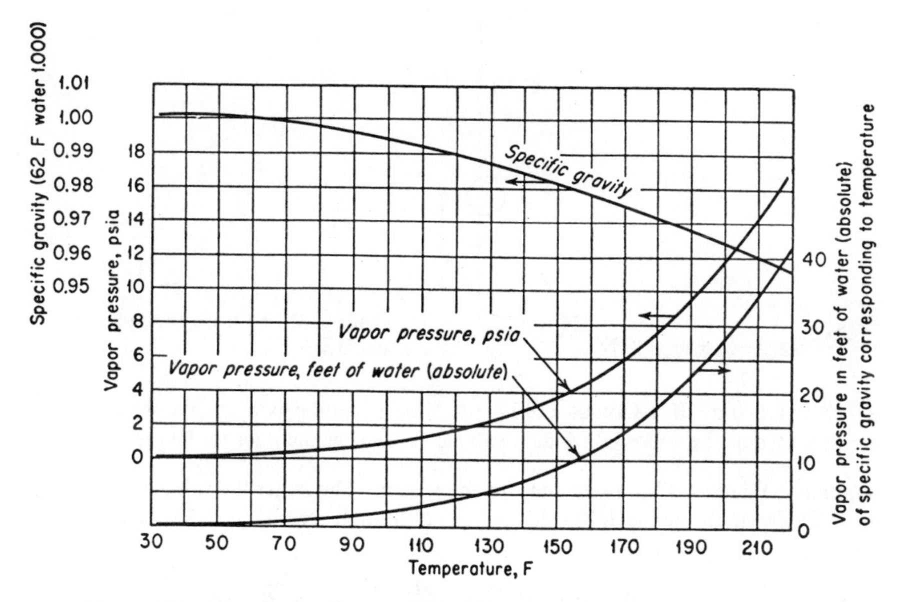

Figure 4-35 Water vapor pressure. (Used with permission, Worthington Division, McGraw-Edison Co.)

$NPSH_A$. In fact, in some installations the pump is placed at the lowest elevation in the system; occasionally, a pump may have to be placed in a sump to meet the requirement of $NPSH_A > NPSH_R$. Pump placement is an important consideration in the layout and specification of a pump–system network since placement can determine whether or not a pump will cavitate.

Example 4-5

Hot water at 200°F is to be pumped vertically upward from a shallow holding pond via a 2-in. ID commercial steel pipe. At a flow rate of 0.25 ft^3/s, the system pressure drop

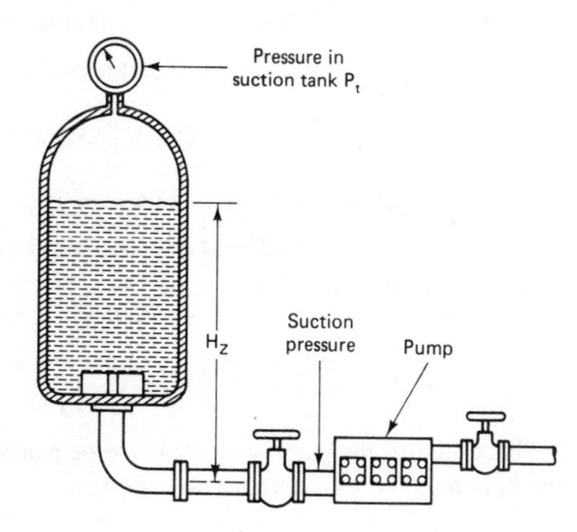

Figure 4-36 System schematic. (Used with permission, File No. 4010, *Plant Engineering*, May 1981.)

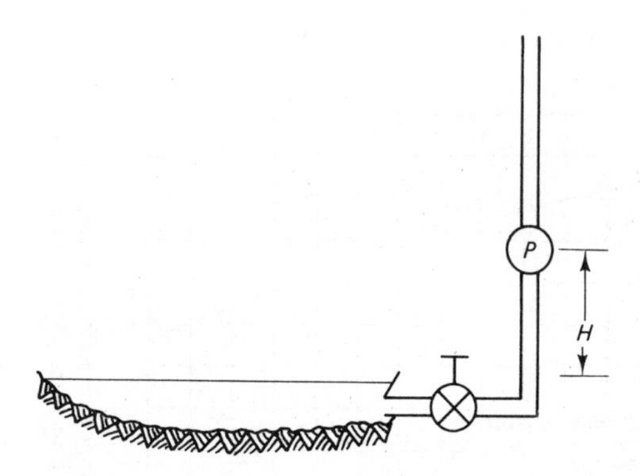

Figure 4-37 System schematic for Example 4-5.

is 200 ft lbf/lbm. If a Goulds $1 \times 1\frac{1}{2} - 8$ 3196 ST pump is available, what location limitations must be followed? A gate valve is located at the inlet to the piping system.

Solution. Figure 4-37 is a schematic of the system. The larger the value of H, the less the NPSH_A at the pump inlet. To prevent cavitation, $\text{NPSH}_R \leq \text{NPSH}_A$. The static pressure at the pump inlet is

$$P_s = P_\infty - K\rho\frac{V^2}{2g_c} - f\frac{L}{D}\rho\frac{V^2}{2g_c} - \rho H\frac{g}{g_c}$$

From Fig. 1-7, $K = 8f_T = 8(0.019) = 0.152$ and the Reynolds number becomes

$$\text{Re}_D = \frac{\rho VD}{\mu} = 60.1\frac{\text{lbm}}{\text{ft}^3}\left(\frac{2}{12}\right)\text{ft}\ 11.46\ \frac{\text{ft}}{\text{s}}\ \frac{\text{ft s}}{0.20 \times 10^{-3}\ \text{lbm}}$$
$$= 5.598 \times 10^5$$

for which $f = 0.0193$. The static pressure at a location H feet above the surface datum becomes

$$P_s = 2116.2 - \underbrace{18.64}_{\substack{\text{valve}\\\text{loss}}} - \underbrace{14.21H}_{\substack{\text{friction}\\\text{loss}}} - \underbrace{60.1H}_{\substack{\text{elevation}\\\text{change}}} \quad (\text{lbf/ft}^2)$$
$$= 2097.56 - 74.31H \quad (\text{lbf/ft}^2)$$
$$= 34.90 - 1.236H \quad (\text{ft lbf/lbm})$$

For the $1 \times 1\frac{1}{2} \times 8$ Goulds pump at 200 ft lbf/lbm and 0.25 ft³ (112.2 gpm) with water at 60°F,

$$\text{NPSH}_R \sim 8\ \text{ft lbf/lbm}$$

Then

$$\text{NPSH}_A = P_s - P_v = 34.90 - 1.236H - 28.75 \quad (\text{ft lbf/lbm})$$
$$= 6.15 - 1.236H$$

To prevent cavitation, we must have

$$\text{NPSH}_A \geq \text{NPSH}_R$$
$$6.150 - 1.236H \geq 8$$
$$H < -1.50\ \text{ft}$$

In this example, the pump would have to be placed 1.5 ft *or more* below the surface of the holding pond to ensure no cavitation.

This completes our discussion of centrifugal pumps and systems. We will next examine a low pressure pump for air, a fan in other words, and we shall use many of the same analysis concepts for fans and systems.

4-9 FANS

Fans are pumps that use air as the working fluid and that have a relatively small increase in head. The concepts that we have developed in this chapter are directly applicable to fans and will form the basis for the development of quantitative procedures. Fan applications are usually classed as heating, ventilating, and air-conditioning (HVAC) considerations and have developed a nomenclature different from that of pumps.

Fan pressure is generally expressed in terms of the pressure rise through the device (4). The fan *total pressure*, P_t, is the difference between the total pressures at the fan outlet and inlet. The fan *velocity pressure* is the dynamic pressure ($\rho \overline{V}^2/2g_c$) computed using the average velocity at the fan outlet. The most used term for fans is the fan *static pressure*, which is defined as the fan total pressure minus the fan velocity pressure. The fan static pressure is not the rise in static pressure across the fan, but it is considered to be the most useful fan pressure. Typical performance characteristics of fans are illustrated in Fig. 4-38. As for the case of pumps, volume flow rate is of primary interest and is usually taken as the independent variable. Maximum efficiency is achieved at a single operating point of the fan and is not usually achieved at

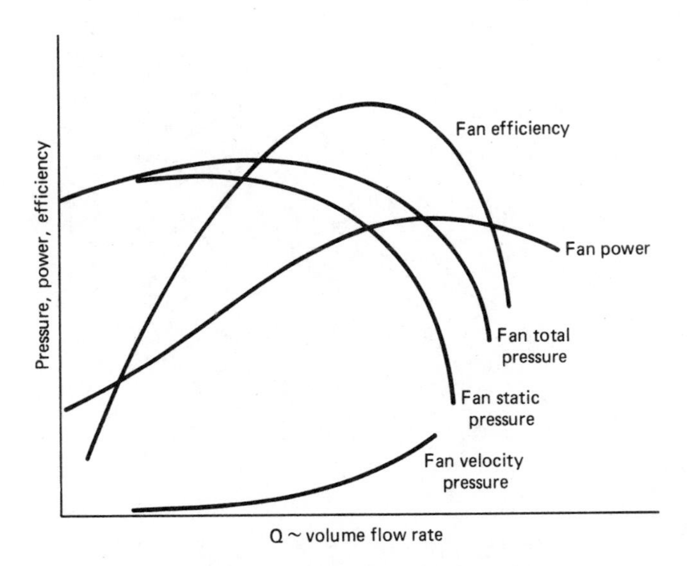

Figure 4-38 Fan characteristic curves.

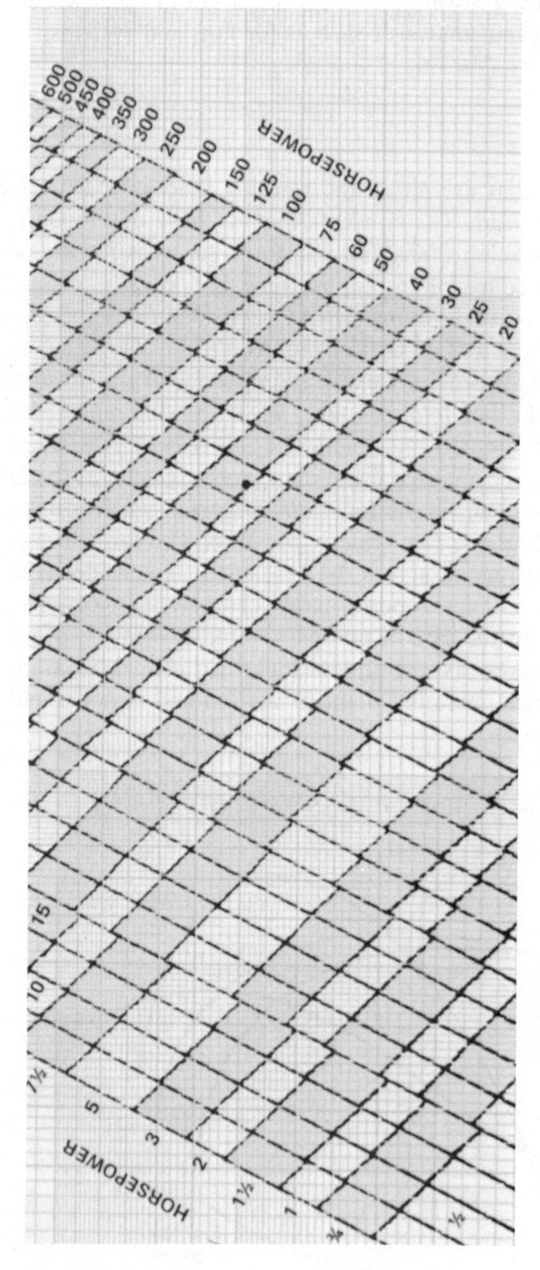

Figure 4-39 Typical fan composite rating chart. (Used with permission, Buffalo Forge Company, Buffalo, N.Y.)

either the maximum of the static or total fan pressures. Fan characteristics can be subtly different from pump characteristics since most pump curves are presented at constant speed (rpm) with effectively different system loads, whereas some fan characteristics are presented with the same load but with varying speed (rpms). In many instances the fan test load consists of only the casing and inlet and outlet ductwork.

Manufacturers' data for fans are typically arranged in a fashion similar to that for pumps; a single composite rating chart is used to identify a suitable fan for the static pressure–capacity requirements, and then detailed performance data are given on that particular fan. Figure 4-39 extracted from the Buffalo Forge Company's ventilating fan manual (10), is typical of fan composite rating curves. On this chart capacity (in CFM) and static pressure (in inches of water) are the abcissa and ordinate, while fan size is the diagonal with positive slope and horsepower the diagonal with negative slope. Thus fan sizes of this family range from the 245, which produces a static fan pressure of 2 in. of water at 1000 CFM, to the 2175, which produces a static fan pressure of 11 in. of water at 220,000 CFM.

A ventilating requirement of 40,000 CFM against a fan static pressure of 4 in. of water would require a 1200 size fan. Performance details for the 1200 Buffalo fan are given in Table 4-6. The requirement can be met by running the fan at about 570 rpms with 32.56 hp required to drive the fan. From Table 4-6 we see that speed (rpm) versus capacity (CFM) for a given static pressure is the basis for data presentation. Manufacturers' performance curves can be extended using the ideas of similarity developed earlier in this chapter. Equations (4-20) through (4-22) are usually called the "fan" laws and can be applied to fans as well as pumps.

Fans are classified as either centrifugal or axial, with the majority being of the centrifugal type. A *centrifugal* fan consists of an impeller running in a casing having a specially shaped contour. Figure 4-40, an illustration of a centrifugal fan, shows clearly the axial inlet and the radial discharge. As with the centrifugal pump, the impeller exit blade angles are backward curved, radial, or forward curved. These are diagrammatically illustrated in Fig. 4-41. The fan pressure (static or total) and the flow rate tend to increase as the exit blade angle increases. Figure 4-42 illustrates typical pressure–flow rate regimes for the various exit blade conditions.

Figures 4-41 and 4-42 provide a study of the various blading arrangements and geometrical considerations. The forward curved blades generally develop the highest pressures and flow rates but only at the expense of being smaller in the radial direction. The small radial depth requires more blades (30 to 60 for forward curved fans as compared to 6 to 16 for backward and radial fans) and allows a larger inlet diameter. Efficiencies for backward curved blading can go as high as 90 percent, as compared to about 75 percent for forward or radial blading. Radial bladed fans are often employed for boiler draft fans.

TABLE 4-6 Detailed Performance Data for the Buffalo 1200 Fan

CFM	OV	2″ SP		$2\frac{1}{4}$″ SP		$2\frac{1}{2}$″ SP		3″ SP		$3\frac{1}{2}$″ SP		4″ SP		$4\frac{1}{2}$″ SP		5″ SP		$5\frac{1}{2}$″ SP		6″ SP	
		rpm	bhp	rpm	bhp	rpm	bhp	rpm	bhp	rpm	bhp	rpm	bhp	rpm	bhp	rpm	bhp	rpm	bhp	rpm	bhp
28,980	1400	402	11.38	419	12.72	436	14.10	469	16.98	501	20.02	532	23.18	562	26.53	590	29.97	618	33.51	645	37.15
31,050	1500	411	12.35	427	13.74	444	15.16	475	18.13	506	21.25	536	24.50	565	27.88	592	31.37	620	35.08	646	38.82
33,120	1600	421	13.41	437	14.84	452	16.31	483	19.36	512	22.56	541	25.90	569	29.37	596	32.96	622	36.65	648	40.56
35,190	1700	431	14.55	446	16.04	461	17.56	491	20.70	519	23.98	547	27.41	574	30.96	600	34.63	626	38.42	651	42.31
37,260	1800	442	15.78	457	17.33	471	18.90	500	22.14	527	25.51	554	29.01	580	32.65	605	36.41	630	40.28	655	44.26
39,330	1900	454	17.09	468	18.70	482	20.33	509	23.67	536	27.14	561	30.73	587	34.45	611	38.29	635	42.25	659	46.31
41,400	2000	465	18.47	479	20.16	493	21.86	519	25.32	545	28.88	570	32.56	594	36.36	618	40.29	642	44.33	664	48.38
43,470	2100	477	19.94	491	21.70	504	23.47	530	27.06	555	30.73	579	34.51	602	38.40	626	42.41	648	46.53	671	50.77
45,540	2200	489	21.49	503	23.34	516	25.18	541	28.90	565	32.69	589	36.58	611	40.57	634	44.66	656	48.87	678	53.18
47,610	2300	502	23.13	515	25.05	528	26.98	552	30.85	576	34.77	599	38.76	621	42.85	643	47.04	664	51.34	685	55.74
49,680	2400	514	24.84	527	26.86	540	28.87	564	32.89	587	36.95	609	41.07	631	45.27	652	49.56	673	53.94	694	58.43
51,750	2500	526	26.64	539	28.75	552	30.85	576	35.03	598	39.24	620	43.49	641	47.81	662	52.20	682	56.69	702	61.27
53,820	2600	539	28.52	552	30.73	564	32.92	588	37.28	610	41.64	631	46.02	652	50.47	672	54.98	692	59.57	712	64.25
55,890	2700	552	30.49	564	32.80	577	35.08	600	39.62	622	44.14	643	48.68	663	53.25	683	57.89	702	62.59	721	67.38
57,960	2800	564	32.55	577	34.96	589	37.34	612	42.06	634	46.75	655	51.45	675	56.16	694	60.93	713	65.75	732	70.65
60,030	2900	577	34.71	590	37.21	602	39.70	624	44.61	646	49.47	666	54.33	686	59.20	705	64.10	724	69.05	742	74.06
62,100	3000	590	36.96	602	39.56	614	42.15	637	47.25	658	52.30	678	57.32	698	62.35	717	67.40	735	72.48	753	77.62
66,240	3200	616	41.76	628	44.57	640	47.35	662	52.85	683	58.28	703	63.66	722	69.02	740	74.38	758	79.76	775	85.16
70,380	3400	643	47.01	654	50.00	666	52.98	687	58.88	708	64.71	727	70.47	746	76.18	764	81.88	781	87.57	798	93.28
74,520	3600	669	52.73	681	55.90	692	59.07	713	65.37	733	71.59	752	77.74	771	83.84	788	89.89	805	95.92	822	101.95

OV ≡ outlet velocity in ft/min.
SP ≡ static pressure.
Used with permission, Buffalo Forge Company, Buffalo, N.Y.

Figure 4-40 Centrifugal fan. (Used with permission, Buffalo Forge Company, Buffalo, N.Y.)

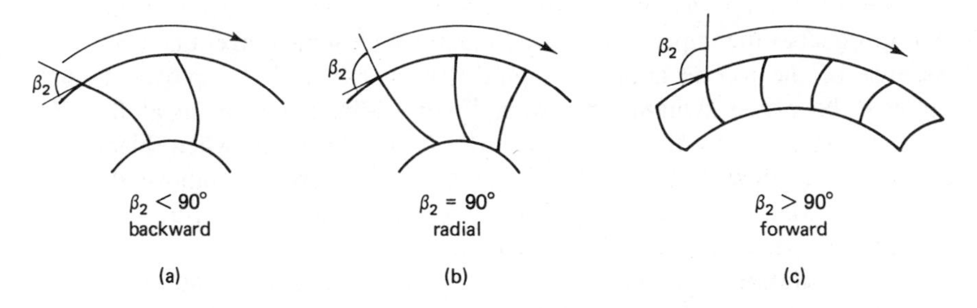

Figure 4-41 Exit blade geometry examples.

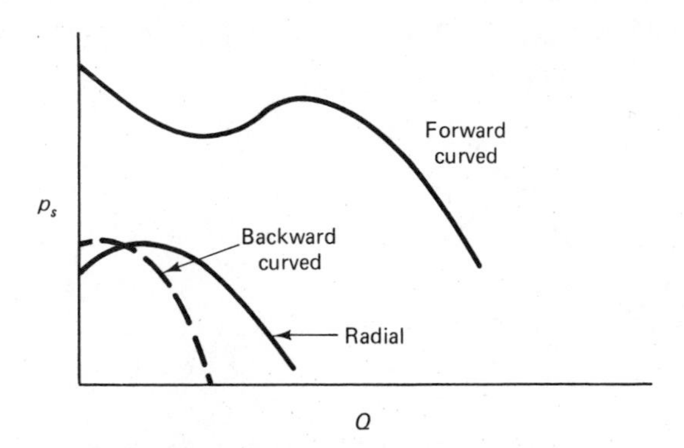

Figure 4-42 Fan characteristic curve examples.

Figure 4-43 Typical axial flow fan. (Used with permission, Chicago Blower Corp., Chicago, Ill.)

Axial flow fans have the flow of air substantially parallel to the axis of the impeller. Figure 4-43 illustrates a typical axial flow fan. The tips of the impeller blades have as fine a clearance as possible with the cylindrical casing. Air approaches the impeller axially and leaves with some tangential component because of the torque applied. A guide vane is typically employed to help recover the excess dynamic pressure. Figure 4-44, a typical axial flow fan characteristic, shows the abrupt change in performance when blade stall occurs. Axial flow fans are typically operated at flow rates above this stall position. Axial fans are not nearly as widely used as centrifugal and are generally restricted to uncontaminated air.

Quantitative techniques for the analysis and design of fan systems are very similar to the concepts examined earlier in this chapter for pump systems. Performance curves for fans in series and parallel, for example, are generated in the same manner as for pumps in series and parallel. Two simplifying assumptions are usually made in the analysis of fan systems: (1) the flow is incompressible, and (2) losses are quadratic in flow rate. The first assumption is nearly satisfied for most fan flows, as both pressure and temperature changes are usually small. Assumption 2 is nearly satisfied as $h_f \propto Q^2$ is quite reasonable for a first-order estimate.

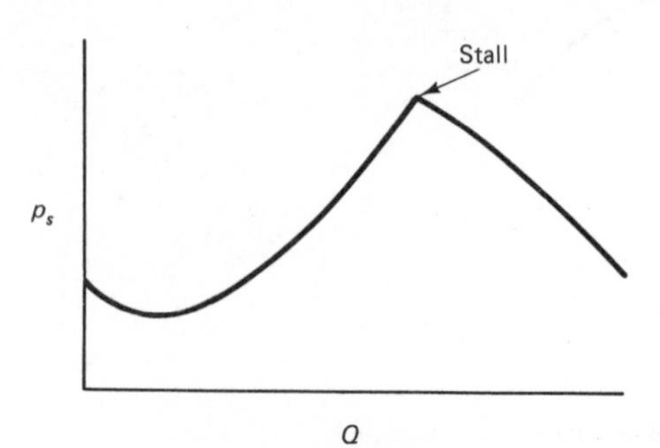

Figure 4-44 Axial flow fan characteristic curve.

As an example of a fan–system operating point problem, consider a supply fan with a duct, an exhaust fan with a duct, and a volume to be ventilated. Schematically, this can be illustrated as in Fig. 4-45. The usual practice (4) in such problems is to consider an effective fan, the fan static pressure minus the duct losses; the fan–system then becomes as in Fig. 4-46. In this sketch $(P-R)$ represents an effective fan. The *effective fan performance* curve is obtained via the procedure illustrated in Fig. 4-47. The *operating point* for this arrangement is determined by plotting the effective fan characteristics with respect to atmospheric pressure; the supply system pressure is positive, and the extract pressure is negative. The intersection point represents the steady-state supply–exhaust flow rate and the room pressure in the absence of infiltration. This process is illustrated as point N in Fig. 4-48.

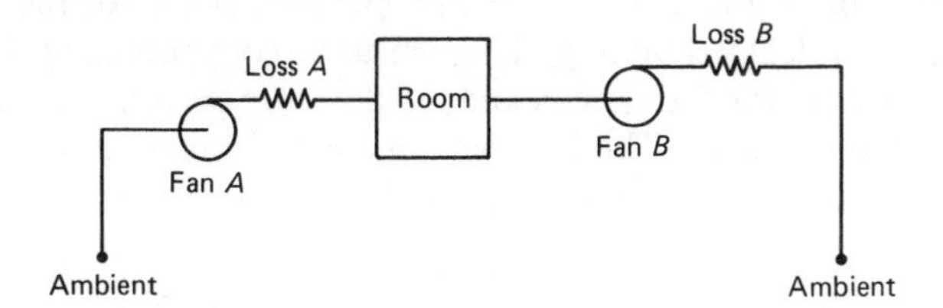

Figure 4-45 Room ventilating schematic with losses.

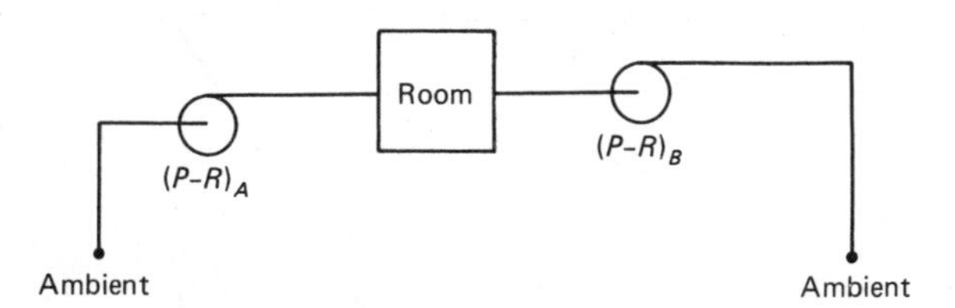

Figure 4-46 Room ventilating schematic using effective performance curves.

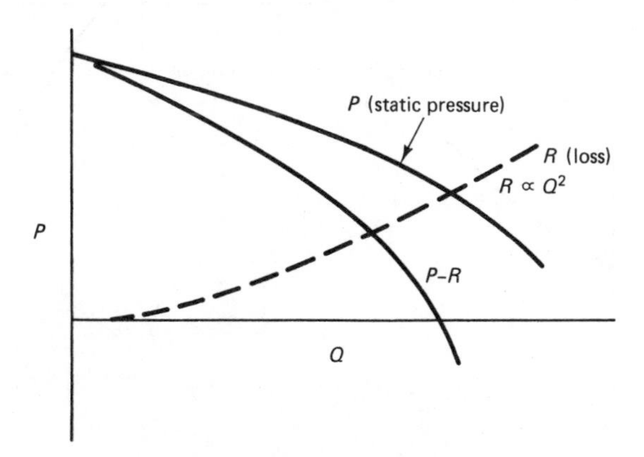

Figure 4-47 Effective fan performance curve.

An analogous procedure can be used to find the operating point of the supply–exhaust–building system when infiltration is present. An exhaust–supply fan arrangement with an operating pressure at atmospheric represents the maximum infiltration possible, an "open" building. Thus the segment a–b represents the maximum *exfiltration* for this supply–exhaust system. Had point N been in the negative (gauge) pressure region, the line a–b would have represented the *infiltration*. The region a–b–N–a represents the *leakage* characteristics of the supply–exhaust fans. If the differences in the supply–exhaust characteristics at given pressures (segment c–d, for example) are plotted as the ordinate (segment $c' - d'$), then a triangular-shaped region N'–d'–b' is generated near the origin. In accordance with assumption 2, we take $p \propto h \propto Q^2$ for the room. This is plotted as line R in Fig. 4-48. Point e, the intersection of R and N'–d'–b' is the operating point for the supply fan, exhaust fan, and leaky building. The pressure corresponding to e is the steady-state pressure for the system, and the flow rate corresponding to e is the exfiltration for the system. The intersections of this pressure with the supply fan and the exhaust fan characteristics are the operating points of the individ-

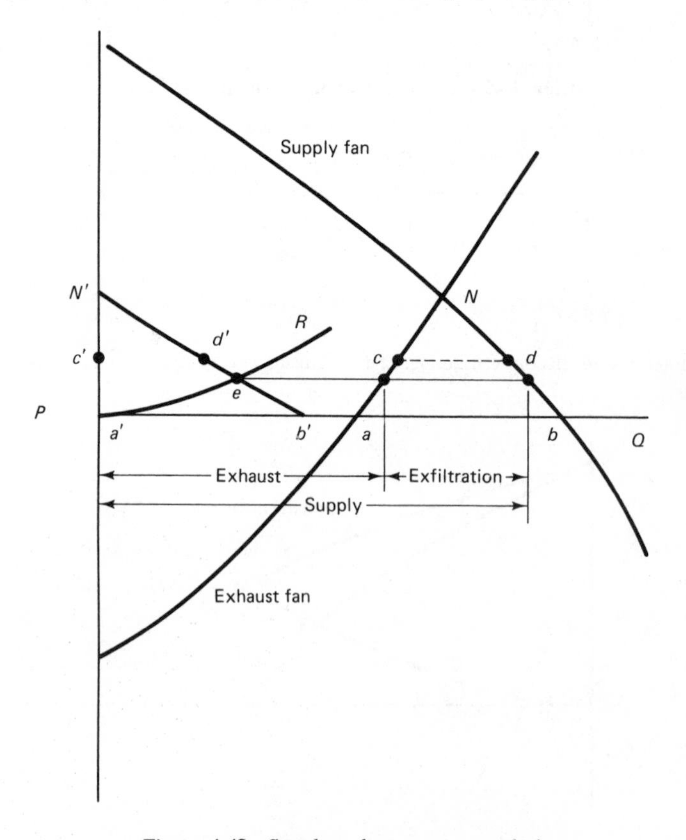

Figure 4-48 Supply–exhaust system analysis.

ual fans. As shown in Fig. 4-48, the supply flow rate is equal to the sum of the exhaust flow rate and the exfiltration.

The preceding represents a rather short examination of fans and fans interacting with systems. Details of HVAC system analysis and design are covered in many basic texts and are really too specialized for coverage here.

4-10 NOZZLES

Nozzles are frequently used discharge devices for both gases and liquids. They come in a wide variety of shapes, sizes, and output patterns, but the salient feature that governs the behavior of all nozzles is the ambient pressure into which they are discharged. Figure 4-49 schematically illustrates the important pressures in a typical nozzle. The pressure at the *nozzle exit plane* is p_e, and the relationship between p_e and p_∞ governs the nozzle efflux for a prescribed $p_i - p_e$, where p_i is the *nozzle inlet pressure*.

If the fluid is considered incompressible, then p_e must *always* be equal to p_∞; that is, the ambient pressure determines the nozzle exit pressure. The discharge velocity is

$$V_e = C_v \sqrt{\frac{2(p_i - p_e)}{\rho} \frac{g_c}{1-(A_e/A_i)^2}} \tag{4-36}$$

where C_v is an experimentally determined velocity coefficient.

If the fluid is compressible, then the pressure in the nozzle exit plane must be greater than or equal to the ambient pressure. For a converging nozzle, such as is shown in Fig. 4-49, the Mach number at the nozzle exit can never exceed unity no matter how low p_∞ becomes. Hence, p_e can never become lower than the static pressure corresponding to Mach 1 in the exit plane. However, if the exit Mach number is less than 1, then the exit pressure is equal to the ambient pressure. When the exit Mach number is equal to 1, the nozzle is said to be choked and the mass flow rate (for a gas with $C_p/C_v = 1.4$) (11) is

$$\dot{m} = 0.532 A_e \frac{\sqrt{T_0}}{p_0} \tag{4-37}$$

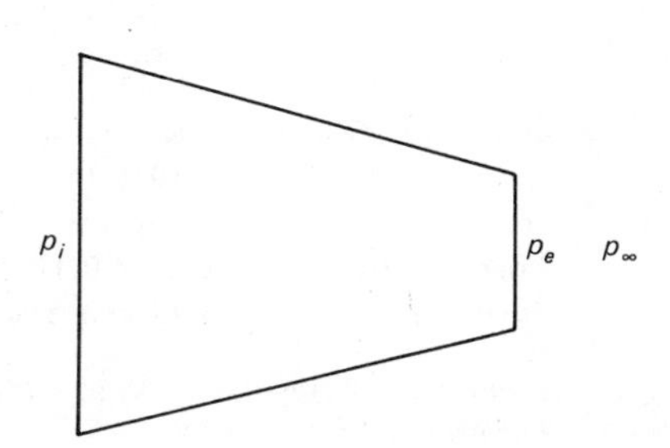

Figure 4-49 Nozzle nomenclature.

where p_0 and T_0 are the stagnation temperature and pressure. The mass flow for situations where p_e is equal to p_∞ is calculated by computing the exit Mach number and exit temperature corresponding to p_e and then using $\rho_e A_e V_e$ to compute the mass flow rate. The flow process through the nozzle is usually taken to be isentropic. We shall not develop the compressible flow study any further, but will examine instead incompressible spray nozzles.

Spray nozzles are widely used in many industries and are classified as either laminar or radial (12). Nozzles usually consist of slotted caps attached to stems on a pipe header. *Laminar nozzles*, shown in Fig. 4-50(a), have slots normal to the nozzle axes, while *radial nozzles*, shown in Fig. 4-50(b), have slots parallel to the nozzle axes. Radial spray nozzles are generally referred to in terms of the spray nozzle pattern, with each pattern being suited to different

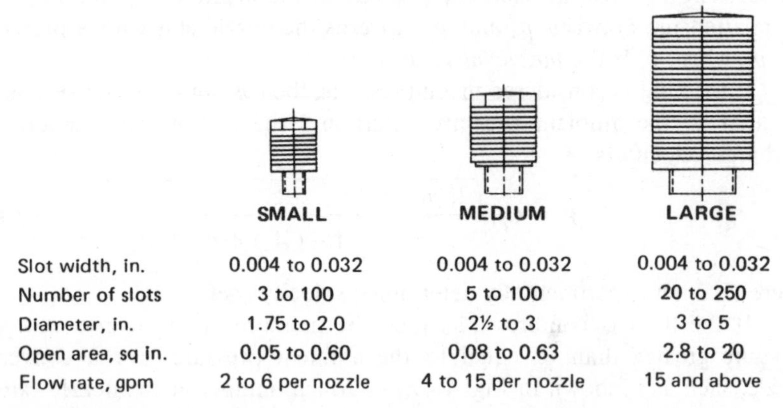

	SMALL	MEDIUM	LARGE
Slot width, in.	0.004 to 0.032	0.004 to 0.032	0.004 to 0.032
Number of slots	3 to 100	5 to 100	20 to 250
Diameter, in.	1.75 to 2.0	2½ to 3	3 to 5
Open area, sq in.	0.05 to 0.60	0.08 to 0.63	2.8 to 20
Flow rate, gpm	2 to 6 per nozzle	4 to 15 per nozzle	15 and above

(a) Design and performance characteristics of laminar distribution nozzles.

(b) Design and performance characteristics of radial distribution nozzles.

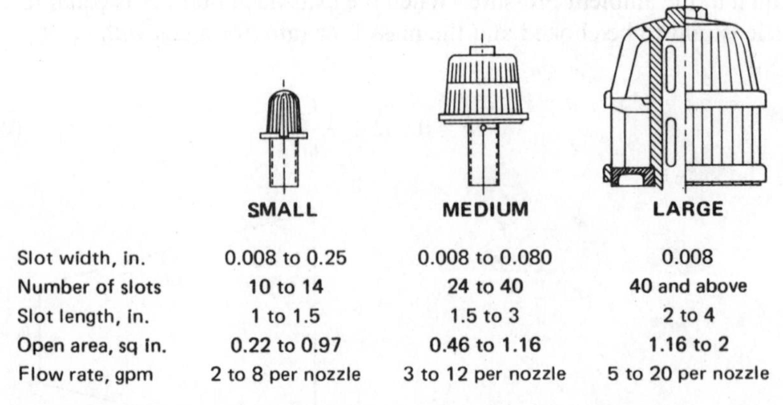

	SMALL	MEDIUM	LARGE
Slot width, in.	0.008 to 0.25	0.008 to 0.080	0.008
Number of slots	10 to 14	24 to 40	40 and above
Slot length, in.	1 to 1.5	1.5 to 3	2 to 4
Open area, sq in.	0.22 to 0.97	0.46 to 1.16	1.16 to 2
Flow rate, gpm	2 to 8 per nozzle	3 to 12 per nozzle	5 to 20 per nozzle

Figure 4-50 Spray nozzle examples. (Used with permission, File No. 7520, *Plant Engineering*, March 1982.)

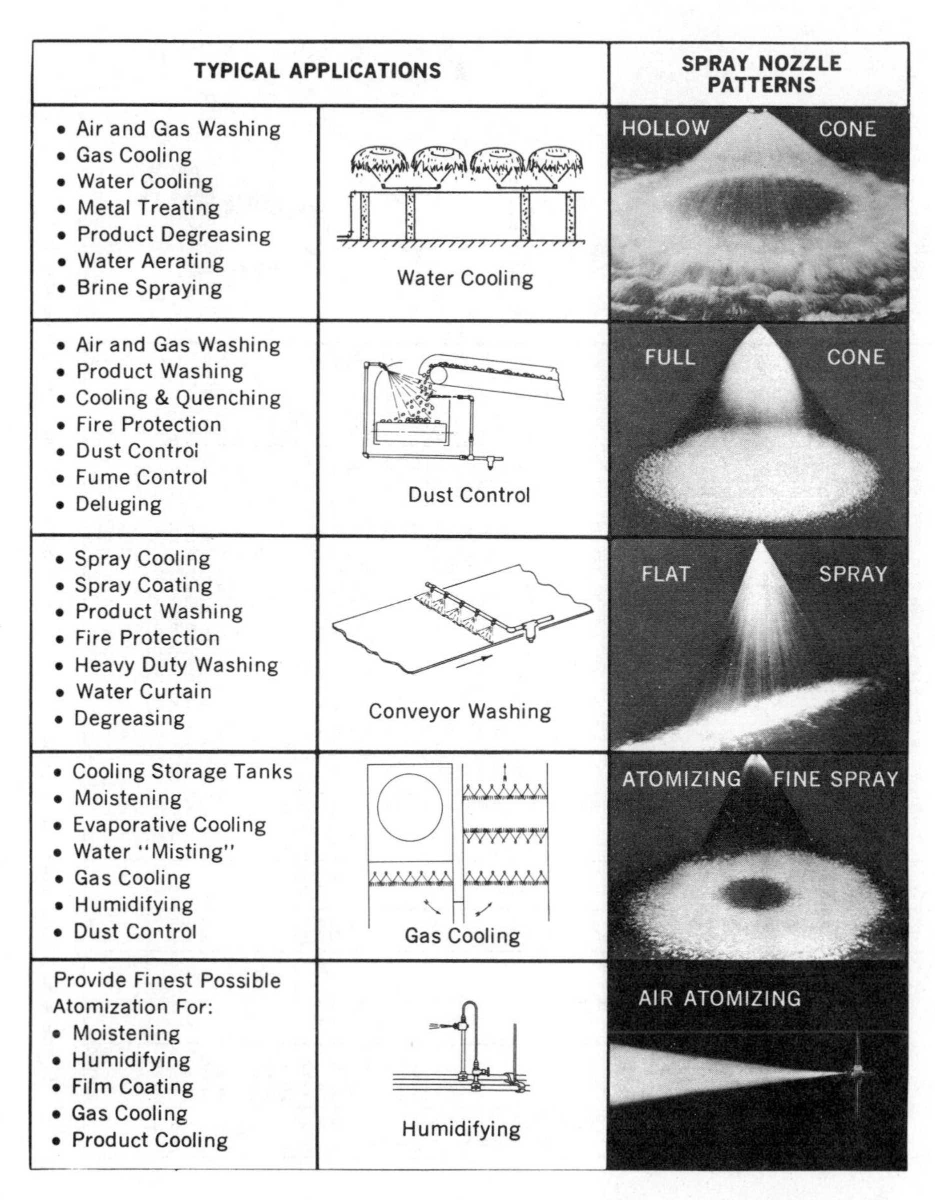

TYPICAL APPLICATIONS		SPRAY NOZZLE PATTERNS
• Air and Gas Washing • Gas Cooling • Water Cooling • Metal Treating • Product Degreasing • Water Aerating • Brine Spraying	Water Cooling	HOLLOW CONE
• Air and Gas Washing • Product Washing • Cooling & Quenching • Fire Protection • Dust Control • Fume Control • Deluging	Dust Control	FULL CONE
• Spray Cooling • Spray Coating • Product Washing • Fire Protection • Heavy Duty Washing • Water Curtain • Degreasing	Conveyor Washing	FLAT SPRAY
• Cooling Storage Tanks • Moistening • Evaporative Cooling • Water "Misting" • Gas Cooling • Humidifying • Dust Control	Gas Cooling	ATOMIZING FINE SPRAY
Provide Finest Possible Atomization For: • Moistening • Humidifying • Film Coating • Gas Cooling • Product Cooling	Humidifying	AIR ATOMIZING

Figure 4-51 Spray nozzle patterns and uses. (Used with permission, Spraying Systems Co., Wheaton, Ill.)

Nozzle No.		Connection		Orifice Diam. Nom. Inches	Maximum Free Passage Diam.* Inches	CAPACITY GPM (gallons per minute) at p.s.i. (pounds per square inch)												SPRAY ANGLE***		
Female Pipe Conn.	Flange Conn.	Pipe Type NPT	Flange Conn.** Inches			3 p.s.i.	5 p.s.i.	7 p.s.i.	10 p.s.i.	15 p.s.i.	20 p.s.i.	30 p.s.i.	40 p.s.i.	60 p.s.i.	80 p.s.i.	100 p.s.i.	150 p.s.i.	7 p.s.i.	20 p.s.i.	60 p.s.i.
1¼ H6		1¼		19/64	¼	4.0	5.1	6.0	7.1	8.5	9.7	11.8	13.5	16.3	18.6	21	25	48°	50°	46°
1¼ H10				⅜	¼	6.8	8.7	10.0	12.0	14.5	16.5	19.9	23	27	31	35	42	64°	67°	61°
1¼ H12				27/64	¼	8.0	10.2	12.0	14.1	17.1	19.4	24	27	33	37	41	50	66°	70°	63°
1¼ H14				31/64	¼	9.3	12.0	14.0	16.5	19.9	23	27	32	38	41	48	58	77°	80°	73°
1¼ H20				19/32	5/16	13.5	17.2	20.0	24	29	33	39	44	53	61	68	82	90°	93°	85°
1½ H10		1½		⅜	11/32	6.8	8.7	10.0	12.0	14.5	16.5	19.9	23	27	31	35	42	48°	50°	46°
1½ H16				½	11/32	10.8	13.7	16.0	18.9	23	26	31	35	43	49	54	66	72°	74°	67°
1½ H20				9/16	11/32	13.5	17.2	20	24	29	33	39	44	53	61	68	82	74°	76°	69°
1½ H30				23/32	13/32	20	26	30	36	43	48	58	66	80	91	102	123	91°	94°	86°
2H17		2		½	7/16	12	15	17	20	24	27	33	38	45	52	63	76	49°	50°	46°
2H30				11/16	7/16	20	26	30	36	43	48	58	66	80	91	102	123	72°	74°	67°
2H35				¾	7/16	24	30	35	42	50	57	68	76	93	108	121	147	75°	77°	70°
2H40				53/64	7/16	27	34	40	47	57	64	77	88	108	124	137	166	78°	80°	73°
2H50				15/16	9/16	34	43	50	59	71	82	99	113	135	156	173	210	83°	85°	78°
2H60				1⅛	9/16	41	51	60	71	85	96	118	140	163	186	203	250	98°	100°	91°
2½ H25		2½		19/32	9/16	17	21	25	30	36	41	49	56	68	78	86	104	49°	50°	46°
2½ H50				⅞	9/16	34	43	50	59	71	81	97	112	135	155	172	208	72°	74°	67°
2½ H60				31/32	9/16	41	51	60	71	85	96	118	140	163	186	203	250	76°	78°	71°
2½ H70				1⅛	9/16	47	60	70	83	100	115	138	158	191	219	243	294	79°	82°	75°
2½ H80				1⅛	11/16	54	69	80	95	115	132	160	183	220	253	280	339	86°	88°	80°
2½ H90				1¼	11/16	61	77	90	108	130	148	180	202	248	282	316	383	95°	97°	88°
3H42		3		¾	11/16	28	36	42	50	60	69	82	94	105	131	145	176	49°	50°	46°
3H80				1 3/32	11/16	54	69	80	95	115	132	160	183	220	253	280	339	81°	84°	76°
3H90				1 3/16	11/16	61	77	90	108	130	148	180	202	248	282	316	383	86°	89°	81°
3H100				1 9/32	11/16	68	86	100	119	143	164	198	228	274	314	349	419	92°	95°	87°
3H120				1⅜	13/16	80	102	120	142	171	194	235	269	324	370	411	500	102°	105°	93°
4H160	4HF160	4	4	1 11/16	¾	108	137	160	189	228	260	311	354	428	487	540	655	87°	90°	73°
4H180	4HF180			1 55/64	⅞	122	154	180	212	257	292	351	403	485	552	615	745	92°	95°	87°
4H200	4HF200			2	1	135	172	200	238	285	325	390	443	533	610	675	820	97°	100°	91°
4H210	4HF210			2 5/32	1	141	180	210	250	302	345	420	480	580	665	740	895	102°	105°	95°
5H250	5HF250	5	5	1⅞	1⅛	168	213	250	298	359	408	490	561	678	775	860	1040	89°	91°	83°
5H280	5HF280			2 3/16	1⅛	188	240	280	330	400	452	546	625	751	860	955	1150	93°	96°	88°
5H320	5HF320			2 11/16	1⅜	215	272	320	378	452	519	621	710	858	980	1100	1330	97°	100°	91°
5H330	5HF330			2 27/32	1⅜	221	281	330	390	469	534	640	732	885	1020	1140	1380	102°	105°	95°
6H350	6HF350	6	6	2 13/32	1⅝	238	301	350	413	499	568	680	775	932	1080	1210	1470	87°	90°	82°
6H400	6HF400			2 23/32	1⅝	270	341	400	470	566	643	770	883	1080	1240	1370	1660	92°	95°	87°
6H450	6HF450			3 1/32	1¾	303	387	450	531	640	730	880	1010	1230	1400	1550	1880	97°	100°	91°
6H480	6HF480			3 7/32	1¾	323	412	480	569	681	780	940	1080	1310	1490	1670	2040	102°	105°	95°
8H500	8HF500	8	8	2¾	1⅞	335	425	500	590	710	815	985	1130	1370	1560	1730	2100	78°	80°	73°
8H600	8HF600			3 5/32	1⅞	359	512	600	708	849	960	1180	1400	1630	1860	2030	2500	86°	88°	80°
8H700	8HF700			3 19/32	1⅞	473	600	700	830	995	1150	1380	1580	1910	2190	2430	2940	92°	95°	87°
8H800	8HF800			4 1/32	2¼	540	685	800	949	1150	1320	1600	1830	2200	2530	2800	3390	102°	105°	95°
8H900	8HF900			4⅞	2¼	610	772	900	1080	1300	1480	1800	2020	2480	2820	3160	3830	106°	110°	100°
	10HF800		10	3 11/32	2½	540	685	800	949	1150	1320	1600	1830	2200	2530	2800	3390	78°	80°	73°
	10HF1000			3 31/32	2½	670	850	1000	1180	1420	1630	1970	2260	2740	3130	3470	4200	86°	89°	81°
	10HF1200			4 51/64	2⅝	810	1030	1200	1420	1720	1970	2380	2720	3290	3750	4170	5020	97°	100°	91°
	10HF1300			5 5/16	2⅝	880	1110	1300	1550	1860	2130	2570	2930	3530	4050	4500	5430	103°	106°	96°

FULL CONE SPRAY PATTERN

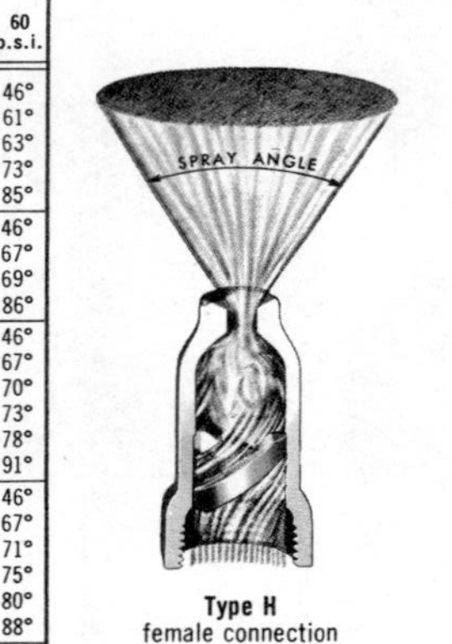

Type H
female connection

All FullJet Nozzles made with internal vanes as shown in cut-away above.

Type HF
flange connection

Figure 4-52 Spray nozzle data. (Used with permission, Spraying Systems Co., Wheaton, Ill.)

10. Indicate the range of stable operation for the pump of Fig. RQ4-10.
11. Why is the infiltration of a structure considered to be a maximum when the internal and external pressures are the same?
12. How could spray nozzle information in Fig. 4-52 be used in applying the energy equation to an analysis for a spraying apparatus?

PROBLEMS

1. A process in a plant requires 300 gal/min at a head increase of 72 ft lbf/lbm. The density of the fluid is 123 lbm/ft^3. A Goulds standard dimension process pump is to be used.
   (a) What two standard Goulds 3196 pumps would be acceptable?
   (b) What 1750 rpm pump would be required?
      1. What would the operating efficiency be?
      2. What size motor would be required? (Assume motors are available in 5-hp increments.)
   (c) Answer 1 and 2 of part (b) for the 3500 rpm pump.
2. A cooling loop for hydraulic fluid (MIL-M-5606) consists of:
   (a) A heat exchanger whose pressure drop versus flow rate curve is given in Fig. P4-2(a).
   (b) 150 ft. of 3 in. schedule 80 pipe + 4 standard elbows ahead of the exchanger.
   (c) 150 ft. of 3 in. schedule 80 pipe + 4 standard elbows after the exchanger.
   (d) 1 gate valve.
   The fluid enters the exchanger at 200°F and leaves at 80°F. The loop leaves and enters the machine at the same level. The outlet reservoir has a sharp entrance, and

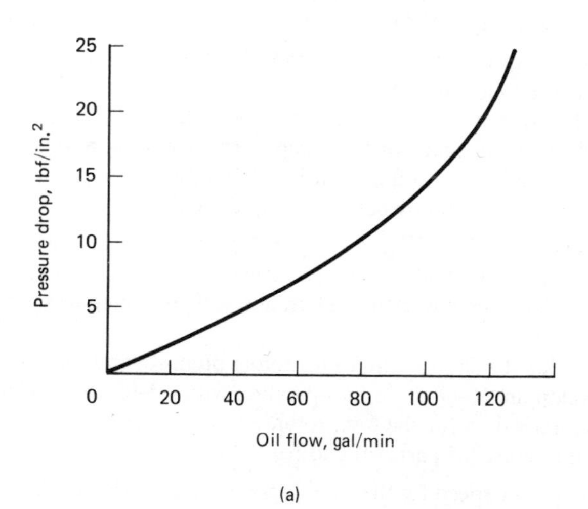

Figure P4-2(a) Heat exchanger pressure loss.

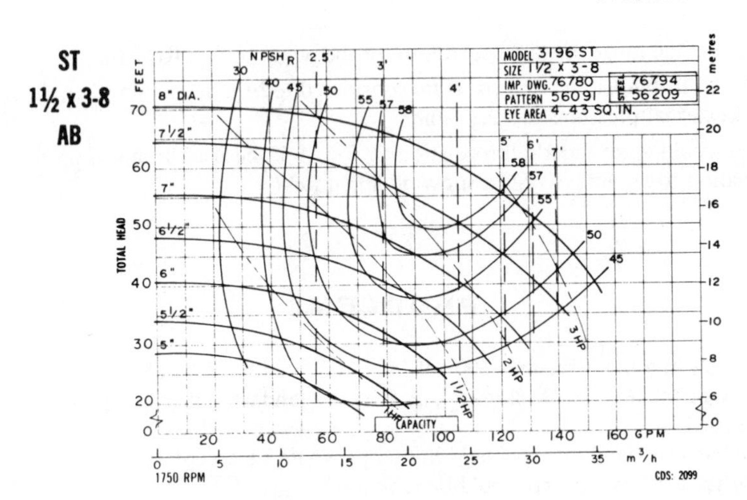

Figure P4-2(b) (Used with permission, Goulds Pumps, Inc., Seneca Falls, N.Y.)

an inlet reservoir is present. The system has a Goulds $1\frac{1}{2} \times 3$-8 pump [characteristics given in Fig. P4-2(b)] that runs at 1750 rpm. The pump has a $7\frac{1}{2}$-in. rotor. For this system:

(a) Plot the H versus Q for the loop with the gate valve fully open.
(b) Plot the H versus Q for the loop with the gate valve one-quarter open.
(c) From the manufacturer's data, plot H versus Q for the pump.
(d) Find the operating point for part (a).
(e) Find the power required for part (d).
(f) Find the operating point for part (b).
(g) Find the power required for part (f).

3. A process in a plant requires 350 gpm and 320 ft lbf/lbm. If a 3500 rpm motor is available, find the following:
   (a) The Goulds 3196 pump required; the rotor diameter required.
   (b) The mechanical efficiency of the pump at the operating point.
   (c) The power required for a fluid density of 100 lbm/ft^3.
   (d) Indicate on a sketch the *important* pump dimensions.

4. **(a)** For a Goulds 2×3-10 (Fig. 4-15) pump operating with water at 70°F and an 8-in. impeller, estimate the pump performance curve at 1750 rpm from the 3500 rpm data. Compare the estimated results with the manufacturer's data for the 1750 rpm case.
   (b) For a Goulds 2×3-10 (with 10-in. rotor) pump operating with water at 3550 rpm, develop an H–Q curve for operation with a 6-in. rotor. Compare with the manufacturer's data for the 6-in. rotor.
   (c) Discuss the results of parts (a) and (b).

5. Compute the specific speed for the Goulds $1 \times 1\frac{1}{2}$-8, 2×3-10, 1×2-10, and 8×10-15 pumps with the maximum rotor diameter at both speeds given. Plot on Figure 4-19 and discuss the results.

6. Several Goulds $1\frac{1}{2} \times 3$-8 pumps are available for use with 1750 rpm motors. The system characteristics are given by $H = 0.002Q^2$ for the parallel systems and $H = 0.020Q^2$ for the series system when Q is in gpm and H is in ft lbf/lbm.
   (a) If two pumps with 8-in. rotors are placed in series, find the equivalent pump curve, the system operating point, the pumps' operating point, and the power required by each pump.
   (b) Work part (a) if one 8-in. rotor pump and one $7\frac{1}{2}$-in. pump in series are used.
   (c) Work part (a) if two 8-in. rotor pumps are in parallel.
   (d) Work part (a) if one 8-in. rotor pump and one $7\frac{1}{2}$-in. rotor pump are in parallel.
   (e) Work part (a) if one 8-in. rotor pump, one $7\frac{1}{2}$-in. rotor pump, and one 7-in. pump are in parallel.

   The pump characteristics are given in Problem 2.
7. Water at 150°F is being pumped from a holding pond to a tank located 80 ft above the surface of the pond. At the required 150 gpm, the system pressure drop is 101 ft lbf/lbm. Where must a Goulds 2×3-10 pump be located to avoid cavitation? The water velocity is 6 ft/s. The pump operates at 1750 rpm.
8. A large process workshop is to be ventilated by a supply fan and system in which the pressure loss is 1.2 in. of water for a flow of 6000 ft^3/min and an extract fan and system for which the pressure loss is 1.7 in. of water for a flow of 5000 ft^3/min. Find the pressure in the workshop and the air volume flow. If there is a loss of air from the space by exfiltration through an aperture, where the pressure loss is 0.5 in. of water for a flow of 1000 ft^3/min, find the pressure in the space, the supply volume flow, and the extract volume flow. The fan characteristic is given in Fig. P4-8.
9. In an industrial manufacturing plant, $1\frac{1}{2}$ pipe type NPT full cone spray nozzles (Fig. 4-52) are to be used to spray 60°F cooling water. Nozzles are to be placed on a pipe along a 10 ft length. A 30 gal/s flow rate is required. Determine the number of nozzles, specify the type, and estimate the pressure required at the pipe inlet. Would the resulting spray patterns be uniform?

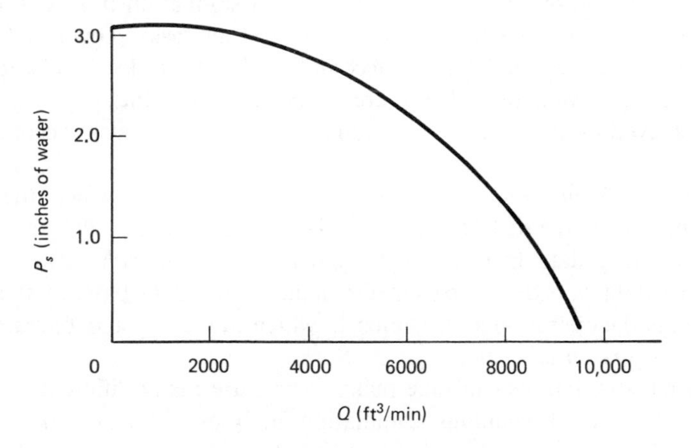

Figure P4-8 Fan characteristics.

5

SYSTEM SIMULATION

5-1 INTRODUCTION AND OVERVIEW

Simulation, and the complementary word modeling, is seen with increasing frequency in journals and textbooks (1–4) of every engineering discipline. Generally, in the literature, *modeling* deals primarily with establishing valid quantitative relationships between real systems and models of real systems, while *simulation* is concerned with implementing via computer codes the models in such a manner that the resulting computations have a high degree of fidelity. Simulation and modeling are a duality. System simulation is often used to denote the output of the modeling and simulation process for a real system. Smith (1) divides this process into the four tasks shown in Fig. 5-1. Task 1 is where the modeling is accomplished, and tasks 2, 3, and 4 are required for the simulation. These are serial tasks, and the only way fidelity can be achieved in the system simulation is to have all four tasks accomplished correctly.

The *degree of validity* of a model is the extent to which the model matches data from the real system. A model is said to be *replicatively valid* if it matches existing data from the real system and *predictably valid* if it can predict the data outside the parameter range of the data base. If the model truly reflects the operation and manner in which the real system operates, then the model is *structurally valid.*

System simulations can take many forms, use many different tools, and connote many varied meanings. Simulation analyses with a timelike independent variable are usually referred to as either discrete or continuous. A *discrete*

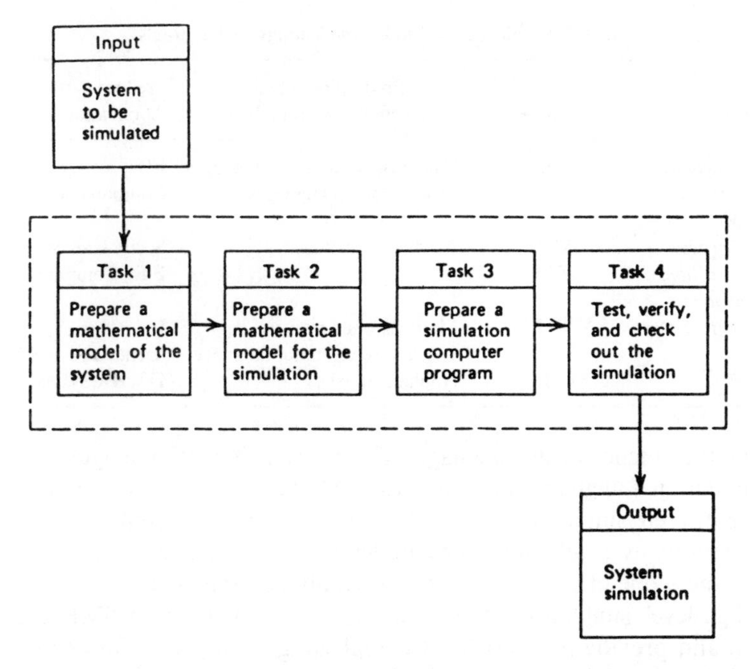

Figure 5-1 The system simulation process. (Used with permission, from J. M. Smith, *Mathematical Modeling and Digital Simulation for Engineers and Scientists*, John Wiley & Sons, Inc., 1977.)

simulation is called for when the dependent variables change by discrete amounts at specified points of the timelike independent variable. These simulations are widely used in the analysis and study of queues and queuing systems. A *continuous* simulation is used when the dependent variables change continuously with the timelike independent variable. Continuous simulations are widely used in dynamical systems analysis and in the analysis and design of control systems. Table 5-1, abstracted from Solomon (4), delineates the wide range of mathematics, applications, and journals in which "simulation" is used. The nature of the input variables determines whether a simulation is deterministic or stochastic. A *deterministic system* simulation has input variables whose values are not uncertain. Stochastic simulation input variables are either completely random or follow some probability distribution. Statistics play a dominant role in analyzing the significance of stochastic simulations.

High-level computer languages have been developed and are routinely used to simulate many discrete and continuous models of real systems. These languages, once mastered, can be used to study an astonishing variety of very sophisticated and complex physical systems. The IBM developed CSMP III (Continuous System Modeling Program) is the most widely used and referenced of languages used for continuous systems. Speckhart and Green (5) examine and explain many applications of CSMP III and are helpful in

TABLE 5-1 Simulation Skills, Applications, and Journals

Journal	Publisher	Predominant Skills Needed to Comprehend	Predominant Applications
AIIE Transactions	AIIE	Calculus, continuous languages	I.E.
Communications of the ACM	ACM	Programming languages	Computer science
Decision Sciences	AIDS	Applied statistics	Social sciences
IEEE Transactions on Computers	IEEE	—	E.E., Computer science
Management Science	TIMS	Math/applied statistics	Industry
Simulation	SCS	Calculus, continuous languages	Science and engineering
Simuletter	ACM/SIGSIM	Applied statistics	Discrete systems

revealing the potency of the language. The discrete simulation language SLAM (6), Simulation Language for Alternative Modeling, is relatively new and is used for discrete simulations (although some network and continuous simulation capability is available). Once understood, these languages enormously enhance the scope of problems that can readily be examined.

High-level simulation languages such as SLAM and CSMP are user oriented and provide routines for accomplishing a variety of common operations—integration of a first-order nonlinear differential equation, for example. The availability of such routines allows the user to rapidly implement and simulate a given model. Output options such as printer plots and free-formatting of specified variables also greatly aid in the rapid developement of simulations.

Simulation and simulation languages are really impressive and undoubtedly represent the future simulation path for many transient models of systems with timelike independent variables; however, we shall leave this category behind as we are interested in simulation of systems in the steady state, a situation for which timelike variables are absent. We are interested in *continuous*, *deterministic*, *steady-state* simulations for fluid and thermal systems.

However, before delving into methodologies for such modeling and simulation, some mathematical tools need to be developed. Virtually all quantitative hardware information is expressed in graphical or tabular forms, modes of presentation that are not appropriate for computer applications. The next section of the chapter presents an examination of some simple curve-fitting procedures. These procedures have very strict limitations, but they are sufficient for many engineering applications.

Two general steady-state simulation procedures used for thermal–fluid systems are the multivariable Newton–Raphson and the Hardy–Cross, which was discussed in Chapter 1. These are next examined in this chapter. The chapter concludes with a generalized procedure for compressible flow calculations.

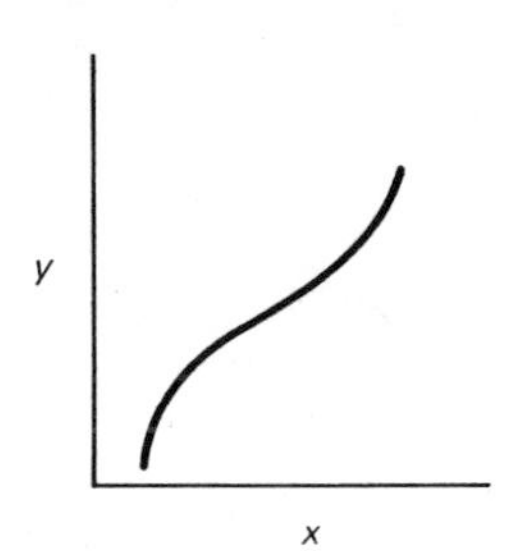

Figure 5-2 Curve with an inflection point.

5-2 PARAMETRIC REPRESENTATION BY CURVE FITTING

Stoecker's book, *Design of Thermal Systems* (7), contains a chapter devoted to curve fitting. Most of the coverage in this section was abstracted from that chapter. Curve fits should never be blindly undertaken; a few plots or sketches of the variables involved will often reveal the functional form of the required curve fit expression. A few of these forms are examined in the following paragraphs.

Polynomials such as

$$y = a_0 + a_1 x + a_2 x^2 + \cdots + a_m x^m \tag{5-1}$$

are often used when there are no indicators that would point to other forms. *Quadratics* ($m = 2$) are frequently used if no inflection point is present. For curves with *inflection* points, such as Fig. 5-2, at least a cubic, a *third-degree polynomial*, is required. When curves are *asymptotic* at large values of the independent variables as in Fig. 5-3, a polynomial with negative exponents

$$y = a_0 + a_1/x + a_2/x^2 \tag{5-2}$$

may represent the curve in a satisfactory manner.

Curves that are *straight lines* when plotted as $\log y$ versus $\log x$ are represented by

$$y = b + ax^m \tag{5-3}$$

This situation is illustrated by Fig. 5-4.

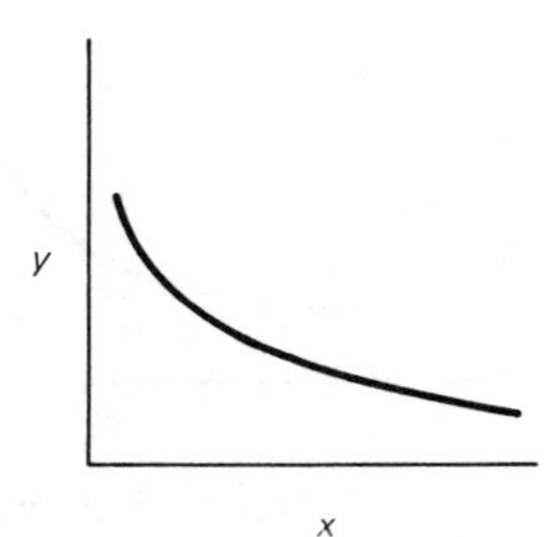

Figure 5-3 Curve with asymptote at large x.

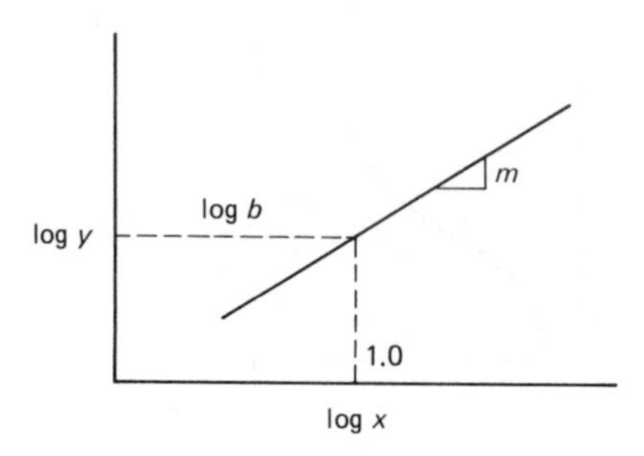

Figure 5-4 Linear relationship on log–log plot.

The S-shaped curve of Fig. 5-5 is represented by the Gompertz equation

$$y = ab^{c^x} \tag{5-4}$$

where a, b, and c are constants and b and c are less than unity.

One of the first tasks in most modeling problems is finding suitable equations to represent the performance characteristics of various components. In many instances, performance curves or tables are available from the manufacturer, and the modeler has "only" to decide on the form of the equation appropriate for representing the data and then accomplish the curve fit for the equation form chosen. Stoecker correctly suggests that manufacturers should present equations for their data, as such a step would save much engineering manpower. We shall examine in some detail polynomial curve fits for one or two independent variables.

A cursory glance at most of the data presented in Chapters 2 through 4 will confirm that a *polynomial* is all that is required to satisfactorily fit much of the data shown. Manufacturers' performance information and data are often presented in graphical form. Pump performance curves are well suited for representation by a quadratic of the form

$$y = a_0 + a_1x + a_2x^2 \tag{5-5a}$$

or

$$H = a_0 + a_1Q + a_2Q^2 \tag{5-5b}$$

Three pairs of information, (H_1, Q_1), (H_2, Q_2), and (H_3, Q_3), chosen in the

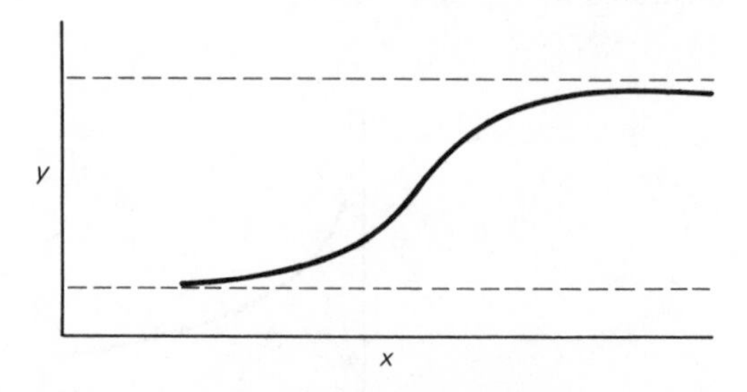

Figure 5-5 Gompertz curve.

range of interest can be substituted into Eq. (5-5b) to yield the linear system

$$H_1 = a_0 + a_1 Q_1 + a_2 Q_1^2$$
$$H_2 = a_0 + a_1 Q_2 + a_2 Q_2^2 \tag{5-5c}$$
$$H_3 = a_0 + a_1 Q_3 + a_2 Q_3^2$$

Equations (5-5c) can then be solved for a_0, a_1, and a_2 by

$$\begin{bmatrix} a_0 \\ a_1 \\ a_2 \end{bmatrix} = \begin{bmatrix} 1 & Q_1 & Q_1^2 \\ 1 & Q_2 & Q_2^2 \\ 1 & Q_3 & Q_3^2 \end{bmatrix}^{-1} \begin{bmatrix} H_1 \\ H_2 \\ H_3 \end{bmatrix} \tag{5-6}$$

The general case of a quadratic curve fit

$$y = a_0 + a_1 x + a_2 x^2 \tag{5-5a}$$

yields for data at (x_1, y_1), (x_2, y_2), and (x_3, y_3) the coefficients

$$\begin{bmatrix} a_0 \\ a_1 \\ a_2 \end{bmatrix} = \begin{bmatrix} 1 & x_1 & x_1^2 \\ 1 & x_2 & x_2^2 \\ 1 & x_3 & x_3^2 \end{bmatrix}^{-1} \begin{bmatrix} y_1 \\ y_2 \\ y_3 \end{bmatrix} \tag{5-7}$$

The pattern is clear as $(m+1)$ points are required for an mth-order polynomial curve fit. We should be wary of using higher-order polynomials where a quadratic or cubic would suffice, since the behavior of the higher-order polynomial between the $m+1$ points may wildly oscillate.

Many situations arise where it is desirable to generate a curve fit for *two* independent variables; that is, $y = f(x, z)$. The output of a pump for different flow rates and speeds (N, rpm) would have the form $H = g(Q, N)$. One satisfactory polynomial representation would have the form

$$H = a_0(N) + a_1(N)Q + a_2(N)Q^2 \tag{5-8}$$

with the coefficients a_i $(i = 0, 1, 2)$ expressed as

$$a_i(N) = b_{i0} + b_{i1}N + b_{i2}N^2 \tag{5-9}$$

The speed dependence in Eq. (5-8) is contained within the coefficients. Figure 5-6 illustrates this situation. The nine points, three at each speed, shown on the sketch are required to determine the coefficients in Eq. (5-8). At speed N_1 a quadratic curve fit using Eq. (5-8) yields

$$H_1 = a_{10} + a_{11}Q + a_{12}Q^2 \tag{5-10a}$$

and at speeds N_2 and N_3 the analogous expressions are

$$H_2 = a_{20} + a_{21}Q + a_{22}Q^2 \tag{5-10b}$$

$$H_3 = a_{30} + a_{31}Q + a_{32}Q^2 \tag{5-10c}$$

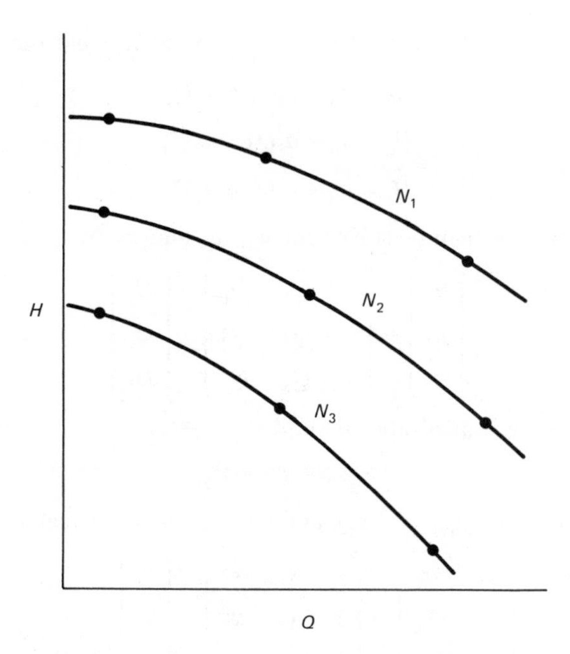

Figure 5-6 H as a function of two independent variables.

The same curve fit procedure is used to evaluate the coefficients in the quadratic expression for $a_0(N)$:

$$a_0(N) = b_{00} + b_{01}N + b_{02}N^2 \tag{5-11}$$

The pairs $(a_{01}, N_1)(a_{02}, N_2)$, and (a_{03}, N_3) are used to evaluate b_{00}, b_{01}, and b_{02}. This process is repeated for $a_1(N)$ to obtain b_{10}, b_{11}, and b_{12} and for $a_2(N)$ to obtain b_{20}, b_{21}, and b_{22}. The final expression for the H as a function of flow rate and speed then becomes

$$\begin{aligned} H = &\left(b_{00} + b_{01}N + b_{02}N^2\right) + \left(b_{10} + b_{11}N + b_{12}N^2\right)Q \\ &+ \left(b_{20} + b_{21}N + b_{22}N^2\right)Q^2 \end{aligned} \tag{5-12}$$

The general case $y = f(x, z)$ becomes

$$\begin{aligned} y = &\left(b_{00} + b_{01}z + b_{02}z^2\right) + \left(b_{10} + b_{11}z + b_{12}z^2\right)x \\ &+ \left(b_{20} + b_{21}z + b_{22}z^2\right)x^2 \end{aligned} \tag{5-13}$$

where the b_{ij}'s are computed as in the preceding. Nine points and six quadratic curve fits are required for this procedure.

Example 5-1

Develop a parametric curve fit for the pump characteristics (8) given in Fig. 5-7. Independent variables are flow rate and speed, and the dependent variable is the increase in head across the pump.

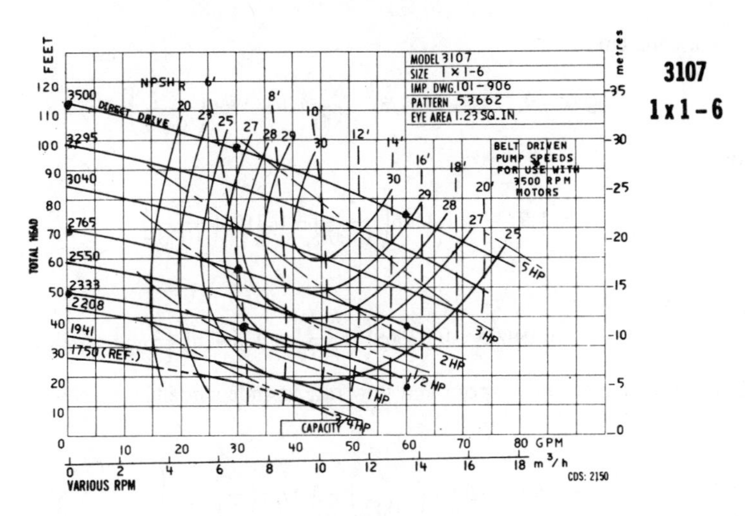

Figure 5-7 Performance data for Goulds model 3107 1 × 1-6 pump. (Used with permission, Goulds Pumps, Inc., Seneca Falls, N.Y.)

Solution. Nine points were chosen from the pump performance characteristics given in Fig. 5-7. These points are indicated by the solid dots. Points were chosen at 3500, 2765, and 2333 rpm and at 0, 30, and 60 gpm. Table 5-2 gives the array of head as a function of speed N and flow rate.

A quadratic curve fit at 3500 rpm yields

$$H(3500 \text{ rpm}) = 113.0 - 0.4167Q - 0.00389Q^2$$

and at 2765 rpm

$$H(2765 \text{ rpm}) = 70.0 - 0.4000Q - 0.00222Q^2$$

and at 2333 rpm

$$H(2333 \text{ rpm}) = 49.0 - 0.2167Q - 0.00500Q^2$$

These expressions give the head versus flow rate characteristics for a given speed; what is needed is a *single* equation with the speed as an explicit parameter. This is accomplished by requiring the coefficients a_i $(i = 1, 2, 3)$ to be quadratic functions of speed; that is,

$$a_i = b_{i0} + b_{i1}N + b_{i2}N^2$$

TABLE 5-2 Pump Performance Array

	gpm		
rpm	0	30	60
3500	113	97	74
2765	70	56	38
2333	49	38	18

Then for a_0 the curve fit is

$$a_0 = -9.7299 + 5.3978\left(\frac{N}{1000}\right) + 8.4766\left(\frac{N}{1000}\right)^2$$

and for a_1

$$a_1 = 1.9739 - 1.4506\left(\frac{N}{1000}\right) + 0.2193\left(\frac{N}{1000}\right)^2$$

and for a_2

$$a_2 = -0.04605 + 0.02869\left(\frac{N}{1000}\right) - 0.00476\left(\frac{N}{1000}\right)^2$$

The final expression for both speed and flow rate as independent variables is

$$\begin{aligned} H = & \left[-9.7299 + 5.3978\left(\frac{N}{1000}\right) + 8.4766\left(\frac{N}{1000}\right)^2\right] \\ & + \left[1.9739 - 1.4506\left(\frac{N}{1000}\right) + 0.2193\left(\frac{N}{1000}\right)^2\right]Q \\ & + \left[-0.04605 + 0.02869\left(\frac{N}{1000}\right) - 0.00476\left(\frac{N}{1000}\right)^2\right]Q^2 \end{aligned}$$

The nine points required for this two-parameter curve fit force the results to pass exactly through each point. If more than $m+1$ points are used in any mth-order polynomial curve fit, then the *method of least squares* is used to ensure the best curve fit in the sense of minimizing the sum of the squares of the deviation from the resulting curve fit and the greater than $m+1$ points. Since most manufacturers' data have had prior smoothing, least squares is not usually necessary for modeling purposes.

Interpolation is a useful alternative to curve fitting when tabular information is available. Many different methods of widely varying accuracy and complexity are available. Here we shall only be concerned with one, the *Lagrange interpolating polynomial.* Previously we talked about using expressions of the form of Eq. (5-5a) for quadratic curve fits. Lagrange interpolation can be to any degree, but in the quadratic form the polynomial expression is

$$\begin{aligned} y = & C_1(x - x_2)(x - x_3) + C_2(x - x_1)(x - x_3) \\ & + C_3(x - x_1)(x - x_2) \end{aligned} \tag{5-14}$$

Equation (5-14) is quadratic and contains three constants. If the available data are at the points (x_1, y_1), (x_2, y_2), and (x_3, y_3), then successive substitution in

Eq. (5-14) yields

$$C_1 = \frac{y_1}{(x_1 - x_2)(x_1 - x_3)} \tag{5-15a}$$

$$C_2 = \frac{y_2}{(x_2 - x_1)(x_2 - x_3)} \tag{5-15b}$$

$$C_3 = \frac{y_3}{(x_3 - x_1)(x_3 - x_2)} \tag{5-15c}$$

For interpolative purposes, the form of Eq. (5-14) makes evaluation of the constants C_1, C_2, and C_3 simple. By using capital pi, Π, to indicate multiplication, the Lagrange interpolating expression can be written in a very compact form as

$$y = \sum_{i=1}^{3} y_i \prod_{\substack{j=1 \\ i \neq j}}^{3} \frac{x - x_j}{x_i - x_j} \tag{5-16}$$

The Lagrange expression is very convenient and compact for computer applications, and the three-point expression as given in Eq. (5-16) is equivalent to forcing a quadratic through three points.

The information and techniques presented herein have barely scratched the surface of knowledge of curve fitting and interpolation. They do, however, provide a convenient way to model hardware performance characteristics; and, while much more sophisticated techniques are available where more fidelity or accuracy is required, the results of the more sophisticated techniques are utilized in the same manner by the two simulation procedures that we will introduce.

5-3 HARDY–CROSS REVISITED

The *Hardy–Cross* method and the *generalized Hardy–Cross* method, both discussed in Chapter 1, are examples of continuous, deterministic, steady-state simulation methods for complex piping networks. The Hazen–Williams expression for the pipe head loss represents the mathematical model for a single pipe; the concepts of nodal mass conservation and uniqueness of pressure represent the mathematical model of the system. The Hardy–Cross method is the simulation procedure used to implement the model concepts.

Now that we have completed Chapters 2–4 we are in a much better position to use the generalized Hardy–Cross method for complex systems containing devices such as pumps and heat exchangers. The mathematical model relating the head loss (gain) and flow rate for any device in line j of loop

i is, from Chapter 1,

$$h_{f_{ij}} = A_{ij} + \sum_{m=1}^{M} B_{ijm} Q^m \tag{1-48}$$

Chapter 4 provides information on the H–Q curves for several pumps, Chapters 2 and 3 provide details and procedures for calculating pressure losses for heat exchangers, and Chapter 1 provides loss coefficients for minor losses in pipe fittings. For use in the generalized Hardy–Cross procedure, such information (head or pressure changes as a function of capacity or flow rate) can be calculated as needed, interpolated from previously calculated values in tabular form, or computed from curve fits based on previously calculated values. Overall the latter approach is preferable both from a computational economy standpoint and from an ease of application standpoint. The previous section developed some simple curve-fitting notions that can be applied to express head loss–capacity information in the form of Eq. (1-48). Thus, devices for inclusion in piping networks can be readily modeled and the Hardy–Cross simulation procedure used to find the steady-state conditions.

The generalized Hardy–Cross simulation method is perhaps the widest used method for steady-state fluid flow networks. It has the advantages of flexibility and ease of application. Wood and Rayes (9) discuss several different techniques for obtaining the steady-state solution to pipe networks. Basically, they suggest that all methods can be categorized as either *loop* or *node*. The Hardy–Cross method used herein is a loop method since flow rates are obtained by satisfying uniqueness of pressure about a given loop. Node methods are based on the heads at nodes throughout the pipe system. The paper of Wood and Rayes presents a concise study of piping network solution techniques and is worth examining prior to undertaking the analysis of a large complex system containing many devices with nonlinear H–Q characteristics. Wood and Rayes noted that, for some networks, convergence difficulties occur. The difficulties seem to occur for large complex networks in which pipes coming into the same node have flow rates that differ by several orders of magnitude. For networks in which the Hardy–Cross procedure experiences convergence difficulties, the multivariable Newton–Raphson technique developed in the next section is often a viable alternative methodology.

The generalized Hardy–Cross technique based on loop equations is fine for fluid systems where mass continuity and uniqueness of pressure must be enforced; however, such a methodology is not easily extendable to system simulations requiring matching of components with complex parametric representations—jet engines, for example. Complex series systems are often simulated by developing and solving nonlinear systems of equations. This method, like the Hardy–Cross technique, requires curve fitting of performance information.

5-4 MULTIVARIABLE NEWTON–RAPHSON SIMULATION METHOD

A more general steady-state simulation method than the Hardy–Cross technique is available. This more general method uses information-flow diagrams to represent each system component, links in an appropriate manner all the component information-flow diagrams, and develops equations to represent the system. Such an approach results in a system of nonliner algebraic equations that must be solved for the steady-state values of the operating parameters. Obviously, in such an approach the number of unknowns must be equal to the number of equations, and the equations must be independent.

As an example of a system simulation using this component approach, consider two pumps in parallel pumping water between two reservoirs with an elevation difference H. Such a system is illustrated in Fig. 5-8. For the purpose of simplicity, assume all pipes are of a diameter D and length L. Using techniques developed in the previous section, the H–Q characteristic of the pumps can be represented as

$$\Delta H_i = a_{0i} + a_{1i}Q_i + a_{2i}Q_i^2 \qquad i = 1,2 \tag{5-17}$$

where the total flow rate Q_T is $Q_1 + Q_2$. Applying the energy equation between points a and b yields the pump head required:

$$\frac{g_c}{g}\Delta H = h + \left(f\frac{L}{D} + K\right)\frac{1}{2g}\frac{Q_T^2}{A^2} \tag{5-18}$$

where $(fL/D + K)$ represents the total major and minor frictional losses. Since the pumps are in parallel, $\Delta H_1 = \Delta H_2$ and the required system of equations is

$$\Delta H = a_{01} + a_{11}Q_1 + a_{21}Q_1^2 \tag{5-19a}$$

$$\Delta H = a_{02} + a_{12}Q_2 + a_{22}Q_2^2 \tag{5-19b}$$

$$\frac{g_c}{g}\Delta H = h + \left(f\frac{L}{D} + K\right)\frac{1}{2g}\frac{Q_T^2}{A^2} \tag{5-19c}$$

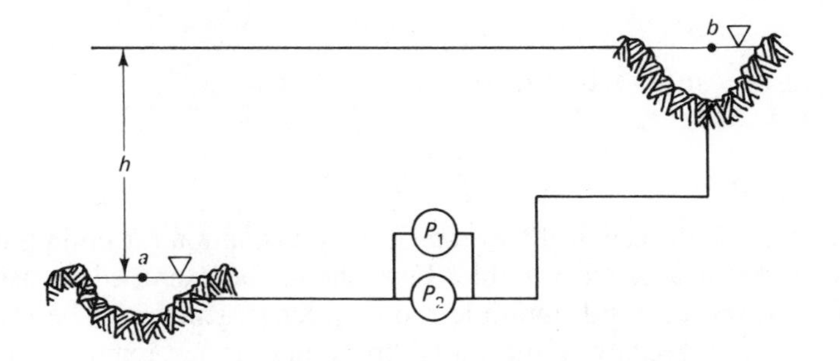

Figure 5-8 Schematic of system with parallel pumps.

$$Q_T = Q_1 + Q_2 \tag{5-19d}$$

Equations (5-19) are nonlinear in the unknowns ΔH, Q_1, Q_2, and Q_T. With four unknowns and four independent equations, the nonlinear algebraic system is solvable; the solution would yield the steady-state operating point in terms of the flow rates Q_1, Q_2, and Q_T, as well as the pump operating points. The question of how to solve such a system must be answered.

Two methods for solving nonlinear systems are generally available: (1) successive substitution and (2) Newton–Raphson iteration. *Successive substitution* is generally the iterative approach followed in the previous chapters where a value (or values) of an unknown variable (or variables) was assumed and an iterative process delineated that allowed corrections to the assumed value (or values) until convergence was achieved. Generally, for a specific simulation problem many successive substitution iterative paths can be developed. Some of the paths will converge rapidly, some slowly, and some diverge. The *Newton–Raphson interative approach* solves the system simultaneously and, if well behaved, generally avoids convergence problems. We shall pursue the Newton–Raphson approach as the major technique for solving systems of nonlinear equations.

The multivariable Newton–Raphson method is closely tied with the Newton–Raphson method for one variable (or unknown). Generally, we wish to find the correct value of x, x_c, such that

$$f(x_c) = 0 \tag{5-20}$$

since in general

$$f(x) = y(x) \neq 0 \tag{5-21}$$

A truncated Taylor series about x_c can be written as

$$y(x) = y(x_c) + y'(x_c)(x - x_c) \tag{5-22}$$

If a trial guess x_t is made for x_c, then

$$y(x_t) = y(x_c) + y'(x_c)(x_t - x_c) \tag{5-23}$$

But since $y(x_c) = 0$ and since x_c is not known, Eq. (5-23) can be approximated by

$$y(x_t) \simeq 0 + y'(x_t)(x_t - x_c) \tag{5-24}$$

Equation (5-24) can be solved to yield an estimate for x_c:

$$x_c \simeq x_t - \frac{y(x_t)}{y'(x_t)} \tag{5-25}$$

Equation (5-25) is the classical Newton–Raphson equation for finding the root of an equation in a single variable. Care should be exercised in using Eq. (5-25), for expressions with multiple roots as convergence may be erratic or may not occur, especially if the initial trial guess is far removed from the correct result.

The multivariable Newton–Raphson methodology (10) is developed, as is the single-variable Newton–Raphson, by considering a Taylor's series expansion. In this case, more than a single independent variable is present, and the more complicated series is

$$
\begin{aligned}
y(x_1, x_2, \ldots, x_n) = y(a_1, a_2, \ldots, a_n) &+ \sum_{j=1}^{n} \left(\frac{\partial y}{\partial x_j} \right)_a (x_j - a_j) \\
&+ \frac{1}{2} \sum_{i=1}^{n} \sum_{j=1}^{n} \left(\frac{\partial^2 y}{\partial x_j \, \partial x_i} \right)_a (x_i - a_i)(x_j - a_j) + \cdots
\end{aligned}
\tag{5-26}
$$

Consider a system of three equations in the independent variables x_1, x_2, and x_3:

$$
\begin{aligned}
f_1(x_1, x_2, x_3) &= 0 \\
f_2(x_1, x_2, x_3) &= 0 \\
f_3(x_1, x_2, x_3) &= 0
\end{aligned}
\tag{5-27}
$$

The correct solution point is (x_{1c}, x_{2c}, x_{3c}), and the initially assumed solution point is (x_{1t}, x_{2t}, x_{3t}). The Taylor series for $f_1(x_{1t}, x_{2t}, x_{3t})$ can be written as

$$
\begin{aligned}
f_1(x_{1t}, x_{2t}, x_{3t}) = f_1(x_{1c}, x_{2c}, x_{3c}) \\
&+ \left(\frac{\partial f_1}{\partial x_1} \right)_c (x_{1t} - x_{1c}) + \left(\frac{\partial f_1}{\partial x_2} \right)_c (x_{2t} - x_{2c}) \\
&+ \left(\frac{\partial f_1}{\partial x_3} \right)_c (x_{3t} - x_{3c}) + \cdots
\end{aligned}
\tag{5-28}
$$

and in similar fashion for $f_2(x_{1t}, x_{2t}, x_{3t})$ and $f_3(x_{1t}, x_{2t}, {}_{3t})$. Making the approximation $(\partial f_1/\partial x_j)_c \sim (\partial f_1/\partial x_j)_t$ and realizing that $f_1(x_{1c}, x_{2c}, x_{3c})$ is zero, Eq. (5-28) becomes

$$
\begin{aligned}
f_1(x_{1t}, x_{2t}, x_{3t}) \approx & \left(\frac{\partial f_1}{\partial x_1} \right)_t (x_{1t} - x_{1c}) \\
&+ \left(\frac{\partial f_1}{\partial x_2} \right)_t (x_{2t} - x_{2c}) + \left(\frac{\partial f_1}{\partial x_3} \right)_t (x_{3t} - x_{3c}) + \cdots
\end{aligned}
\tag{5-29a}
$$

and f_2 and f_3 appear as

$$
\begin{aligned}
f_2(x_{1t}, x_{2t}, x_{3t}) \approx & \left(\frac{\partial f_2}{\partial x_1} \right)_t (x_{1t} - x_{1c}) \\
&+ \left(\frac{\partial f_2}{\partial x_2} \right)_t (x_{2t} - x_{2c}) + \left(\frac{\partial f_2}{\partial x_3} \right)_t (x_{3t} - x_{3c}) + \cdots
\end{aligned}
\tag{5-29b}
$$

$$f_3(x_{1t}, x_{2t}, x_{3t}) \approx \left(\frac{\partial f_3}{\partial x_1}\right)_t (x_{1t} - x_{1c}) + \left(\frac{\partial f_3}{\partial x_2}\right)_t (x_{2t} - x_{2c}) + \left(\frac{\partial f_3}{\partial x_3}\right)_t (x_{3t} - x_{3c}) + \cdots \tag{5-29c}$$

Equations (5-29) constitute a *linear* system of algebraic equations for $(x_{it} - x_{ic})$, $i = 1, 2, 3$, which can be written as

$$\begin{bmatrix} \frac{\partial f_1}{\partial x_1} & \frac{\partial f_1}{\partial x_2} & \frac{\partial f_1}{\partial x_3} \\ \frac{\partial f_2}{\partial x_1} & \frac{\partial f_2}{\partial x_2} & \frac{\partial f_2}{\partial x_3} \\ \frac{\partial f_3}{\partial x_1} & \frac{\partial f_3}{\partial x_2} & \frac{\partial f_3}{\partial x_3} \end{bmatrix} \begin{bmatrix} x_{1t} - x_{1c} \\ x_{2t} - x_{2c} \\ x_{3t} - x_{3c} \end{bmatrix} = \begin{bmatrix} f_1 \\ f_2 \\ f_3 \end{bmatrix} \tag{5-30}$$

So that

$$\begin{bmatrix} x_{1t} - x_{ic} \\ x_{2t} - x_{2c} \\ x_{3t} - x_{3c} \end{bmatrix} = \begin{bmatrix} \frac{\partial f_i}{\partial x_j} \\ i = 1, 2, 3, \\ j = 1, 2, 3 \end{bmatrix}^{-1} \begin{bmatrix} f_1 \\ f_2 \\ f_3 \end{bmatrix} \tag{5-31}$$

and the new corrected values for the next iterative pass become

$$\begin{aligned} x_{1c\,\text{new}} &= x_{1t\,\text{old}} - (x_{1t} - x_{1c}) \\ x_{2c\,\text{new}} &= x_{2t\,\text{old}} - (x_{2t} - x_{2c}) \\ x_{3c\,\text{new}} &= x_{3t\,\text{old}} - (x_{3t} - x_{3c}) \end{aligned} \tag{5-32}$$

When $(x_{it} - x_{ic})$, $i = 1, 2, 3$, are sufficiently close to zero the procedure has converged and x_{1c}, x_{2c}, and x_{3c} are the required solution to the nonlinear system.

Stoecker (7) gives a seven-step procedure for implementing this technique:

1. Rewrite all the equation in the form $f_i(x_1, x_2, x_3) = 0$, $i = 1, 2, 3$.
2. Assume values of x_{1t}, x_{2t}, and x_{3t}.
3. Calculate $f_i(x_{1t}, x_{2t}, x_{3t})$ for $i = 1, 2, 3$.
4. Compute $\partial f_i / \partial x_j$ at (x_{1t}, x_{2t}, x_{3t}) for $i = 1, 2, 3$ and $j = 1, 2, 3$.
5. Solve the set of linear equations [Eqs. (5-30)].
6. Compute the new values from Eqs. (5-32).
7. Using the values from step 6 in step 2, test for convergence, and if not converged, repeat steps 2–7.

This procedure is best illustrated by an example problem:

Example 5-2

Consider the system shown schematically in Fig. 5-8 with $D = 12$ in., $L = 10{,}000$ ft, $K = 543$, $\varepsilon/D = 0.001$, and $h = 300$ ft. Two 3×6-13 pumps with 13-in. rotors are in parallel. The pump characteristics are given in Fig. 5-9. Find the flow rate if water at 70°F is the pumped fluid.

Solution. At 70°F the physical properties of water are density, $\rho = 62.3$ lbm/ft^3; kinematic viscosity, $\nu = 1.06 \times 10^{-5}$ ft^2/s; and dynamic viscosity, $\mu = 0.658 \times 10^{-3}$ lbm/ft s. The pump characteristic for the 3×6-13 (13-in. rotor) pump shown in Fig. 5-9 is next curve fitted using the previously discussed method, with the result

$$\Delta H = 740.0 + 40.579Q - 39.210Q^2 \tag{5-33a}$$

Equations (5-19) for the particular values given in the problem statement then become

$$\Delta H = 740.0 + 40.579Q_1 - 39.210Q_1^2 \tag{5-33b}$$

$$\Delta H = 740.0 + 40.579Q_2 - 39.210Q_2^2 \tag{5-33c}$$

$$\Delta H = 18.72Q_T^2 + 300.0 \tag{5-33d}$$

$$Q_T = Q_1 + Q_2 \tag{5-33e}$$

where f is taken as 0.02, Q_1 is the flow rate through pump 1, Q_2 is the flow rate through pump 2, Q_T is the total flow rate, and ΔH is the change in head. The value of the friction factor should be verified when convergence is obtained. Following the steps

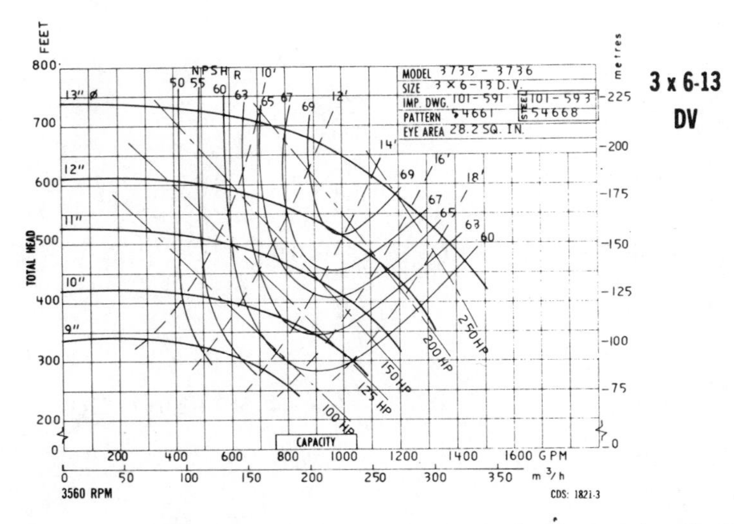

Figure 5-9 Pump characteristics for Example 5-2. (Used with permission, Goulds Pumps, Inc., Seneca Falls, N.Y.)

recommended by Stoecker:

Step 1:

$$F1 = \Delta H - 18.72 Q_T^2 - 300.0 \tag{5-34a}$$

$$F2 = Q_T - Q_1 - Q_2 \tag{5-34b}$$

$$F3 = \Delta H - 740.0 - 40.579 Q_1 + 39.21 Q_1^2 \tag{5-34c}$$

$$F4 = \Delta H - 740.0 - 40.579 Q_2 + 39.21 Q_2^2 \tag{5-34d}$$

Step 2: Initial estimates:

$$\Delta H = 575.0 \text{ ft} \tag{5-35a}$$

$$Q_T = 4.00 \text{ ft}^3/\text{s} \tag{5-35b}$$

$$Q_1 = 2.00 \text{ ft}^3/\text{s} \tag{5-35c}$$

$$Q_2 = 2.00 \text{ ft}^3/\text{s} \tag{5-35d}$$

Step 3: Evaluate Eqs. (5-34) using the conditions of step 2; for all but the first iterative pass use the current values of ΔH, Q_T, Q_1, and Q_2.

Step 4: The partial derivatives are

$$\frac{\partial f_1}{\partial x_1} = 1.0 \qquad \frac{\partial f_1}{\partial x_2} = -2.0 * 18.72 Q_T$$

$$\frac{\partial f_1}{\partial x_3} = 0.0 \qquad \frac{\partial f_1}{\partial x_4} = 0.0$$

$$\frac{\partial f_2}{\partial x_1} = 0.0 \qquad \frac{\partial f_2}{\partial x_2} = 1.0$$

$$\frac{\partial f_2}{\partial x_3} = -1.0 \qquad \frac{\partial f_2}{\partial x_4} = -1.0 \tag{5-36}$$

$$\frac{\partial f_3}{\partial x_1} = 1.0 \qquad \frac{\partial f_3}{\partial x_2} = 0.0$$

$$\frac{\partial f_3}{\partial x_3} = -40.579 + 2.0 * 39.21 Q_1 \qquad \frac{\partial f_3}{\partial x_4} = 0.0$$

$$\frac{\partial f_4}{\partial x_1} = 1.0 \qquad \frac{\partial f_4}{\partial x_2} = 0.0$$

$$\frac{\partial f_4}{\partial x_3} = 0.0 \qquad \frac{\partial f_4}{\partial x_4} = -40.579 + 2.0 * 39.21 Q_2$$

$$F1 = f_1 \qquad x_1 = \Delta H$$

$$F2 = f_2 \qquad x_2 = Q_T$$

$$F3 = f_3 \qquad x_3 = Q_1$$

$$F4 = f_4 \qquad x_4 = Q_2$$

Step 5: Then using the expression from step 4, the corrections are computed as

$$\begin{bmatrix} \Delta(\Delta H) \\ \Delta(Q_T) \\ \Delta(Q_1) \\ \Delta(Q_2) \end{bmatrix} = \begin{bmatrix} \dfrac{\partial f_i}{\partial x_j} \\ i=1,2,3,4 \\ j=1,2,3,4 \end{bmatrix}^{-1} \begin{bmatrix} F1 \\ F2 \\ F3 \\ F4 \end{bmatrix} \tag{5-37}$$

Step 6: The new values of the solution are

$$\begin{aligned} \Delta H &= \Delta H\,(\text{step 2}) - \Delta(\Delta H) \\ Q_T &= Q_T\,(\text{step 2}) - \Delta(Q_T) \\ Q_1 &= Q_1\,(\text{step 2}) - \Delta(Q_1) \\ Q_2 &= Q_2\,(\text{step 2}) - \Delta(Q_2) \end{aligned} \tag{5-38}$$

These are the current values of the iterates.

Step 7: Test for convergence by checking the values of $F1$, $F2$, $F3$, and $F4$ using the results of step 6. Repeat steps 2 to 7 until convergence.

For the stated problem, the converged solution is:

$$\begin{aligned} \Delta H &= 646.04 \text{ ft} \\ Q_T &= 4.299 \text{ ft}^3/\text{s} \\ Q_1 &= 2.15 \text{ ft}^3/\text{s} \\ Q_2 &= 2.15 \text{ ft}^3/\text{s} \end{aligned} \tag{5-39}$$

The approach to convergence is illustrated in Table 5-3. The rapidity of convergence is typical of this method, although the relatively close initial guesses are partly responsible. The computer coding is simple and straightforward.

Figure 5-10 is a reproduction of the FORTRAN listing for this problem. The variables are for the most part self-explanatory: QT for Q_T, Q1 for Q_1, Q2 for Q_2, and DELH for ΔH. Advantage is taken of the FORTRAN FUNCTION DEFINITION capability to encode F1, F2, F3, and F4 directly. Subroutine Gauss is a Gaussian elimination code taken from a standard scientific subroutine package; A is the matrix of coefficients, B the function value, and V the returned solution. The partial derivatives from step 4 are in the A array. Instead of algebraically evaluating these terms, the defined functions F1, F2, F3, and F4 could be used to numerically evaluate the partial

TABLE 5-3 Approach to Convergence

Iterate	ΔH	Q_T	Q_1	Q_2
1	575.0	4.000	2.000	2.000
2	646.2	4.312	2.156	2.156
3	646.04	4.299	2.150	2.150

derivatives for each iteration. For example,

$$A(1,2) = \frac{\partial f_1}{\partial x_2} = \frac{\text{F1(DELH, QT + DELQT)} - \text{F1(DELH, QT)}}{\text{DELQT}} \tag{5-40}$$

where DELQT is a small change in QT, would approximate the $\partial f_1/\partial x_2$. This numerical evaluation may be easier to accomplish than the actual partial differentiation for very complex equations.

```
C
C       MULTIPLE NEWTON-RAPHSON METHOD
C       EXAMPLE PROBLEM FOR ME 4333
C
        DIMENSION A(4,4),AINV(4,4),B(4),V(4)
C       FUNCTION DEFINITIONS
        F1(DELH,QT) = DELH - 18.72*QT*QT - 300.0
        F2(QT,Q1,Q2) = QT - Q1 - Q2
        F3(DELH,Q1) = DELH -740.0 - 40.579*Q1 + 39.210*Q1*Q1
        F4(DELH,Q2) = DELH -740.0 - 40.579*Q2 + 39.210*Q2*Q2
C
C       INITIAL CONDITIONS
        ITER = 0
        ITERMX = 200
        WRITE(5,300)
300     FORMAT (///,10X,'READ IN VALUES OF EPS AND LUN.',//)
        ACCEPT * ,EPS,LUN
        DELH = 575.0
        QT = 4.0
        Q1 = 2.0
        Q2 = 2.0
        B(1) = F1(DELH,QT)
        B(2) = F2(QT,Q1,Q2)
        B(3) = F3(DELH,Q1)
        B(4) = F4(DELH,Q2)
        WRITE(LUN,200)ITER,DELP,Q1,Q2,QT,B(1),B(2),B(3),B(4)
C
C       ITERATION STARTS
C
100     ITER = ITER + 1
        DELHO = DELH
        QTO = QT
        Q1O = Q1
        Q2O = Q2
        A(1,1) = 1.0
        A(1,2) = -18.72*2.0*QT
        A(1,3) = 0.0
        A(1,4) = 0.0
        A(2,1) = 0.0
        A(2,2) = 1.0
        A(2,3) = -1.0
        A(2,4) = -1.0
        A(3,1) = 1.0
        A(3,2) = 0.0
        A(3,3) = -40.579 + 2.0*39.210*Q1
        A(3,4) = 0.0
        A(4,1) = 1.0
        A(4,2) = 0.0
        A(4,3) = 0.0
        A(4,4) = -40.579 + 2.0*39.210*Q2
        CALL GAUSS(4,A,B,V)
```

Figure 5-10 Computer program listing for Newton–Raphson simulation program.

```
      DELH = DELHO - V(1)
      QT = QTO - V(2)
      Q1 = Q10 - V(3)
      Q2 = Q20 - V(4)
C     CHECK ON CONVERGENCE
      B(1) = F1(DELH,QT)
      B(2) = F2(QT,Q1,Q2)
      B(3) = F3(DELH,Q1)
      B(4) = F4(DELH,Q2)
      WRITE (LUN,200) ITER,DELH,QT,Q1,Q2,B(1),B(2),B(3),B(4)
200   FORMAT (/,5X,'ITERATION = ',I3,5X,'DELH = ',F8.2,5X,'QT = ',
     1 F7.3,5X,'Q1 = ',F7.3,5X,'Q2 = ',F7.3,/,20X,'F1 = ',F8.3,5X,
     2 'F2 = ',F8.3,5X,'F3 = ',F8.3,5X,'F4 = ',F8.3/)
      DO 150 I = 1,4
150   IF (ABS(B(I)) .GT. EPS) GO TO 110
      STOP
C
C     NEXT PASS
C
110   IF (ITER .LT. ITERMX) GO TO 100
      WRITE (LUN,210)
210   FORMAT (10X,10(2H* ),'ITERATION COUNT EXCEEDED',10(2H *))
      STOP
      END
```

Figure 5-10 (continued)

In addition to simplifying the coding by numerically evaluating the required partial derivatives, an examination of Eqs. (5-23) indicates that further simplification might be possible by combining Eqs. (5-33d) and (5-33e) in order to eliminate Q_T and reduce the system to three equations and three unknowns. Generally, we would expect that further algebra might reduce the number of equations in any simultaneous formulation. Such a procedure is *not* recommended since the chance for algebraic errors exists and since values for most parameters are needed. Computer debugging efforts for a system of $2n$ simple equations are less than for a system of n complex equations, especially if the values of the remaining n parameters are required anyway. Thus far we have studied only examples concerned with simulation of fluid networks or with systems in which the pressures or heads and flow rates are of interest. Many thermal systems are concerned also with thermal energy transfer, and an examination of such an application is in order.

As an illustration of a more complex system, consider the typical gas turbine schematically illustrated in Fig. 5-11. The system indicated in the figure is typical of a compressor–combustor–turbine engine with power output to a shaft; it could represent a turboprop aircraft engine or an industrial gas turbine. Air is taken in at the inlet at temperature T_1 and pressure P_1 and compressed to a temperature T_2 and pressure P_2. In the combustor, fuel is injected and burned, raising the temperature to T_3 and reducing the pressure to P_3. Heat addition to a compressible fluid in a constant area results in a decrease in static and total pressure. This is usually called Rayleigh flow and is a good model for the process in the combustor. The air mass flow is $\dot{m}_a$ and the fuel flow rate $\dot{m}_f$, so the mass flow entering the turbine is $\dot{m}_a + \dot{m}_f$. The

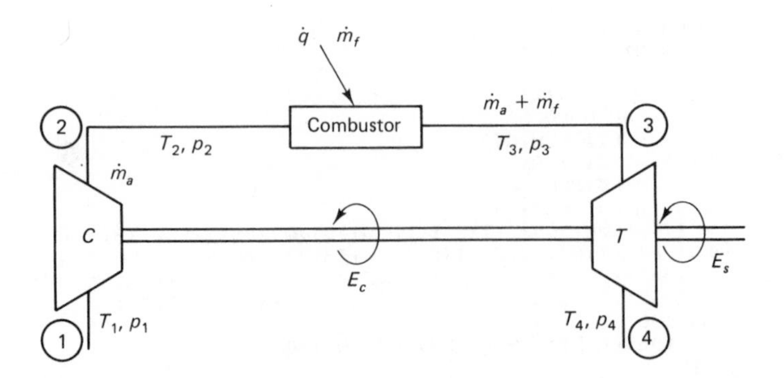

Figure 5-11 Gas turbine schematic. (Used with permission, from W. F. Stoecker, *Design of Thermal Systems*, 2nd ed., McGraw-Hill Book Company, 1980.)

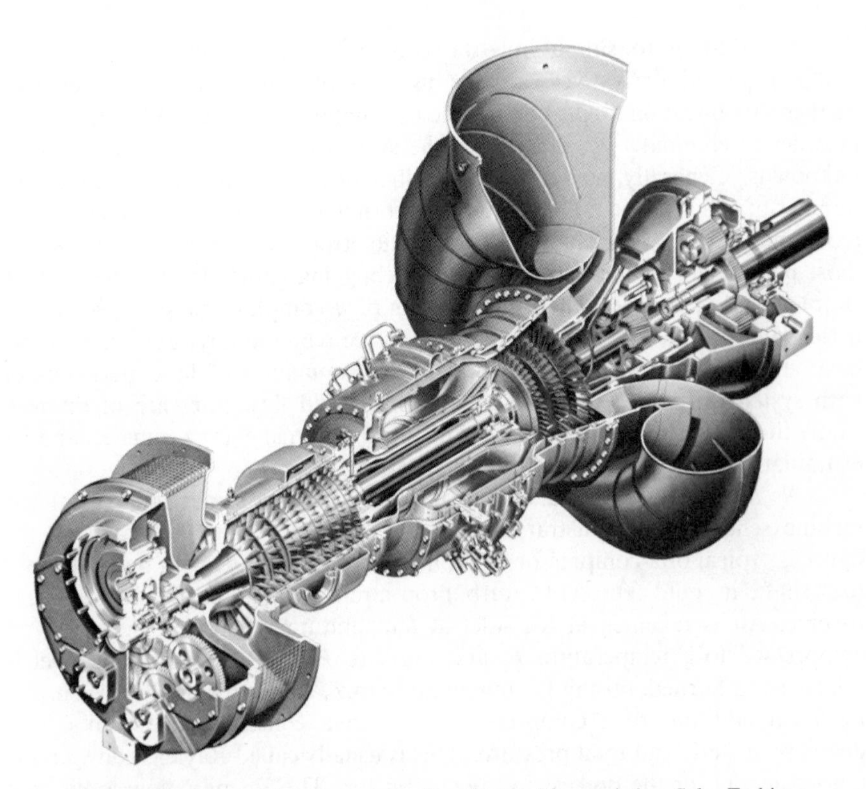

Figure 5-12 Gas turbine cutaway drawing. (Used with permission, Solar Turbines Incorporated.)

flow is expanded through the turbine where part of the energy extracted is used to drive the compressor (E_c) and the remainder is output to the shaft (E_s). The combusted air–fuel mixture leaves the turbine with a temperature T_4 and a pressure P_4. In an industrial gas turbine, energy would be recovered from the turbine discharge, while in a turboprop the turbine discharge would be expanded through a nozzle and a reaction thrust produced. Figure 5-12 presents a cutaway drawing of a gas turbine arrangement. Before simulating such a system, the various components must be modeled.

The compressor is basically a pump, and its characteristics are modeled and presented in the same fashion as for the pump. Figure 5-13 illustrates a typical presentation of compressor characteristics at a given speed. The two presentations, mass flow versus discharge pressure (P_2) and discharge pressure versus compressor power required, model the compressor in sufficient detail. In many instances, the compressor power required is presented in terms of the isentropic compression efficiency η_c and appears as

$$E_c = \dot{m}C_p \frac{T_1}{\eta_c}\left[\left(\frac{P_2}{P_1}\right)^{(\gamma-1)/\gamma} - 1\right] \tag{5-41}$$

Using the data representation as in Fig. 5-13, three expressions are required to model the compressor:

$$P_2 = a_{p0} + a_{p1}\dot{m}_a + a_{p2}\dot{m}_a^2 \tag{5-42a}$$

$$E_c = a_{e0} + a_{e1}P_2 + a_{e2}P_2^2 \tag{5-42b}$$

$$E_c = \dot{m}_a C_p (T_2 - T_1) \tag{5-43}$$

Equations (5-42a) and (5-42b) model the pressure–power characteristics, and Eq. (5-43) is the energy balance for the compressor. The combustor is next examined.

The combustor can be modeled as a simple temperature change by assuming that all the added energy goes to increasing the temperature. An energy balance yields

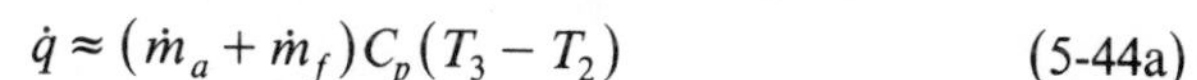

$$\dot{q} \approx (\dot{m}_a + \dot{m}_f)C_p(T_3 - T_2) \tag{5-44a}$$

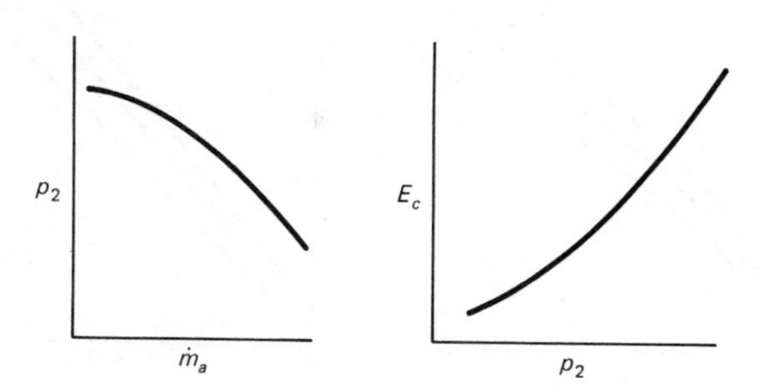

Figure 5-13 Functional representation of compressor characteristics.

and the pressure P_3 is taken as

$$P_3 = P_3(\dot{m}_a, \dot{m}_f, P_2, T_2) \tag{5-44b}$$

A constant value of C_p has been assumed. Combustors are usually designed to minimize the pressure drop, and often P_3 is taken as equal to P_2. In the same vein, preliminary estimates are often generated neglecting the $\dot{m}_f$, since in any reasonable engine $\dot{m}_f/\dot{m}_a \ll 1$. The turbine modeling is considered next.

The turbine is modeled (11) in a manner analogous to the compressor. Typical turbine characteristics are presented in Fig. 5-14. The issue is complicated somewhat by the effects of turbine inlet temperature, which makes representation of the characteristics more complex. Turbine data are usually represented as a function of two independent variables, T_3 (turbine inlet temperature) and P_3 (turbine inlet pressure). Thus

$$\dot{m}_T = a_{m0}(T_3) + a_{m1}(T_3)P_3 + a_{m2}(T_3)P_3^2 \tag{5-45a}$$

$$E_T = a_{t0}(T_3) + a_{t1}(T_3)P_3 + a_{t2}(T_3)P_3^2 \tag{5-45b}$$

where $a_{m0}(T_3)\ldots a_{t2}(T_3)$ are quadratic functions in T_3. An energy balance yields

$$E_t = E_c + E_s \tag{5-45c}$$

The data representation for the turbine, like the compressor, is sometimes simplified by defining an efficiency, in this case the isentropic expansion efficiency η_t. The use of η_t eliminates the E_t versus inlet pressure P_3 plot, since E_t becomes

$$E_t = \dot{m}\eta_t C_p T_3 \left[1 - \left(\frac{1}{P_3/P_2}\right)^{(\gamma-1)/\gamma}\right] \tag{5-46}$$

This completes the component modeling.

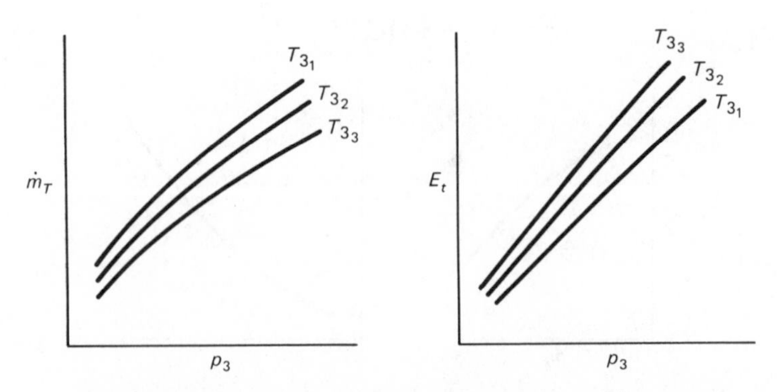

Figure 5-14 Functional representation of turbine characteristics.

Assembling all the equations for the components gives

$$P_2 = a_{P0} + a_{P1}\dot{m}_a + a_{P2}\dot{m}_a^2 \tag{5-42a}$$

$$E_c = a_{e0} + a_{e1}P_2 + a_{e2}P_2^2 \tag{5-42b}$$

$$E_c = \dot{m}_a C_p (T_2 - T_1) \tag{5-43}$$

$$\dot{q} = (\dot{m}_a + \dot{m}_f) C_p (T_3 - T_2) \tag{5-44a}$$

$$P_3 = P_3(\dot{m}_a, \dot{m}_f, P_2, T_2) \tag{5-44ab}$$

$$\dot{m}_T = \dot{m}_f + \dot{m}_a \tag{5-47}$$

$$\dot{m}_T = a_{m0}(T_3) + a_{m1}(T_3)P_3 + a_{m2}(T_3)P_3^2 \tag{5-45a}$$

$$E_t = a_{t0}(T_3) + a_{t1}(T_3)P_3 + a_{t2}(T_3)P_3^2 \tag{5-45b}$$

$$E_t = E_c + E_s \tag{5-45c}$$

The unknowns are P_2, $\dot{m}_a$, E_c, T_2, $\dot{q}$, T_3, P_3, $\dot{m}_T$, E_t, and E_s. There are ten unknowns in the nine equations used to describe the gas turbine system. One of the unknowns must be specified if the system is to have a solution. All this is saying is that the type of problem must be specified: a $\dot{q}$ specification will yield the steady-state solution for a given fuel flow, while an E_s specification will yield the $\dot{q}$ (or equivalently the fuel flow rate) required. Other variations are possible, but these are the most feasible. Once the problem is specified, the multivariable Newton–Raphson procedure can be utilized to solve the nonlinear system.

The preceding indicates the general procedure that should be followed in posing the steady-state simulation for a complex system.

5-5 GENERALIZED ONE-DIMENSIONAL COMPRESSIBLE FLOW

Many energy systems use a compressible fluid as the working medium. Most undergraduate-level textbooks and courses in compressible fluid flow consider only the so-called *simple flow* (13): a steady one-dimensional flow in which a single driving potential is responsible for the changes in the fluid properties. Isentropic flow with only area change, heat addition (or loss) in the absence of friction and area change, and friction in the absence of heat addition (or loss) and area change are all examples of simple flows. For perfect gases (thermal and caloric), the simple flows can be solved in closed form; and for imperfect (real) gases, relatively simple numerical approaches can be used to obtain solutions for specific problems.

TABLE 5-4 Equations for Simple One-Dimensional Compressible Flows

	Simple Area Change	Simple Frictional Flow	Simple Diabatic Flow	Simple Mass Addition, $y=0$
$\frac{A}{A^*}$	$\frac{1}{M}\left[\left(\frac{2}{\gamma+1}\right)\psi\right]^{(\gamma+1)/2(\gamma-1)}$	1	1	1
$\frac{T}{T^*}$	1	1	$\frac{2(\gamma+1)M^2\psi}{(1+\gamma M^2)^2}$	1
$\frac{4\bar{f}L^*}{D}$	0	$\frac{1-M^2}{\gamma M^2}+\left(\frac{\gamma+1}{2\gamma}\right)\ln\left\{M^2\left[\left(\frac{2}{\gamma+1}\right)\psi\right]^{-1}\right\}$	0	0
$\frac{\dot{m}}{\dot{m}^*}$	1	1	1	$\frac{M\sqrt{2(\gamma+1)\psi}}{1+\gamma M^2}$
$\frac{p}{p^*}$	$\left[\left(\frac{2}{\gamma+1}\right)\psi\right]^{-\gamma/(\gamma-1)}$	$\frac{1}{M}\left[\left(\frac{2}{\gamma+1}\right)\psi\right]^{-1/2}$	$\frac{\gamma+1}{1+\gamma M^2}$	$\frac{\gamma+1}{1+\gamma M^2}$
$\frac{\rho}{\rho^*}$	$\left[\left(\frac{2}{\gamma+1}\right)\psi\right]^{-1/(\gamma-1)}$	$\frac{1}{M}\left[\left(\frac{2}{\gamma+1}\right)\psi\right]^{1/2}$	$\frac{1+\gamma M^2}{(\gamma+1)M^2}$	$\frac{2\psi}{1+\gamma M^2}$
$\frac{t}{t^*}$	$\left[\left(\frac{2}{\gamma+1}\right)\psi\right]^{-1}$	$\left[\left(\frac{2}{\gamma+1}\right)\psi\right]^{-1}$	$\frac{(\gamma+1)^2M^2}{(1+\gamma M^2)^2}$	$\left[\left(\frac{2}{\gamma+1}\right)\psi\right]^{-1}$
$\frac{V}{V^*}$	$M\left[\left(\frac{2}{\gamma+1}\right)\psi\right]^{-1/2}$	$M\left[\left(\frac{2}{\gamma+1}\right)\psi\right]^{-1/2}$	$\frac{(\gamma+1)M^2}{1+\gamma M^2}$	$M\left[\left(\frac{2}{\gamma+1}\right)\psi\right]^{-1/2}$
$\frac{P}{P^*}$	1	$\frac{1}{M}\left[\left(\frac{2}{\gamma+1}\right)\psi\right]^{(\gamma+1)/2(\gamma-1)}$	$\frac{\gamma+1}{1+\gamma M^2}\left[\left(\frac{2}{\gamma+1}\right)\psi\right]^{\gamma/(\gamma-1)}$	$\frac{\gamma+1}{1+\gamma M^2}\left[\left(\frac{2}{\gamma+1}\right)\psi\right]^{\gamma/(\gamma-1)}$
$\frac{\mathscr{F}}{\mathscr{F}^*}$	$\frac{1+\gamma M^2}{M\sqrt{2(\gamma+1)\psi}}$	$\frac{1+\gamma M^2}{M\sqrt{2(\gamma+1)\psi}}$	1	1
$\frac{s-s^*}{c_p}$	0	$\ln\left\{M^{(\gamma-1)/\gamma}\left[\left(\frac{2}{\gamma+1}\right)\psi\right]^{-(\gamma+1)/2\gamma}\right\}$	$\ln\left\{M^2\left(\frac{\gamma+1}{1+\gamma M^2}\right)^{(\gamma+1)/\gamma}\right]$	$\ln\left\{\left[\left(\frac{2}{\gamma+1}\right)\psi\right]^{-1}\left(\frac{1+\gamma M^2}{\gamma+1}\right)^{(\gamma-1)/\gamma}\right\}$

$\psi = 1+\frac{\gamma-1}{2}M^2$

Used with permission, from M. J. Zucrow and J. D. Hoffman, *Gas Dynamics*, John Wiley & Sons, Inc., 1976.

Table 5-4 lists the resulting closed-form expressions for the variables of interest for the simple flows of area change, friction (Fanno), diabatic flow (Rayleigh), or simple mass addition. The assumption of a perfect gas was invoked to obtain these results. The starred quantities (A^* and T^*, for example) refer to conditions at the location where M, the Mach number, equals 1. Table 5-4 is provided for completeness and to illustrate the consequence of various simple flows. A detailed discussion of simple compressible flows is not given. Unfortunately, many of the problems of interest to us in the design or analysis of energy system components are *combinations* of several simple flows.

Combinations of simple one-dimensional flows are usually called *generalized one-dimensional* flows. A flexible model for generalized compressible flows is the *influence coefficient* method introduced by Shapiro (13), studied by Beans (15), and discussed by Zucrow and Hoffman (14). We shall examine this method, as its flexibility makes it a good general model for a variety of situations encountered in energy systems.

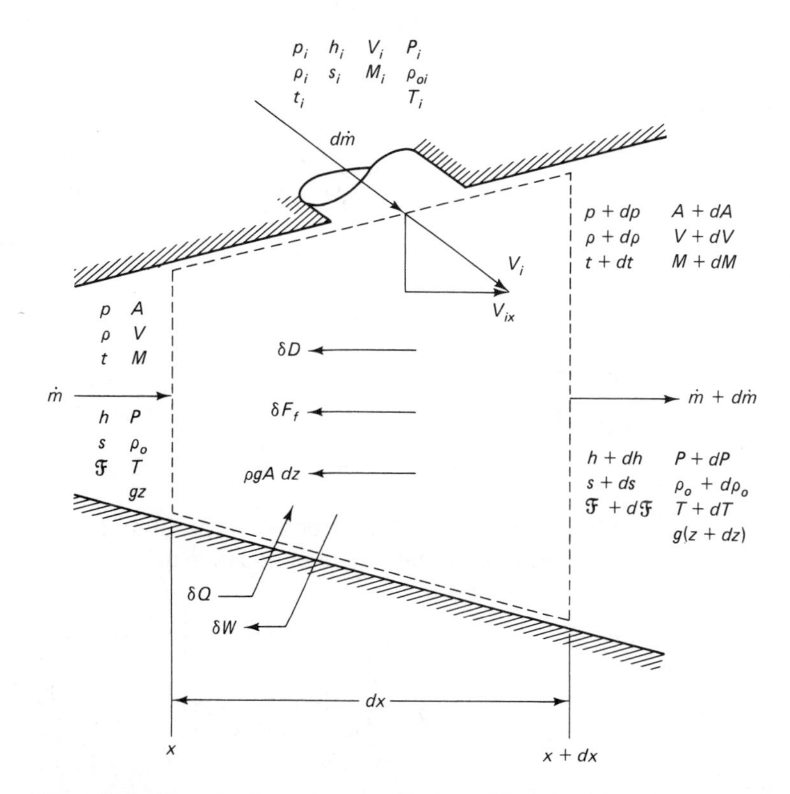

Figure 5-15 Control volume for generalized one-dimensional flow model. (Used with permission, from M. J. Zucrow and J. D. Hoffman, *Gas Dynamics*, John Wiley & Sons, Inc., 1976.)

The influence coefficient method is developed by application of the concepts of conservation of mass, momentum, and energy to a differential control volume with simultaneous area change, friction, heat transfer, and mass addition (or deletion). This control volume is shown in Fig. 5-15. An examination of the figure shows the simultaneous effects: (1) area change (A to $A + dA$ in a distance dx), (2) friction (δF_f, frictional force), (3) heat transfer and work (δQ and δW), (4) drag (δd), and (5) mass addition ($d\dot{m}$ with injection properties h_i, T_i, ρ_i, etc.). Properties of interest at the inlet and exit are p (static pressure), P (total pressure), t (static temperature), T (total temperature), A (area), ρ (density), M (Mach number), V (velocity), h (enthalpy), ρ_0 (stagnation or total density), s (entropy), z (elevation), and F (impulse). Obviously, the aforementioned are not all independent, but they are all of interest. The conservation laws [Eqs. (1-1) to (1-4)] are next applied with the following results (details are given in Shapiro or Zucrow and Hoffman):

Conservation of Mass:

$$\frac{d\dot{m}}{\dot{m}} = \frac{d\rho}{\rho} + \frac{dV}{V} + \frac{dA}{A} \tag{5-48}$$

Conservation of Momentum:

$$\frac{dp}{p} + \gamma M^2 \frac{dM}{M} + \frac{\gamma M^2}{2}\frac{dt}{t} + \frac{\gamma M^2}{2}\left[\frac{4f}{D}\,dx + \frac{2\delta d}{\gamma M^2 p A}\right] + \gamma M^2 (1-y)\frac{d\dot{m}}{\dot{m}} = 0 \tag{5-49}$$

where $f =$ Fanning friction factor and $y = V_{i_x}/V_i$.

Conservation of Energy:

$$\frac{\delta Q - \delta W - dH_i}{c_p t} = \left(1 + \frac{\gamma - 1}{2} M^2\right)\frac{dT}{T} \tag{5-50}$$

The following expressions, developed in differential form from the stated concept, complete the specification of the system of equations:

Equation of State:

$$\frac{dp}{p} = \frac{d\rho}{\rho} + \frac{dt}{t} \tag{5-51}$$

Mach Number:

$$\frac{dM}{M} = \frac{dV}{V} - \frac{1}{2}\frac{dt}{t} \tag{5-52}$$

Static and Stagnation (Total) Temperature:

$$\frac{dT}{T} = \frac{dt}{t} + \frac{(\gamma - 1)M^2}{1 + \frac{\gamma - 1}{2}M^2} \frac{dM}{M} \tag{5-53}$$

Static and Stagnation (Total) Pressure:

$$\frac{dP}{P} = \frac{dp}{p} + \frac{\gamma M^2}{1 + \frac{\gamma - 1}{2}M^2} \frac{dM}{M} \tag{5-54}$$

Impulse Function $[\mathscr{F} = A(p + V^2)]$:

$$\frac{d\mathscr{F}}{\mathscr{F}} = \frac{dp}{p} + \frac{dA}{A} + \frac{2\gamma M^2}{1 + \gamma M^2} \frac{dM}{M} \tag{5-55}$$

Equation of State (Entropy):

$$\frac{ds}{c_p} = \frac{dt}{t} - \frac{\gamma - 1}{\gamma} \frac{dp}{p} \tag{5-56}$$

Equations (5-48), (5-49), and (5-51) through (5-56) comprise a set of eight equations for the dependent variables dp/p, $d\rho/\rho$, dt/t, dV/V, dP/P, $d\mathscr{F}/\mathscr{F}$, and ds/c_p expressed in terms of the four independent variables or driving potentials dA/A, dT/T, $d\dot{m}/\dot{m}$, and $(4f\,dx/d + 2\delta D/\gamma M^2 pA)$. These equations are linear in the dependent variables and the driving potentials, which are taken as known quantities.

The solution of this linear system results in eight differential expressions, one for each of the dependent variables, which are functions of the local Mach number M and the driving potentials. The solution is given in tabular form in Table 5-5. The entries are called *influence coefficients* and represent the partial derivative of each flow property with respect to each driving potential. The influence of each driving potential on the change in each flow property is determined by each influence coefficient.

Consider the first row in Table 5-5 as defining the differential equation

$$\frac{1}{M}\frac{dM}{dx} = -\frac{\psi}{1 - M^2}\frac{1}{A}\frac{dA}{dx} + \frac{\gamma M^2 \psi}{2(1 - M^2)}\left[\frac{4f}{D} + \frac{2}{\gamma M^2 pA}\frac{\delta d}{dx}\right]$$
$$+ \frac{(1 + \gamma M^2)\psi}{2(1 - M^2)}\frac{1}{T}\frac{dT}{dx} + \frac{\psi\left[(1 + \gamma M^2) - y\gamma M^2\right]}{1 - M^2}\frac{1}{\dot{m}}\frac{d\dot{m}}{dx} \tag{5-57}$$

TABLE 5-5 Influence Coefficients for a Perfect Gas

Change in Flow Property	Driving Potentials			
	$\frac{dA}{A}$	$\left[\left(\frac{4f\,dx}{\mathscr{D}}\right) + \frac{2\,\delta d}{\gamma M^2 pA}\right]$	$\frac{dT}{T}$	$\frac{d\dot{m}}{\dot{m}}$
$\frac{dM}{M}$	$-\frac{\psi}{1-M^2}$	$\frac{\gamma M^2 \psi}{2(1-M^2)}$	$\frac{(1+\gamma M^2)\psi}{2(1-M^2)}$	$\frac{\psi\left[(1+\gamma M^2) - y\gamma M^2\right]}{1-M^2}$
$\frac{dp}{p}$	$\frac{\gamma M^2}{1-M^2}$	$-\frac{\gamma M^2\left[1+(\gamma-1)M^2\right]}{2(1-M^2)}$	$-\frac{\gamma M^2 \psi}{1-M^2}$	$-\frac{\gamma M^2[2\psi(1-y)+y]}{1-M^2}$
$\frac{d\rho}{\rho}$	$\frac{M^2}{1-M^2}$	$-\frac{\gamma M^2}{2(1-M^2)}$	$-\frac{\psi}{1-M^2}$	$-\frac{\left[(\gamma+1)M^2 - y\gamma M^2\right]}{1-M^2}$
$\frac{dt}{t}$	$\frac{(\gamma-1)M^2}{1-M^2}$	$-\frac{\gamma(\gamma-1)M^4}{2(1-M^2)}$	$\frac{(1-\gamma M^2)\psi}{1-M^2}$	$-\frac{(\gamma-1)M^2\left[(1+\gamma M^2) - y\gamma M^2\right]}{1-M^2}$
$\frac{dV}{V}$	$-\frac{1}{1-M^2}$	$\frac{\gamma M^2}{2(1-M^2)}$	$\frac{\psi}{1-M^2}$	$\frac{\left[(1+\gamma M^2) - y\gamma M^2\right]}{1-M^2}$
$\frac{dP}{P}$	0	$-\frac{\gamma M^2}{2}$	$-\frac{\gamma M^2}{2}$	$-\gamma M^2(1-y)$
$\frac{d\mathscr{F}}{\mathscr{F}}$	$\frac{1}{1+\gamma M^2}$	$-\frac{\gamma M^2}{2(1+\gamma M^2)}$	0	$\frac{y\gamma M^2}{1+\gamma M^2}$
$\frac{ds}{c_p}$	0	$\frac{(\gamma-1)M^2}{2}$	ψ	$(\gamma-1)M^2(1-y)$

$\psi = 1 + \frac{\gamma-1}{2}M^2$

Used with permission, from M. J. Zucrow and J. D. Hoffman, *Gas Dynamics*, John Wiley & Sons, Inc., 1976.

where

$$\psi = 1 + \frac{\gamma - 1}{2} M^2$$

Once the variations of the driving potentials are known, Eq. (5-57) can be integrated to give the distribution of Mach number in the x direction. This is an initial-value problem as the Mach number M must be known to start the process. The remaining seven differential equations specified by Table 5-5 can be integrated in a similar fashion for the values of the other seven dependent variables.

Such integrations are not necessary, as simple algebraic expressions that allow evaluation of the remaining dependent variables can be developed. These useful expressions, labeled "integral relations" by Shapiro, are obtained by forming ratios of various equations. These equations are

$$\frac{t_2}{t_1} = \frac{T_2}{T_1} \frac{1 + \frac{\gamma - 1}{2} M_1^2}{1 + \frac{\gamma - 1}{2} M_2^2} \tag{5-58}$$

$$\frac{V_2}{V_1} = \frac{M_2}{M_1} \sqrt{\frac{t_2}{t_1}} \tag{5-59}$$

$$\frac{p_2}{p_1} = \frac{\dot{m}_2}{\dot{m}_1} \frac{A_1}{A_2} \frac{M_1}{M_2} \sqrt{\frac{t_2}{t_1}} \tag{5-60}$$

$$\frac{\rho_2}{\rho_1} = \frac{\dot{m}_2}{\dot{m}_1} \frac{A_1}{A_2} \frac{V_1}{V_2} \tag{5-61}$$

$$\frac{\mathscr{F}_2}{\mathscr{F}_1} = \frac{p_2}{p_1} \frac{A_2}{A_1} \frac{1 + \gamma M_2^2}{1 + \gamma M_1^2} \tag{5-62}$$

$$s_2 = s_1 + c_p \ln \frac{t_2}{t_1} - R \ln \frac{p_2}{p_1} \tag{5-63}$$

$$\frac{P_2}{P_1} = \frac{p_2}{p_1} \left[\frac{1 + \frac{\gamma - 1}{2} M_2^2}{1 + \frac{\gamma - 1}{2} M_1^2} \right]^{\gamma/(\gamma - 1)} \tag{5-64}$$

With these equations, a simple procedure for solving any generalized one-dimensional steady-state compressible flow can be used.

The first step in solving a generalized problem is to define the initial (and boundary) conditions and develop expressions for the required driving potentials. Equation (5-57), a first-order nonlinear ordinary differential equation, is then numerically integrated using any appropriate numerical integration algorithm. Because of its simplicity and accuracy, a Runge–Kutta method is often used. The integration of Eq. (5-57) yields $M(x)$, which in conjunction with

Eqs. (5-58) through (5-64) can be used to obtain the values of the remaining dependent variables as a function of x.

Many variations of the Runge–Kutta procedure are available in the literature, the most common being the following "4-5" (fourth-order algorithm, error of fifth order) version:

$$y_{n+1} = y_n + \frac{h}{6}(m_1 + 2m_2 + 2m_3 + m_4) \tag{5-65}$$

which solves the equation

$$\frac{dy}{dx} = f(x, y) \tag{5-66}$$

and where h is the step size in the independent variable with

$$m_1 = f(x_n, y_n) \tag{5-67}$$

$$m_2 = f\left(x_n + \frac{h}{2}, y_n + \frac{h}{2}m_1\right) \tag{5-68}$$

$$m_3 = f\left(x_n + \frac{h}{2}, y_n + \frac{h}{2}m_2\right) \tag{5-69}$$

$$m_4 = f(x_n + h, y_n + hm_3) \tag{5-70}$$

Example 5-3

Air enters a combustor at $T = 1400°\text{R}$ and $P = 335$ psia with an inlet Mach number of 0.2. Energy is added to the air such that the exit stagnation temperature is 2200°R. The combustor is 2 ft long and tapers from a diameter of 1 ft to a diameter of 10 in. The friction factor is estimated to be $f = 0.005$. Neglect the effects of fuel mass flow rate and assume the properties throughout are those of air. Solve for t, T, p, P, and M as a function of x.

Solution. This does not represent a simple flow, so influence coefficients will be used. Since $\delta D = 0$ and $d\dot{m} = 0$, Eq. (5-57) reduces to

$$\frac{1}{M}\frac{dM}{dx} = -\frac{\psi}{1 - M^2}\frac{1}{A}\frac{dA}{dx} + \frac{\gamma M^2 \psi}{2(1 - M^2)}\frac{4f}{D} + \frac{(1 + \gamma M^2)\psi}{2(1 - M^2)}\frac{1}{T}\frac{dT}{dx}$$

A linear change in the total temperature gives

$$\frac{dT}{dx} = 400.0°\text{R/ft}$$

so that $T(x) = 1400 + 400x$. The diameter as a function of x is $D(x) = 1.0 - x/12$, which leads to

$$A(x) = \frac{\pi}{4}\left(1 - \frac{x}{12}\right)^2$$

and

$$\frac{dA}{dx} = -\frac{\pi}{24}\left(1 - \frac{x}{12}\right)$$

Then

$$\frac{1}{A}\frac{dA}{dx} = -\frac{1}{6(1 - x/12)}$$

The differential equation then becomes

$$\frac{dM}{dx} = M\left[\frac{\psi}{1-M^2}\frac{1}{6(1-x/12)} + \frac{\gamma M^2\psi}{2(1-M^2)}\frac{4f}{1-x/12} + \frac{(1+\gamma M^2)\psi}{2(1-M^2)}\frac{400}{1400+400x}\right]$$

Integration of this equation with the initial condition of $M = 0.2$ at $x = 0.0$ leads to the Mach number distribution as a function of x, and $M(x)$ can then be used with Eqs. (5-58) through (5-64) to obtain the distributions of the other desired properties as functions of x.

Figure 5-16 is the output for an influence coefficient model Runge–Kutta simulation computer program for this problem. PT is the total pressure, PS is the static pressure, TT is the total temperature, and TS is the static temperature. Units are as indicated in the parentheses. During heating, the Mach number increased from 0.20 to 0.398, while the total pressure decreased from 335 to 325.7 psia. Had friction not been present, the PT at $x = 2.0$ ft would have been 326.6 psia. Thus most of the pressure drop is due to heat addition—the Rayleigh effect. This is a poor design as an exit Mach number of 0.398 is getting dangerously high. The total temperature, a driving potential that is specified, increases from 1400° to 2200°R.

The computer program required is straightforward; functions are defined for the total temperature, the area, and the dM/dx relation and are then used in the Runge–Kutta procedure given by Eqs. (5-65) through (5-70). This is a complex situation, yet the influence coefficient method was simple to set up

I	X(FT)	MACH	PT(PSI)	PS(PSI)	TT(R)	TS(R)
1	0.000	0.200	335.0	325.8	1400.0	1388.9
2	0.100	0.207	334.7	324.9	1440.0	1427.8
3	0.200	0.214	334.4	324.0	1480.0	1466.6
4	0.300	0.221	334.1	323.0	1520.0	1505.3
5	0.400	0.228	333.8	321.9	1560.0	1543.9
6	0.500	0.236	333.5	320.8	1600.0	1582.4
7	0.600	0.244	333.2	319.7	1640.0	1620.7
8	0.700	0.252	332.8	318.4	1680.0	1658.9
9	0.800	0.261	332.5	317.1	1720.0	1696.9
10	0.900	0.270	332.1	315.8	1760.0	1734.8
11	1.000	0.279	331.7	314.3	1800.0	1772.5
12	1.100	0.288	331.3	312.7	1840.0	1809.9
13	1.200	0.298	330.9	311.1	1880.0	1847.1
14	1.300	0.309	330.4	309.3	1920.0	1884.1
15	1.400	0.319	330.0	307.4	1960.0	1920.8
16	1.500	0.331	329.5	305.4	2000.0	1957.2
17	1.600	0.343	328.9	303.3	2040.0	1993.2
18	1.700	0.355	328.4	301.0	2080.0	2028.8
19	1.800	0.368	327.8	298.5	2120.0	2064.0
20	1.900	0.382	327.2	295.9	2160.0	2098.7
21	2.000	0.397	326.6	293.0	2200.0	2132.8

Figure 5-16 Combustor Mach number and property distribution.

and use. The problem is a good example of a differential-equation-based model and simulation.

Influence coefficient models work quite well as long as a sonic point ($M = 1$) is not encountered. An examination of Eq. (5-57) reveals that for a Mach number of 1 the numerator vanishes and the equation is singular. Hence, if the solution is to proceed continuously from subsonic to supersonic flow, as in a converging–diverging nozzle, some overt action must be taken to avoid the singularity about the sonic point. Zucrow and Hoffman discuss several strategies treating the singular point. We shall herein be concerned with only one of these, that due to Beans (15).

Beans suggested that equations such as Eq. (5-57) could be written as

$$\frac{dM}{dx} = M\psi \frac{G(x)}{1 - M^2} \tag{5-71}$$

When $M = 1.0$, the numerator tends to zero and, unless $\lim_{M \to 1} G(x)$ is also zero, the value of dM/dx is unbounded. Beans used this line of thought to infer that $\lim_{M \to 1} G(x) = 0$. Then at the sonic point

$$\frac{G(x)}{1 - M^2} = \frac{0}{0} \tag{5-72}$$

and l'Hôpital's rule can be used to evaluate Eq. (5-71) at the sonic point. Hence

$$\left(\frac{dM}{dx}\right)^* = \lim_{M \to 1} \frac{dM}{dx} = \frac{(dG/dx)^*}{(-2M\,dM/dx)^*} \tag{5-73}$$

For a given system model, Eq. (5-73) becomes

$$\left[\left(\frac{dM}{dx}\right)^*\right]^2 + \frac{\gamma + 1}{4}\frac{dG}{dx} = 0 \tag{5-74}$$

However, $(dG/dx)^*$ contains terms with $(dM/dx)^*$ so that Eq. (5-74) is actually quadratic in $(dM/dx)^*$. The quadratic solution of Eq. (5-74) yields a differential equation, different from Eq. (5-71), valid about the sonic point. Equation (5-71) is integrated to near the sonic point; then the expression from Eq. (5-74) is used.

Such a procedure destroys the initial-value nature of the problem since $G(x)$ is zero at the sonic point. The general form of $G(x)$ is

$$G(x) = \frac{1}{A}\frac{dA}{dx} + \frac{\gamma M^2}{2}\left(\frac{4f}{D} + \frac{2}{\gamma MpA}\frac{\delta d}{dx}\right) + \frac{1 + \gamma M^2}{2}\frac{1}{T}\frac{dT}{dx} + \frac{(1 + \gamma M^2) - y\gamma M^2}{\dot{m}}\frac{d\dot{m}}{dx} \tag{5-75}$$

*Evaluated at $M = 1$.

so that at the sonic point

$$G(x) = 0 = \frac{1}{A}\frac{dA}{dx} + \frac{\gamma}{2}\left(\frac{4f}{D} + \frac{2}{\gamma p A}\frac{\delta d}{dx}\right) + \frac{1+\gamma}{2}\frac{1}{T}\frac{dT}{dx} + \frac{1+\gamma-\gamma y}{\dot{m}}\frac{d\dot{m}}{dx} \tag{5-76}$$

With values of the driving potentials given, Eq. (5-76) yields the x location of the sonic point. The properties cannot be determined since the inlet conditions are not known. The inlet Mach number is then determined by integrating Eq. (5-57) [and near the sonic point the expression for $(dM/dx)^*$] backward with a negative f to the inlet. Once the inlet Mach number is known, the inlet static conditions can be determined and the entire converging–diverging nozzle solution generated by integrating forward and evaluating the physical properties at each step. Details are given by Beans (15).

Converging–diverging nozzles often contain shock waves in the supersonic portion. The method of influence coefficients can be used to model shock waves in ducts. A shock wave is treated as a discontinuity, so the properties change abruptly. Figure 5-17 schematically represents these changes for a stationary shock wave. The flow is from the left to the right, is supersonic ahead of the shock wave, and is subsonic behind the shock wave. The variables are related by the "jump conditions" across a shock wave:

$$M_2^2 = \frac{M_1^2 + \dfrac{2}{\gamma - 1}}{\dfrac{2\gamma}{\gamma - 1}M_1^2 - 1} \tag{5-77}$$

$$\frac{p_2}{p_1} = \frac{2\gamma}{\gamma - 1}M_1^2 - \frac{\gamma - 1}{\gamma + 1} \tag{5-78}$$

$$\frac{\rho_2}{\rho_1} = \frac{(\gamma + 1)M_1^2}{2 + (\gamma - 1)M_1^2} \tag{5-79}$$

$$\frac{t_2}{t_1} = \frac{1 + \dfrac{\gamma - 1}{2}M_1^2}{1 + \dfrac{\gamma - 1}{2}M_2^2} \tag{5-80}$$

When a shock wave is encountered, the integration of Eq. (5-57) is continued up to the shock wave and the required properties are computed. Equations (5-77) through (5-80) are then used to compute the properties downstream of the shock wave, and the integration of Eq. (5-57) is started again.

As explored within this section, the influence coefficient method is a potent one for modeling of "nonsimple" compressible flows. The combination of a differential equation model and a Runge–Kutta numerical integration for

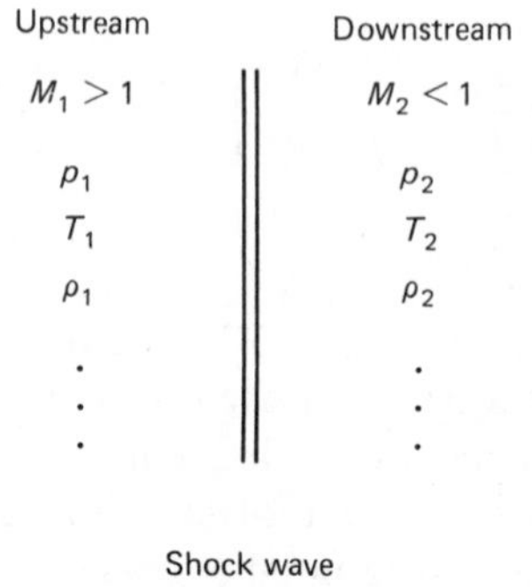

Upstream | Downstream
$M_1 > 1$ | $M_2 < 1$
p_1 | p_2
T_1 | T_2
ρ_1 | ρ_2

Shock wave

Figure 5-17 Stationary shock wave nomenclature.

the simulation is a powerful general technique that can be applied to many thermal and fluid sciences situations. Since x increases monotonically in this simulation, continuous modeling programs such as CSMP III can be used for the integrations by calling x the "time."

5-6 CONCLUDING REMARKS

This chapter has been but a brief introduction to modeling and simulation. The concepts of curve fitting, the generalized Hardy–Cross method, and the use of the multivariable Newton–Raphson technique for solving systems of equations are extremely versatile and can be used to model and simulate an extensive range of different energy systems. The particular systems examined here, a parallel pump–piping arrangement and a gas turbine engine, were used only to illustrate the procedures involved and were not meant to suggest limitations of the method.

The importance of modeling and simulation in the design process should not be overlooked. Complex systems whose components interact with each other in highly nonlinear fashions can be understood much better if simulations with some degree of predictive validity can be generated. In many instances the off-design behavior of a system is important. Optimum design of complex systems and components, a subject not examined in this text, depends heavily on modeling that has good fidelity across a wide range of parameters.

An extensive literature is devoted to the modeling and simulation of physical systems. For techniques beyond those previewed here and for more details on system modeling, this literature should be consulted. References 1 to 3, 7, 11, and 16 of this chapter represent useful additional reading.

REFERENCES

1. Smith, J. M., *Mathematical Modeling and Digital Simulation for Engineers and Scientists*. New York: Wiley, 1977.

2. Wellstead, P. E., *Introduction to Physical System Modeling*. New York: Academic Press, 1979.
3. Dym, C. L., and Ivey, E. S., *Principles of Mathematical Modeling*. New York: Academic Press, 1979.
4. Solomon, S. L., "Building Modelers: Teaching the Art of Simulation," Proceedings of the 12th Annual Simulation Symposium, Annual Simulation Symposium, Tampa, Florida, 1979.
5. Speckhart, F. H., and Green, W. L., *A Guide to Using CSMP—The Continuous System Modeling Program*. Englewood Cliffs, N.J.: Prentice-Hall, 1976.
6. Pritsker, A. A. B., and Pegden, C. D., *Introduction to Simulation and SLAM*. New York: Wiley, 1979.
7. Stoecker, W. F., *Design of Thermal Systems*, 2nd ed. New York: McGraw-Hill, 1980.
8. *Goulds Pump Manual*, 3rd ed., Goulds Pump, Inc., Seneca Falls, N.Y.
9. Wood, D. J., and Rayes, A. G., "Reliability of Algorithms for Pipe Network Analysis," *J. Hydraulics Division, ASCE*, Vol. 107, No. HY10, October 1981, pp. 1145–1161.
10. Conte, S. D., and DeBoor, Carl, *Elementary Numerical Analysis*, 2nd ed. New York: McGraw-Hill, 1972.
11. Harmon, R. T. C., *Gas Turbine Engineering*. New York: Wiley, 1982.
12. Shepherd, D. G., *Principles of Turbomachinery*. New York: Macmillan, 1956.
13. Shapiro, A. H., *The Dynamics and Thermodynamics of Compressible Fluid Flow*, Vol. I. New York: Ronald Press, 1953.
14. Zucrow, M. J., and Hoffman, J. D., *Gas Dynamics*. New York: Wiley, 1976.
15. Beans, E. W., "Computer Solution to Generalized One-Dimensional Flow," *J. Spacecraft and Rockets*, Vol. 7, No. 12, December 1970, pp. 1460–1464.
16. Palm, W. J., III. *Modeling, Analysis, and Control of Dynamic Systems*. New York: Wiley, 1983.

REVIEW QUESTIONS

1. What is simulation?

2. What is modeling?

3. In a system simulation with 14 unknowns, would it be desirable to use algebra to eliminate 7 equations so that the smaller system of 7 equations might be solved? Explain.

4. If a pump with $H = 200 - 0.01Q^2$ is placed in line 2–loop 1, what are the values of A_{ij} and B_{ijm} for use in the generalized Hardy–Cross method?

5. What minimum-order polynomial is required to fit a curve with an inflection point?

6. Which of the four steps of the system simulation process as given in Fig. 5-1 is the most important if good fidelity is to be achieved?

7. Why is the Newton–Raphson approach generally preferred over successive substitution approaches?

8. What advantages does the multivariable Newton–Raphson method possess over the Hardy–Cross method for general simulation problems?
9. What is a simple flow?
10. Why is the influence coefficient method so useful for compressible flow?
11. For any nonsimple compressible flow, must the sonic point always be located at a minimum area? Explain your answer.
12. How might Eq. (5-44b) be evaluated?

PROBLEMS

1. Two pumps in *series* are used to pump water from a low reservoir to a high reservoir. A schematic representation is shown in Fig. P5-1.

$$\text{For pump 1:} \quad H_1 = a_1 - b_1 Q - c_1 Q^2$$
$$\text{For pump 2:} \quad H_2 = a_2 - b_2 Q - c_2 Q^2$$

   (a) Write the energy equation between stations a and b. State all assumptions.
   (b) Find the equations needed to determine the steady-state operating point using the multivariable Newton–Raphson approach.
   (c) In what other ways could the operating point for this system be determined?

2. The following data were taken in a free convection heat transfer experiment:

Gr Pr	$\overline{\text{Nu}}$	Gr Pr	$\overline{\text{Nu}}$
10^5	10	10^9	100
10^6	20	10^{10}	250
10^7	36	10^{11}	580
10^8	56	10^{12}	1400

   (a) Using only three points, find $\text{Nu} = a + b(\text{Gr Pr}) + c(\text{Gr Pr})^2$.
   (b) Using a log–log plot, find two curves of the form $\text{Nu} = a(\text{Gr Pr})^b$ that are much better fits to the data than the results of part (a).
   (c) In log–log coordinates, plot the data, the curve from part (a), and the curves from part (b).
   (d) Discuss the implications.

3. Air initially at $P = 117.0$ psi, $T = 560°\text{R}$, and $M_0 = 0.16$ flows through a copper

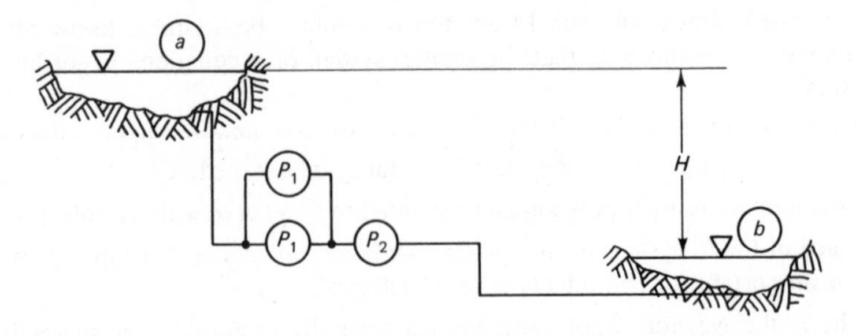

Figure P5-1

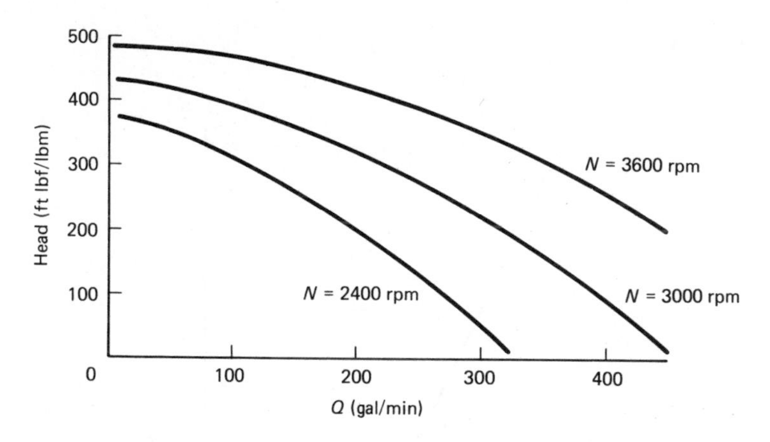

Figure P5-5

pipe whose internal surface is maintained at 1500°R. The pipe ID is 1 in. and $\varepsilon/D = 0.0004$. Assuming the flow to be one dimensional, find p, m, T, P_t, and T_t as a function of the pipe length. Does a maximum length exist for these initial conditions?

4. Work Example 1-4 if a Goulds pump 2 × 3-10 A60 (Fig. 4-15) is placed in line 2 of loop 2.

5. Develop a curve fit for $H = H(N, Q)$ for the pump characteristic of Fig. P5-5.

6. The gas turbine system in Fig. P5-6(a) is to be simulated. Compressor and turbine characteristics are given in Figs. P5-6(b) and (c). Use air as the working fluid. T_1 is 100°F and 210 gal/h of JP-4 (heat content 130,000 Btu/gal) fuel is used. Assuming $p_2 = p_3$ and neglecting $\dot{m}_f$ effects on the mass flow rate, complete the following in order:

(a) Generate curve fits for $p_2 = p_2(\dot{m})$ and $E_c(\dot{m})$.
(b) Generate curve fits for $\dot{m} = \dot{m}(p_3, T_3)$ and $E_t = E_t(p_3, T_3)$.
(c) Develop the system of equations required to model the system.
(d) Develop a multiple-variable Newton–Raphson computer program to simulate the system.
(e) List the steady-state value of all variables.
(f) What is the turbine *exit* temperature? If the unit is an industrial gas turbine, examine the potential for energy recovery from the exhaust gases.

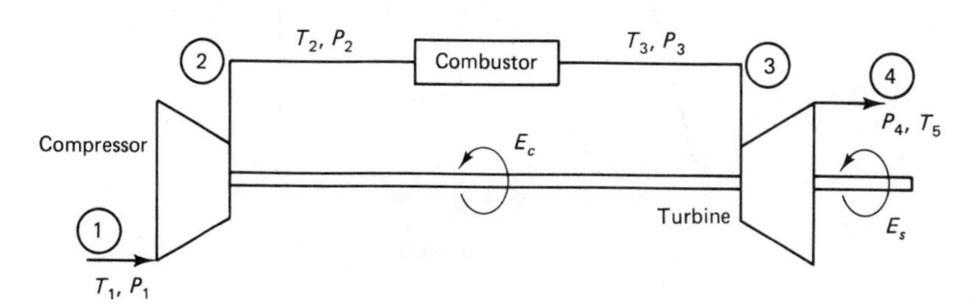

Figure P5-6(a) System schematic.

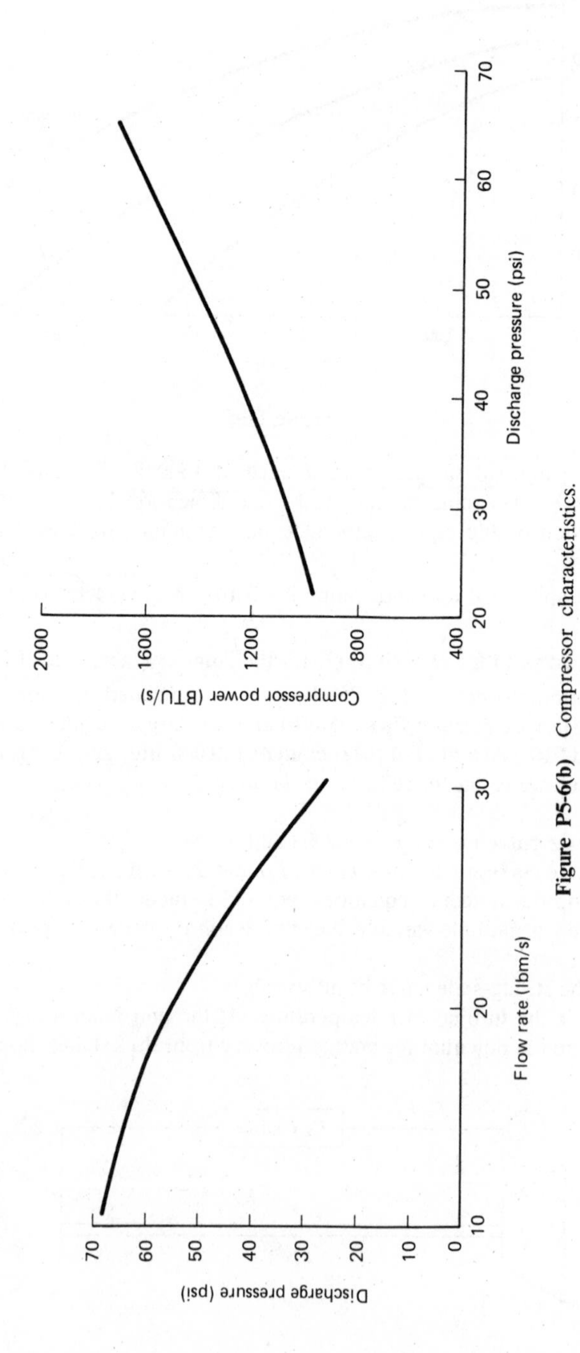

Figure P5-6(b) Compressor characteristics.

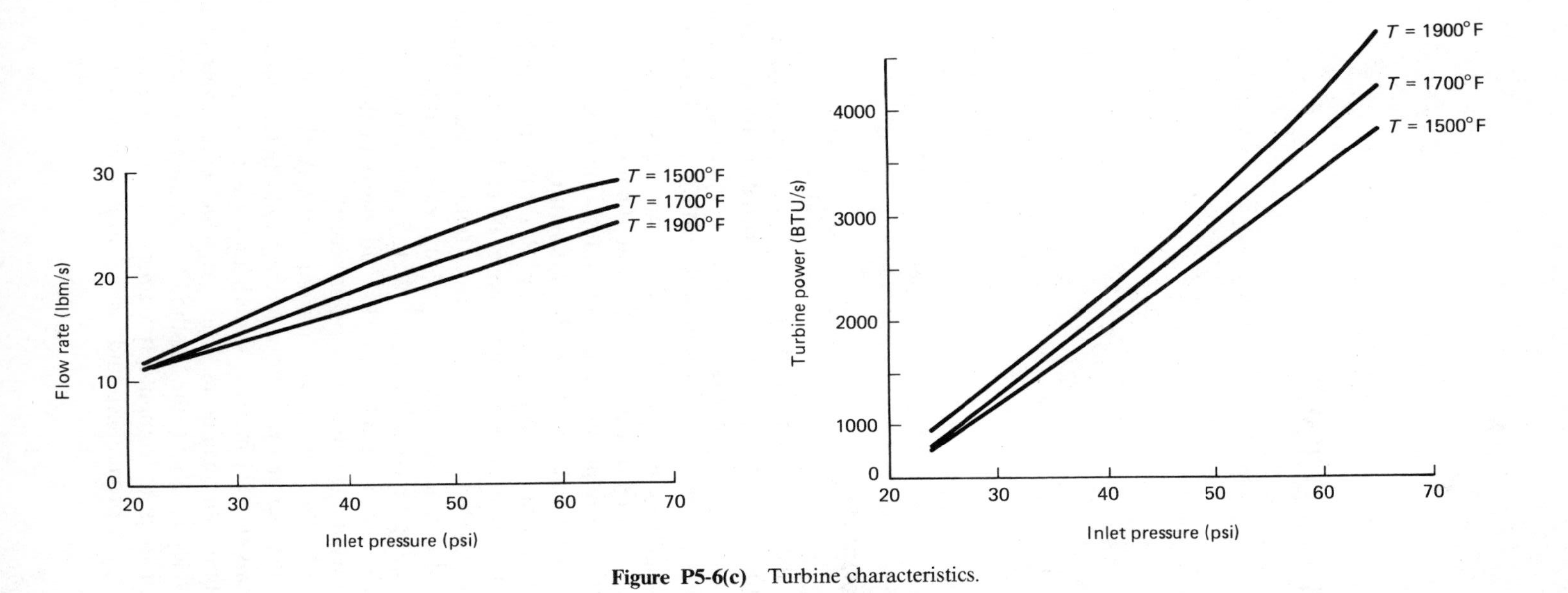

Figure P5-6(c) Turbine characteristics.

6

FLUID TRANSIENTS

6-1 INTRODUCTION

Thus far we have been concerned only with steady-state calculations in fluid networks. However, fluid transients are important in the startup or shutdown or modulation of fluid networks; and if devices such as valves are opened or closed rapidly, traveling wave phenomena called *waterhammer* can result in excessive local pressures. Analysis of fluid transients is usually divided into two categories: (1) rigid theory and (2) waterhammer. *Rigid theory* is developed with the assumptions of an absolutely incompressible fluid and an absolutely inelastic pipe and is useful in estimating unsteady flow behavior for situations where changes occur gradually. Rigid theory is rather easily developed and is a mathematically simple model of reality. *Waterhammer*, on the other hand, is developed from the correct notions that a liquid has a finite speed of sound (i.e., is not absolutely incompressible), that the pipe is elastic, and that a fluid–pipe coupling exists. Waterhammer analysis is accurate for very rapid changes in fluid networks and uses much more sophisticated mathematical techniques than does rigid theory.

Nowhere in fluid mechanics has the digital computer had more impact than in waterhammer analysis. Prior to the publication of Streeter and Wylie's *Hydraulic Transients* (1) in 1967, waterhammer analysis was understood and practiced by only a few engineers. This text, which emphasized computer-based solutions developed about the method of characteristics and which gave FORTRAN listings of computer programs for virtually every major topic covered, has had a pronounced and lasting positive effect on this subject; and the methods and techniques it espoused have become commonplace. Ten years later, books by Fox (2), Wylie and Streeter (3), and Jaeger (4) continued the trend, as have later publications by Watters, (5) and Chaudhry (6). In a matter of 15 years, waterhammer analysis has moved from the realm of a difficult, poorly understood subject to a subject with widespread understanding and a well-developed literature.

This chapter will deal briefly with both rigid theory and waterhammer. The intent is only to convey the rudiments—enough to allow us to understand a few salient features and applications of each topic. As in the previous chapters, what is presented herein is not new; information contained in the works of Wylie and Streeter, Fox, and Watters forms the bulk of the material presented here. For more extended applications, the previously referenced books are recommended.

6-2 RIGID THEORY

Rigid theory treats the fluid as an absolutely incompressible medium. This assumption implies an unbounded speed of sound in the fluid so that pressure changes are propagated instantaneously throughout the system. Within this framework, the elastic properties of the pipe walls are of no consequence, and conservation of linear momentum [Eq. (1-2)] leads to the governing equation.

Consider the control volume, taken along a streamline, as illustrated in Fig. 6-1. From Chapter 1, conservation of momentum in the integral form is

$$\sum \overline{F} = \frac{\partial}{\partial t}\int_{cv} \frac{\rho \overline{V}}{g_c}\, d\,\text{vol} + \int_{cs} \frac{\overline{V}\rho\overline{V}}{g_c} \cdot d\overline{A} \tag{1-2}$$

Application of Eq. (1-2) to the control volume illustrated in Fig. 6-1 yields

$$-\frac{1}{\gamma}\frac{\partial P}{\partial s} - \frac{dz}{ds} - \frac{f}{D}\frac{V^2}{2g} = \frac{1}{g}\frac{dV}{dt} \tag{6-1}$$

where γ is $\rho g/g_c$, s is the distance along the streamline, V is the average

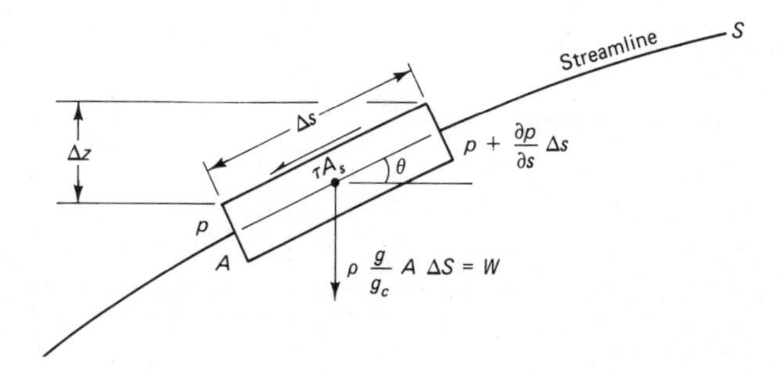

Figure 6-1 Unsteady flow derivation control volume.

instantaneous velocity, D is the hydraulic diameter, f is the Darcy–Weisbach friction factor, and z is the change in elevation. Rigid theory assumptions mean that at any instant of time the velocity V is the same at all locations with the same hydraulic diameter D or that the velocity varies only with time for a given pipe of hydraulic diameter D. Equation (6-1) is the fundamental equation governing rigid theory or rigid water column theory, as it is sometimes called.

A typical application is illustrated in Fig. 6-2, a horizontal pipe of length L and diameter D located H below the free surface. Integration of Eq. (6-1) along the length of the pipe from station 1 to station 2 yields

$$-\frac{1}{\gamma}\int_L \frac{\partial P}{\partial s}\,ds - \int_L \frac{dz}{ds}\,ds - \int_L \frac{fV^2}{2gD}\,ds = \int_L \frac{1}{g}\frac{dV}{dt}\,ds \tag{6-2}$$

If the additional assumption is made that the unsteady friction factor at an instantaneous value V is the same as for steady flow at the same velocity V, then Eq. (6-2) becomes

$$\frac{P_1}{\gamma} - \frac{P_2}{\gamma} - \frac{fL}{D}\frac{V^2}{2g} = \frac{L}{g}\frac{dV}{dt} \tag{6-3}$$

No matter what the situation from (1) to (2), the pressure at (1), P_1, is always γH; so H is P_1/γ and

$$H - \frac{P_2}{\gamma} - \frac{fL}{D}\frac{V^2}{2g} = \frac{L}{g}\frac{dV}{dt} \tag{6-4}$$

If the valve is open at time $t = 0$, then, for all $t > 0$, P_2 is zero pressure (gauge) and

$$H - \frac{fL}{D}\frac{V^2}{2g} = \frac{L}{g}\frac{dV}{dt} \tag{6-5}$$

Separating the variables and integrating Eq. (6-5) yields

$$\int_0^t dt' = \frac{L}{g}\int_0^V \frac{dV'}{H - \dfrac{fL}{D}\dfrac{V^2}{2g}} \tag{6-6}$$

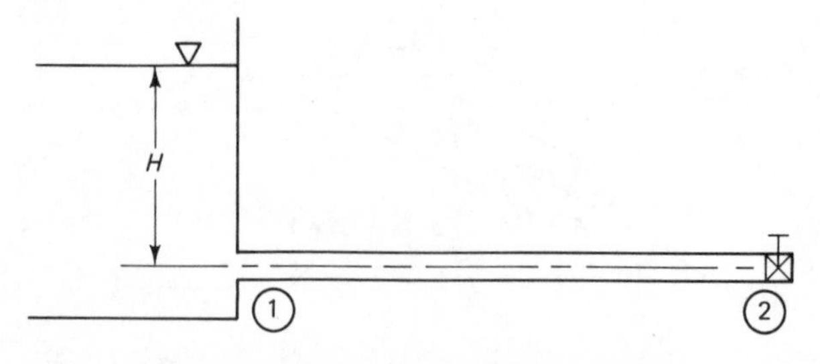

Figure 6-2 Rigid theory illustration schematic.

which gives

$$t = \sqrt{\frac{LD}{2gfH}} \ln \frac{\sqrt{\frac{2gHD}{fL}} + V}{\sqrt{\frac{2gHD}{fL}} - V} \tag{6-7}$$

The steady-state velocity for the situation portrayed in Fig. 6-2 is

$$V_0 = \sqrt{\frac{2gHD}{fL}} \tag{6-8}$$

with which Eq. (6-7) reduces to

$$t = \frac{LV_0}{2gH} \ln \frac{V_0 + V}{V_0 - V} \tag{6-9}$$

Equation (6-9) indicates that a steady flow V_0 is reached only as $t \to \infty$; Fig. 6-3 graphically demonstrates the behavior of Eq. (6-9). Steady flow is usually assumed to occur at $V = 0.99V_0$ for which the time to steady state is

$$t_{ss} = 2.65 \frac{LV_0}{gH} \tag{6-10}$$

Another situation of interest is shown in Fig. 6-4, the *slow* closing of a valve in a constant diameter line connecting two reservoirs. The valve closing is specified to follow a schedule such that dV/dt is maintained constant from the initiation of valve closure at $t = 0$ to the completion of closure at $t = t_c$. The solution to this problem is obtained in two parts, sections 1 to 2 and sections 3 to 4. Denoting the decrease in velocity by $\dot{V}$, Eq. (6-3) for section 1 to 2 becomes

$$\frac{P_2}{\gamma} - \frac{P_1}{\gamma} + \frac{fL}{D}\frac{V^2}{2g} + \frac{L}{g}\dot{V} = 0 \tag{6-11}$$

So

$$\frac{P_2}{\gamma} = H_1 - \frac{fL}{D}\frac{V^2}{2g} - \frac{L}{g}\dot{V} \tag{6-12}$$

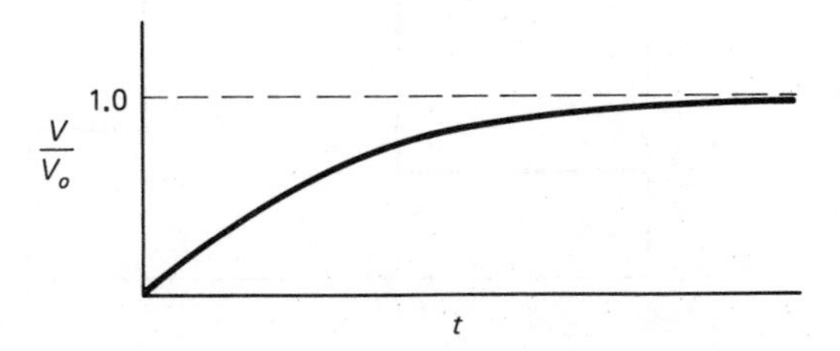

Figure 6-3 Asymptotic approach of V to V_0.

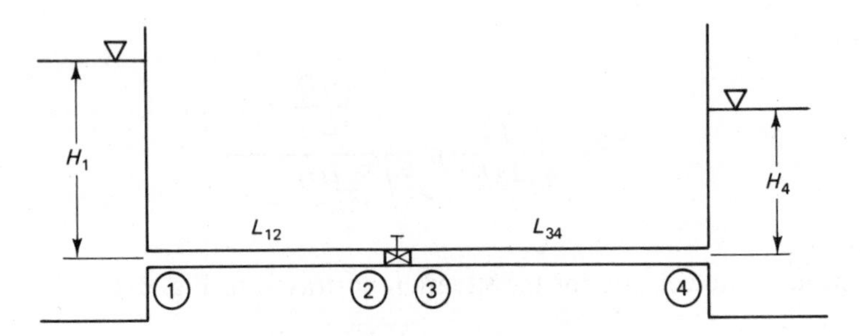

Figure 6-4 Closing of a valve in a line.

But, since $\dot{V} < 0$,

$$\frac{P_2}{\gamma} = \left(H_1 + \frac{L}{g}|\dot{V}|\right) - \frac{fL}{D}\frac{V^2}{2g} \tag{6-13}$$

The maximum pressure at station 2 occurs when $V = 0$, that is, at the instant of full closure of the valve. After valve closure is complete, the pressure at station 2 then "tails" off to a value γH_1. This behavior is illustrated in Fig. 6-5.

The behavior of the section downstream of the valve, section 3 to 4, is analyzed in a similar manner. The basic expression for this section during valve closure is

$$\frac{P_4}{\gamma} - \frac{P_3}{\gamma} + \frac{fL}{D}\frac{V^2}{2g} + \frac{L}{g}\dot{V} = 0 \tag{6-14}$$

which can be written as

$$\frac{P_3}{\gamma} = \left(H_4 - \frac{L}{g}|\dot{V}|\right) + \frac{fL}{D}\frac{V^2}{2g} \tag{6-15}$$

Thus, on the downstream side of the valve, the greatest pressure at station 3

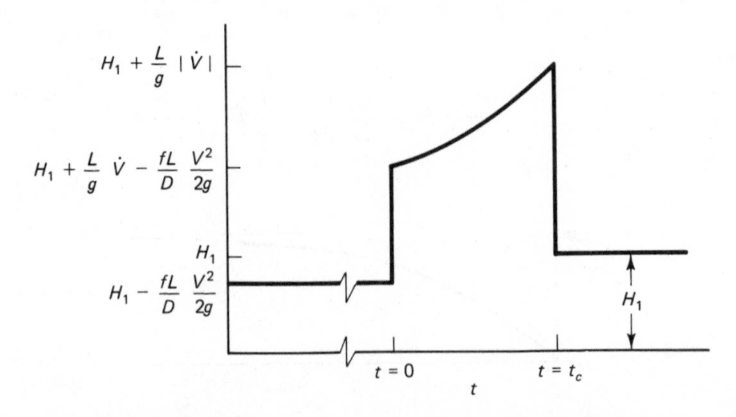

Figure 6-5 Behavior of the head at station 2 for valve closing.

occurs just prior to the initiation of valve closure; the minimum pressure at the station occurs at complete valve closure at time t_c. After t_c the pressure tails up to γH_4. This behavior is illustrated in Fig. 6-6.

Equation (6-3), which has been utilized in the previous illustrations, forms the basis for applying rigid theory. In situations where the rate of change of velocity, $\dot{V}$, is not known, an auxiliary equation for the pressure drop across the valve must be provided. This additional equation is of the form

$$\frac{P_3 - P_2}{\gamma} = K(t)\frac{V^2}{2g} \tag{6-16}$$

Equation (6-16) in conjunction with Eq. (6-12) typically requires a numerical solution. Values of KC_d for several valve positions and types are given in Fig. 6-7.

Some example problems using the previously developed equations are in order.

Example 6-1

For the situation depicted in Fig. 6-2, $L = 10{,}000$ ft, $D = 1$ ft, $H = 80$ ft, and the fluid is water at 70°F. The pipe is commercial steel. Find the time required to reach steady-state flow if the friction factor is constant at its steady state value.

Solution. The steady state velocity is

$$V_0 = \sqrt{\frac{2gHD}{fL}} = \left[(2)\left(32.174\frac{\text{ft}}{\text{s}^2}\right)(80\text{ ft})\frac{1\text{ ft}}{(10{,}000\text{ f})\text{ft}}\right]^{1/2}$$

$$= \frac{0.7175}{\sqrt{f}}\text{ ft/s}$$

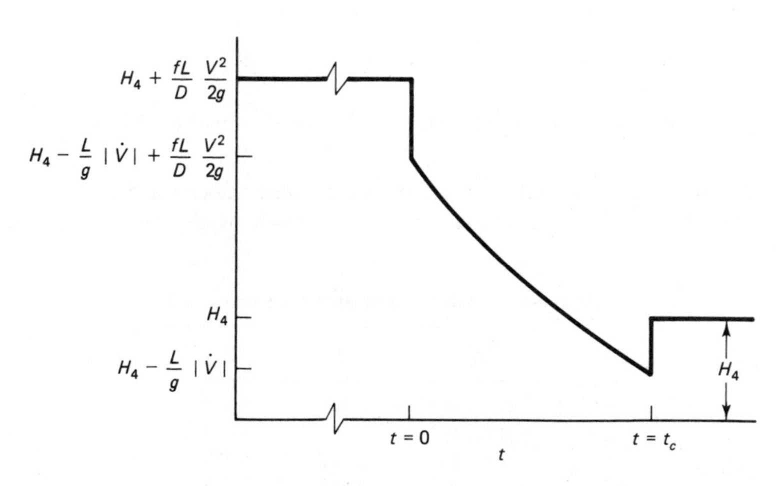

Figure 6-6 Behavior of the head at station 3 for valve closing.

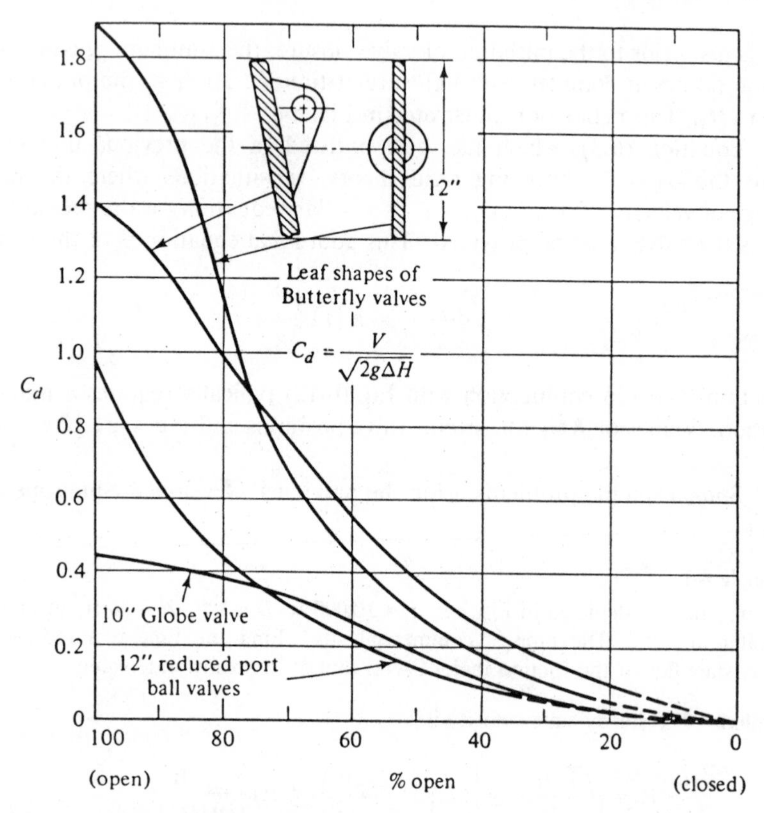

Figure 6-7 Discharge coefficient as a function of valve position. (Used with permission, from E. B. Wylie and V. L. Streeter, *Fluid Transients*, FEB Press, P.O. Box 2431, Ann Arbor, Mich. 48106.)

The Reynolds number is

$$\mathrm{Re}_D = \frac{VD}{\nu} = (V\ \mathrm{ft/s})(1\ \mathrm{ft})\frac{1}{1.06\times 10^{-5}}\frac{\mathrm{s}}{\mathrm{ft}^2} = 9.434\times 10^4\ \mathrm{V}$$

and Table 6-1 can be generated. Thus, the steady-state velocity is 5.96 ft/s and the steady-state friction factor is 0.0146. The time to reach steady state is given by Eq.

TABLE 6-1 Steady-State Solution for Example 6-1

V	Re_D	f	V_0
1	9.434×10^4	0.019	5.205
5	4.717×10^5	0.0152	5.820
6	5.660×10^5	0.0145	5.959
5.96	5.620×10^5	0.0146	5.940

(6-10) and for this example is

$$t_{ss} = 2.65\frac{LV_0}{gH} = 2.65\frac{(10{,}000\text{ ft})(5.96\text{ ft/s})}{80\text{ ft}}\frac{\text{s}^2}{32.174\text{ ft}}$$
$$= 61.36\text{ s}$$

Example 6-2

Conditions are the same as in Example 6-1 except at time t_1 a pump with an increase in head of 100 ft lbf/lbm is placed in the line. Estimate the time required to reach a steady state.

Solution. For times $t < t_1$, $V = 5.96$ ft/s. The situation is as in Fig. 6-8. The driving potential then is effectively 180 ft lbf/lbm so that H_{new} is 180 ft and

$$V_{0\text{new}} = \frac{1.076}{\sqrt{f}}$$

Proceeding as in Example 6-1, we find $V_{0\text{new}} = 9.00$ ft/s with the steady-state friction factor of 0.0143. The time required to reach $V_{0\text{old}}$ under a head of 180 ft lbf/lbm is

$$t_1 = \frac{(10{,}000)(9)}{2g(180)}\ln\frac{9+5.46}{9-5.46}$$
$$= 12.38\text{ s}$$

and the time required to reach $V_{0\text{new}}$

$$t_{c\,\text{new}} = 2.65\frac{(10{,}000)(9)}{(180g)} = 41.18\text{ s}$$

Since the velocity was 5.96 ft/s when the pump was brought on line, the time required to reach 9 ft/s is

$$\Delta t = t_{c\,\text{new}} - t_1 = 41.18\text{ s} - 12.38\text{ s} = 28.80\text{ s}$$

Example 6-3

The situation as given in Fig. 6-4 has $H_1 = 100$ ft, $H_4 = 50$ ft, $L_{12} = L_{34} = 2000$ ft, $D = 1$ ft, and $V = 5$ ft/s. This corresponds to $f = 0.03217$. Find the maximum pressure at the valve if the valve is closed in 100 s such that V varies linearly.

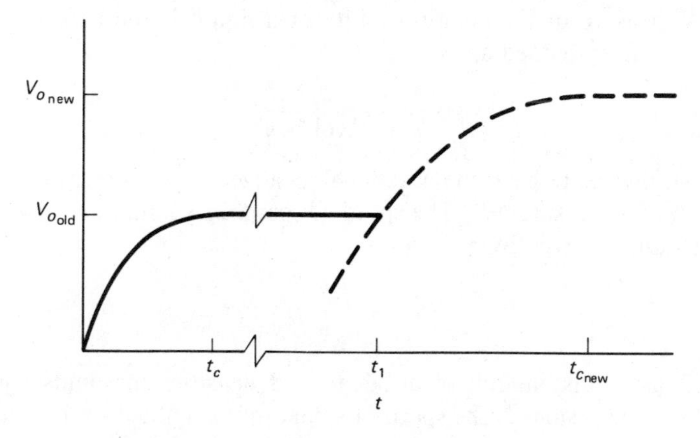

Figure 6-8 Velocity-time behavior for Example 6-2.

Solution. The change in velocity, $\dot{V}$, is

$$\dot{V} = -\frac{5\ \text{ft/s}}{100\ \text{s}} = -0.05\ \text{ft/s}^2$$

From Eq. (6-13), the valve pressure at station 2 is

$$\frac{P_2}{\gamma} = \left(H_1 + \frac{L}{g}|\dot{V}|\right) - \frac{fL}{D}\frac{V^2}{2g}$$

$$= \left(100 + \frac{2000}{32.174}|-0.05|\right) - \frac{(0.03217)(2000)}{1}\frac{V^2}{(2)(32.174)}$$

$$= (103.11 - 1.0V^2)\ \text{ft}$$

Figure 6-5 and the preceding show the maximum upstream valve pressure head to be 103.11 ft lbf/lbm. Downstream of the valve the pressure from Eq. (6-15) is

$$\frac{P_3}{\gamma} = \left(H_4 - \frac{L}{g}|\dot{V}|\right) + \frac{fL}{D}\frac{V^2}{2g}$$

$$= (46.89 + 1.0V^2)\ \text{ft}$$

During valve closure, the maximum head is 71.89 ft lbf/lbm, but just prior to valve closure the head at station 3 is 75 ft lbf/lbm. Thus, for this problem, the maximum pressure head at the valve is 103.11 ft lbf/lbm and occurs on the upstream side at the instant of valve closure.

6-3 WATERHAMMER

Implicit in the coverage in the previous section were the assumptions of an inelastic fluid and an inelastic pipe. In this section, these restrictions are removed and the resulting analysis, *waterhammer*, is examined. All liquids, even the ones usually labeled "incompressible," are compressible to some extent. A measure of the compressibility of a liquid is the bulk modulus of elasticity K. K is defined as

$$K = -\frac{\Delta P}{\Delta\text{vol}/\text{vol}} \tag{6-17}$$

and physically represents the fractional change in volume caused by the imposition of a pressure ΔP. The speed of sound in a liquid in the absence of any constraints is given by

$$a_s = \sqrt{\frac{K}{\rho}g_c} \tag{6-18}$$

Table 6-2 lists bulk moduli of elasticity and densities for fluids commonly involved in pipe systems. The speed of sound in unconfined fresh water, using Eq. (6-18) and Table 6-2, is calculated as 4856 ft/s.

TABLE 6-2 Density and Bulk Modulus of Various Fluids

Liquid	Density (lbm/ft^3)	Bulk Modulus ($lbf/in.^2$)
Benzene	54.94	1.523×10^5
Ethyl alcohol	49.32	1.914×10^5
Glycerin	78.66	6.293×10^5
Kerosene	50.19	1.914×10^5
Mercury	847.13	37.990×10^5
Oil	56.18	2.175×10^5
Fresh water	62.36	3.176×10^5
Sea water	63.99	3.29×10^5

If we had fresh water in a pipe and measured the speed with which a pressure pulse would propagate, we would measure a value considerably less than 4856 ft/s; and, in fact, we would find that the pipe material influenced the speed of propagation.

The reason for the influence lies in the coupling between the elastic fluid in the pipe and the elastic walls of the pipe. Consider the case, as pictured in Fig. 6-9, of a pressure head change ΔH propagating at some speed a (not a_s) along a pipe. The elastic coupling between the fluid and pipe, the *pipe bulge*, causes the head change ΔH and the bulge to propagate at a speed a. The relationship between the head change and the velocity change ΔV is derived by Watters (5) (and others) and is

$$\Delta H = \frac{a}{g} \Delta V \tag{6-19}$$

It is apparent from Eq. (6-19) that the speed of propagation, a, plays an important role in waterhammer analysis.

References 1 to 6 all contain derivations of the waterhammer wave speed, a. The results presented here are taken from Chaudhry (6). The general expression for the waterhammer wave velocity is

$$a = \sqrt{\frac{K}{\rho[1 + (K/E)c]} g_c} \tag{6-20}$$

where E is Young's modulus for the pipe (or conduit) wall and c is a nondimensional parameter depending on the conduit's elastic properties and

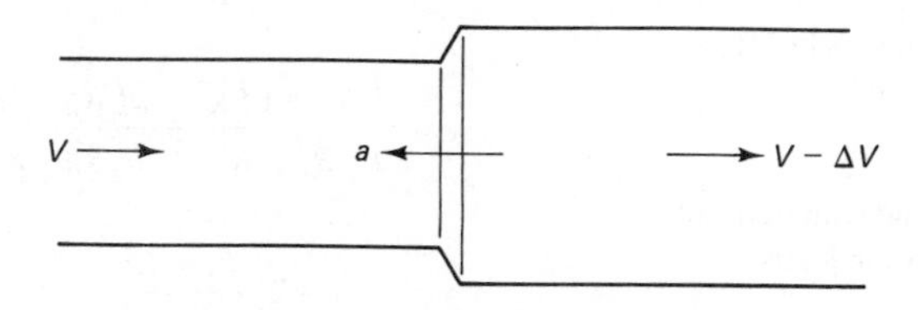

Figure 6-9 Waterhammer nomenclature.

restraints. Table 6-3 lists a few of the more important forms of c. The symbol μ is Poisson's ratio. Table 6-4 lists moduli of elasticity and Poisson's ratios for a number of common pipe materials.

A phenomenological examination of the progression of events with waterhammer wave passage is in order. Consider the sequence of events schematically portrayed in Fig. 6-10. At time $t = 0$ the pipe is flowing at a velocity V when the valve is suddenly closed. Friction effects are neglected. The sudden closing of the valve forces the water velocity at the valve to suddenly become zero. This can only happen if the head at the valve is suddenly increased by an amount ΔH.

This increase in pressure at the valve results in the swelling of the pipe near the valve (shown greatly exaggerated in Fig. 6-10) and the propagation of a pressure wave moving at a speed a [given by Eq. (6-20)] toward the inlet. Behind the moving wave the velocity is zero, and the head is increased by an amount ΔH. At $t = L/a$ the wave has reached the pipe inlet, the velocity in the pipe is everywhere zero, and the head is $H + \Delta H$.

TABLE 6-3 Forms of c for Waterhammer Wave Speed

Description	Form of c	Eq.
1. Rigid conduits	$c = 0$	(6-21)
2. Thin-walled elastic conduits (e = wall thickness)		
a. Conduit anchored against longitudinal movement	$c = \frac{D}{e}(1 - \mu^2)$	(6-22)
b. Conduit anchored at upper end	$c = \frac{D}{e}(1.25 - \mu)$	(6-23)
c. Conduit with frequent expansion joints	$c = \frac{D}{e}$	(6-24)
3. Thick-walled elastic conduits (R_0 = external radii; R_i = internal radii)		
a. Conduit anchored against longitudinal movement	$c = 2(1 + \mu)\frac{R_0^2 + R_i^2}{R_0^2 - R_i^2} - \frac{2\mu R_i^2}{R_0^2 - R_i^2}$	(6-25)
b. Conduit anchored at upper end	$c = 2\left[\frac{R_0^2 + 1.5R_i^2}{R_0^2 - R_i^2} + \frac{\mu(R_0^2 - 3R_i^2)}{R_0^2 - R_i^2}\right]$	(6-26)
c. Conduit with frequent expansion joints	$c = 2\left(\frac{R_0^2 + R_i^2}{R_0^2 - R_i^2} + \mu\right)$	(6-27)

TABLE 6-4 Moduli of Elasticity and Poisson's Ratios for Pipe Material

Material	E	μ
Steel	$E = 30 \times 10^6$ psi	$\mu \approx 0.30$
Ductile cast iron	$E = 24 \times 10^6$ psi	$\mu \approx 0.28$
Copper	$E = 16 \times 10^6$ psi	$\mu \approx 0.30$
Brass	$E = 15 \times 10^6$ psi	$\mu \approx 0.34$
Aluminum	$E = 10.5 \times 10^6$ psi	$\mu \approx 0.33$
PVC	$E \approx 4 \times 10^5$ psi	$\mu \approx 0.45$
Fiberglass reinforced plastic (FRP)	$E_2 = 4.0 \times 10^6$ psi	$\mu_2 = 0.27$–0.30
	$E_1 = 1.3 \times 10^6$ psi	$\mu_1 = 0.20$–0.24
Asbestos cement	$E \approx 3.4 \times 10^6$ psi	$\mu \approx 0.30$
Concrete	$E = 57{,}000\sqrt{f_c'}$	$\mu \approx 0.30$

$f_c' =$ 28-day strength.
Used with permission, from G. Z. Watters, *Modern Analysis and Control of Unsteady Flow in Pipelines*, Ann Arbor Science Publishers, Inc., 1979.

Thus, at $t = L/a$, the fluid in the pipe has zero velocity and is pressurized to $H + \Delta H$ while the head just outside the inlet is H.

Under these conditions, flow from the pipe to the reservoir is initiated as the distended pipe's strain energy ejects the fluid. A waterhammer wave propagates from the inlet (starting at $t = L/a$) to the valve (arriving at $t = 2L/a$), decreasing the head from $H + \Delta H$ to H while accelerating the fluid to a velocity V. Time $t = 2L/a$ finds the pipe at a head of H and the fluid moving with a velocity V toward the inlet.

Since at $t = 2L/a$ no source of fluid is available at the valve, the waterhammer wave (which is an expansion wave) must reflect from the valve (as an expansion wave) to force the velocity to zero. This wave propagates from the valve (starting at $t = 2L/a$) toward the inlet (arriving at $t = 3L/a$), decreasing the head to $H - \Delta H$, shrinking the pipe diameter, and establishing zero velocity within the pipe.

The lower head within the pipe at time $t = 3L/a$ induces a flow from the reservoir into a pipe. This is accomplished by a waterhammer wave propagating from the inlet (starting at $t = 3L/a$) to the valve (arriving at $t = 4L/a$), increasing the pressure from $H - \Delta H$ to H, and accelerating the flow toward the valve with a velocity V. Thus, at $t = 4L/a$ the velocity within the pipe is V (moving toward the valve) and the head is H. But these are the conditions at $t = 0$!

The instantaneous valve closure waterhammer wave response has a period of $4L/a$ and in the absence of friction would repeat indefinitely. All the pressure and velocity changes occur because of the passage of the wave; and, as a result, the time required for wave passage determines the pressure (head) and velocity at a given location in the pipe.

What would happen if the valve were closed in n increments, each increasing the head by $\Delta H/n$? If the closure were accomplished in less time

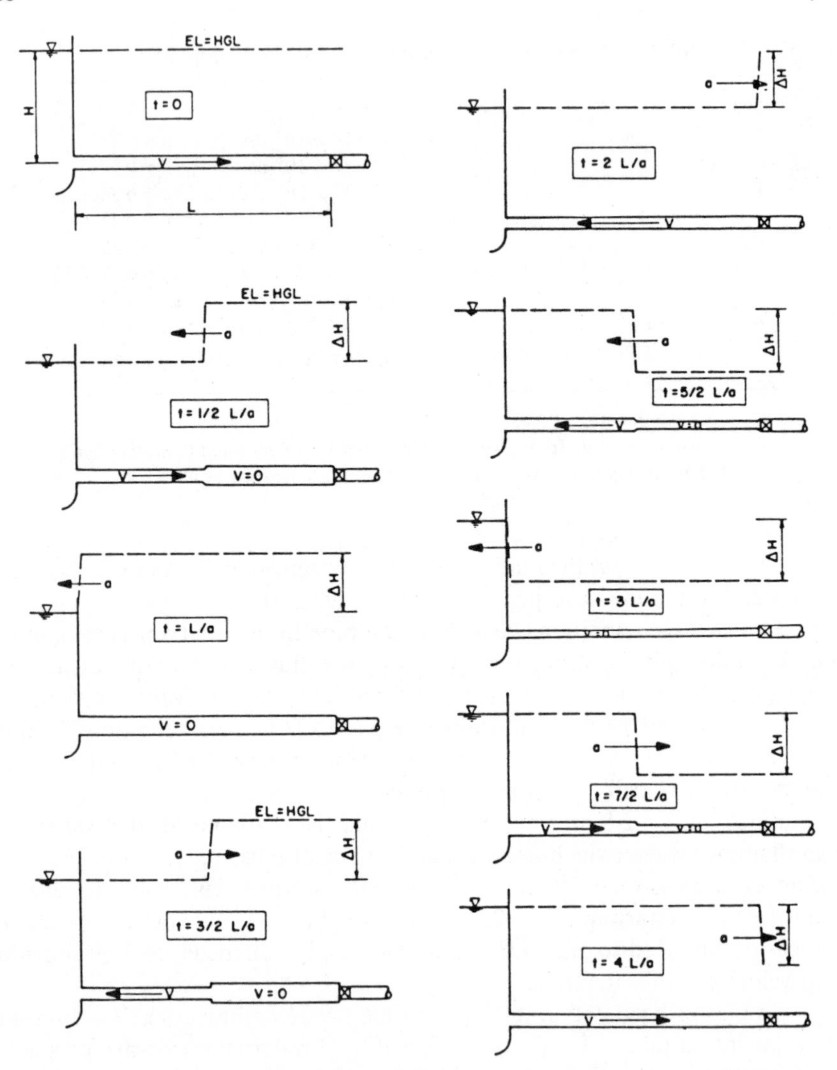

Figure 6-10 Waterhammer wave propagation sequence. (Used with permission, from G. Z. Watters, *Modern Analysis and Control of Unsteady Flow in Pipelines*, Ann Arbor Science Publishers, Inc. 1979.)

than $2L/a$, the results would be essentially unchanged as the head at the valve would still build up to $H + \Delta H$, since no "relief" from the reservoir could arrive prior to $t = 2L/a$. A generalization of this says that any valve closure protocol taking less time than $2L/a$ will always result in the full waterhammer pressure. In reality, valve closing times much *greater* than $2L/a$ may be required to prevent high waterhammer wave pressures from occurring.

The preceding is also useful in establishing general guidelines for applicability of the rigid theory discussed in the previous section. If events occur slowly enough to permit the gradual relief of the pressure head, then rigid theory may be applied. If L/a is taken as the characteristic time involved, then rigid theory can be applied if the time scale of the event is much longer than L/a. This corresponds to situations in which "small" finite pressure waves are able to traverse the pipe many times while the event (valve closing or opening, for example) is occurring. Such traverses will prevent the waterhammer pressure wave from building to large magnitudes, and the assumptions of rigid theory will be satisfied.

Wylie and Streeter list 11 events (or boundary conditions) that may require a waterhammer analysis:*

1. Changes in valve settings, accidental or planned.
2. Starting or stopping of pumps.
3. Changes in power demand in turbines.
4. Action of reciprocating pumps.
5. Changing elevation of a reservoir.
6. Waves on a reservoir.
7. Turbine governor hunting.
8. Vibration of impellers or guide vanes in pumps, fans, or turbines.
9. Vibration of deformable appurtenances such as valves.
10. Draft-tube instabilities due to vortexing.
11. Unstable pump or fan characteristics.

A full waterhammer analysis requires the solution of two coupled partial differential equations:

$$\frac{dV}{dt} + \frac{g_c}{\rho}\frac{\partial P}{\partial s} + g\frac{dz}{ds} + \frac{f}{2D}V|V| = 0 \tag{6-28}$$

$$\frac{a^2}{g_c}\frac{\partial V}{\partial s} + \frac{1}{\rho}\frac{dP}{dt} = 0 \tag{6-29}$$

The method of characteristics is the preferred method of solving these equations. This approach is readily adopted to the digital computer and is developed and applied in detail in references 1 to 6. The fidelity of the method of characteristics is shown in Fig. 6-11 (Wylie and Streeter). In this figure a pressure-regulating valve is closed, and the resulting calculated values of pressure head and flow rate are compared with measured results. Most of the complex nuances are faithfully reproduced, although both magnitudes and times exhibit decreasing accuracy as the event unfolds. The potency of this

*Used with permission from E. B. Wylie and V. L. Streeter, *Fluid Transients*, FEB Press, P.O. Box 2431, Ann Arbor, Mich. 48106.

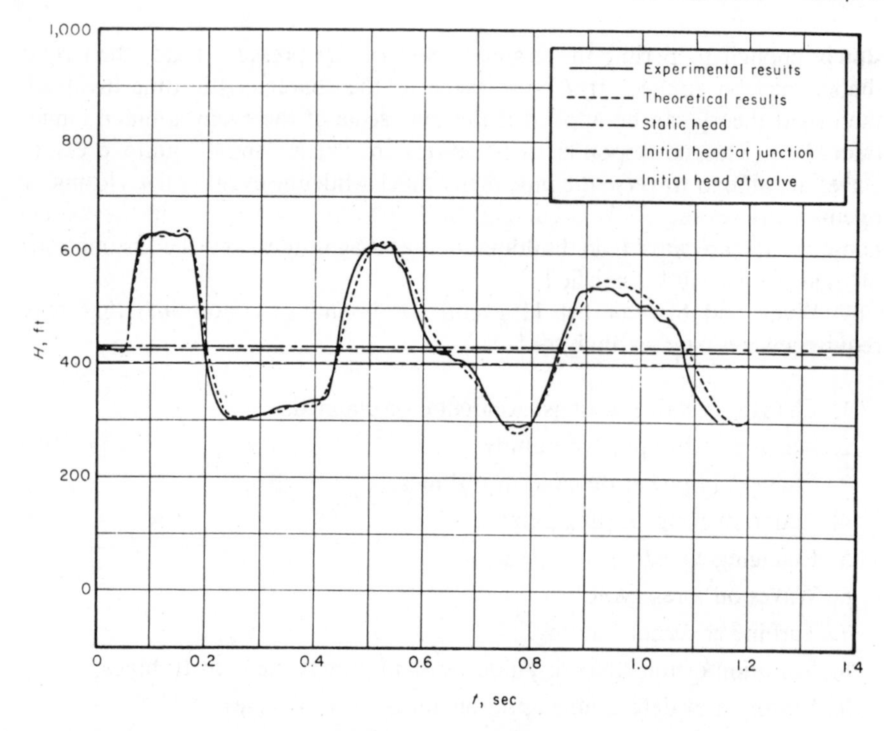

Figure 6-11 Measured and computed method of characteristics results. (Used with permission, from E. B. Wylie and V. L. Streeter, *Fluid Transients*, FEB Press, P.O. Box 2431, Ann Arbor, Mich. 48106.)

method is obvious, and any further study of waterhammer should be based on the method of characteristics.

6-4 WATERHAMMER ANALYSIS

Defining dz/dx as $\sin\theta$, the two governing partial differential equations for waterhammer analysis introduced in the previous section can be written (with x replacing s) as

$$L_1 = \frac{\partial V}{\partial t} + \frac{g_c}{\rho}\frac{\partial P}{\partial x} + g\sin\theta + \frac{f|V|V}{2D} = 0 \tag{6-30}$$

and

$$L_2 = \frac{\partial P}{\partial t} + \frac{\rho a^2}{g_c}\frac{\partial V}{\partial x} = 0 \tag{6-31}$$

The method of characteristics (MOC) approach is the generally preferred solution technique for this system of equations. Virtually all partial differential equation textbooks have sections examining in detail (and rigor) the MOC. We shall take a much more pragmatic view of this method. Additional information is also readily available in References 1–6 of this chapter. Most of these references formulate the equations in terms of Q (flow rate) and H (head) or V (velocity) and H (head). We shall continue to use V and P as the dependent variables.

Equations (6-30) and (6-31) form a coupled system for the velocity V and the pressure P; thus, any two independent linear combinations of these equations would also form a coupled system of two equations relating V and P. Hence, for two *independent* values of λ, Eqs. (6-30) and (6-31) can be written as

$$L = L_1 + \lambda L_2 \tag{6-32}$$

with the result

$$L = \left(\frac{\partial V}{\partial x} \lambda \frac{\rho a^2}{g_c} + \frac{\partial V}{\partial t} \right) + \lambda \left(\frac{\partial P}{\partial x} \frac{g_c}{\rho \lambda} + \frac{\partial P}{\partial t} \right) + g \sin\theta + \frac{fV|V|}{2D} \tag{6-33}$$

For $V = V(x, t)$, we can write

$$\frac{dV}{dt} = \frac{\partial V}{\partial x} \frac{dx}{dt} + \frac{\partial V}{\partial t} \tag{6-34}$$

Then if

$$\frac{dx}{dt} = \frac{\lambda \rho a^2}{g_c} \tag{6-35a}$$

and

$$\frac{dx}{dt} = \frac{g_c}{\rho \lambda} \tag{6-35b}$$

the equation for L [Eq. (6-33)] becomes the ordinary differential equation

$$L = \frac{dV}{dt} \pm \frac{g_c}{\rho a} \frac{dP}{dt} + g \sin\theta + \frac{fV|V|}{2D} = 0 \tag{6-36}$$

with

$$\lambda = \pm \frac{g_c}{\rho a} \quad \text{and} \quad \frac{dx}{dt} = \pm a \tag{6-37}$$

The terms $dx/dt = \pm a$ are called the *characteristics* of Eq. (6-33) since they specify the relationship between x and t for which Eqs. (6-30) and (6-31) reduce to the ordinary differential equation, Eq. (6-36).

The basis of the method of characteristics is that by moving along the characteristics, $dx/dt = \pm a$, Eq. (6-36) can be simply integrated to recover the solution $V = V(x, t)$ and $P = P(x, t)$. The MOC is directly applicable to partial differential equations that are *hyperbolic*. These are, in general, initial-value problems where a known solution at time t (for all x) is advanced to

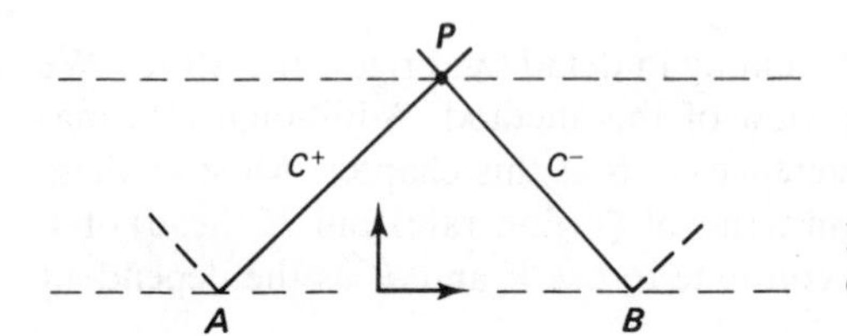

Figure 6-12 Left- and right-running characteristic nomenclature.

time $t+\Delta t$ (for all x). Let us then consider the situation as depicted in Fig. 6-12. At points A and B (time t) the velocities and pressures are known. We seek to find the velocity V_P and pressure P_P at point P (time $t+\Delta t$). Adopting the usual nomenclature, we take C^+ to be the right-running characteristic, $dx/dt = +a$, and C^- to be the left-running characteristic, $dx/dt = -a$. Along C^+,

$$\frac{dV}{dt} + \frac{g_c}{\rho a}\frac{dP}{dt} + g\sin\theta + \frac{fV|V|}{2D} = 0 \tag{6-38}$$

which upon multiplication by $\rho a\,dt$ and integration from point A to point P yields

$$\rho a(V_P - V_A) + g_c P_P - g_c P_A + \rho g\sin\theta\,\Delta x + \rho\frac{f\Delta x}{2D}V_A|V_A| = 0 \tag{6-39}$$

with $a\,\Delta t = \Delta x$. Here V is taken constant as V_A for the integration. Equation (6-39) has V_P and P_P as unknowns. Grouping the known quantities together as

$$C_A = \rho g\sin\theta\,\Delta x + \rho\,\Delta x\frac{f}{2D}|V_A|V_A - \rho a V_a - g_c P_A \tag{6-40a}$$

then Eq. (6-39) becomes

$$\rho a V_P + g_c P_P + C_A = 0 \tag{6-41a}$$

In a similar fashion we define along C^-

$$C_B = \rho g\sin\theta\,\Delta x + \rho\,\Delta x\frac{f}{2D}|V_B|V_B - \rho a V_B - g_c P_B \tag{6-40b}$$

and

$$\rho a V_P - g_c P_P + C_B = 0 \tag{6-41b}$$

Equations (6-41a) and (6-41b) form a system of two equations for V_P and P_P; solving, we obtain

$$P_P = \frac{C_B - C_A}{2g_c} \tag{6-42a}$$

$$V_P = -\frac{C_A + C_B}{2\rho a} \tag{6-42b}$$

Hence, given values of V and P at A and B, we can find P_P and V_P at time $t+\Delta t$ by using Eqs. (6-42a) and (6-42b). This general procedure will suffice for all *interior* points.

Consider the general computational grid for a single pipe as shown in Fig. 6-13. At time level t, C_B and C_A can be evaluated at all points, 1 to N. Then Eqs. (6-42a) and (6-42b) will yield the new pressures $PN(I)$ and the new velocities $VN(I)$ for points 2 to $N-1$. Then we only need to obtain the solution at points 1 and N, the beginning and ending points of the pipe.

Since at points 1 and N only a single characteristic "runs" from the known solution at time t, additional information must be known. This is where the boundary conditions are imposed. Consider $PN(1)$ and $VN(1)$ (pressure and velocity at point 1 at time $t+\Delta t$). The C^- characteristic from point 2, time t, is

$$\rho a VN(1) - g_c PN(1) + CB(2) = 0 \tag{6-43a}$$

In Eq. (6-43a), $CB(2)$ is Eq. (6-40b) evaluated at point 2, time t, and is known. Likewise, at point N, the C^+ characteristic from $N-1$ at time t becomes

$$\rho a VN(N) + g_c PN(N) + CA(N-1) = 0 \tag{6-43b}$$

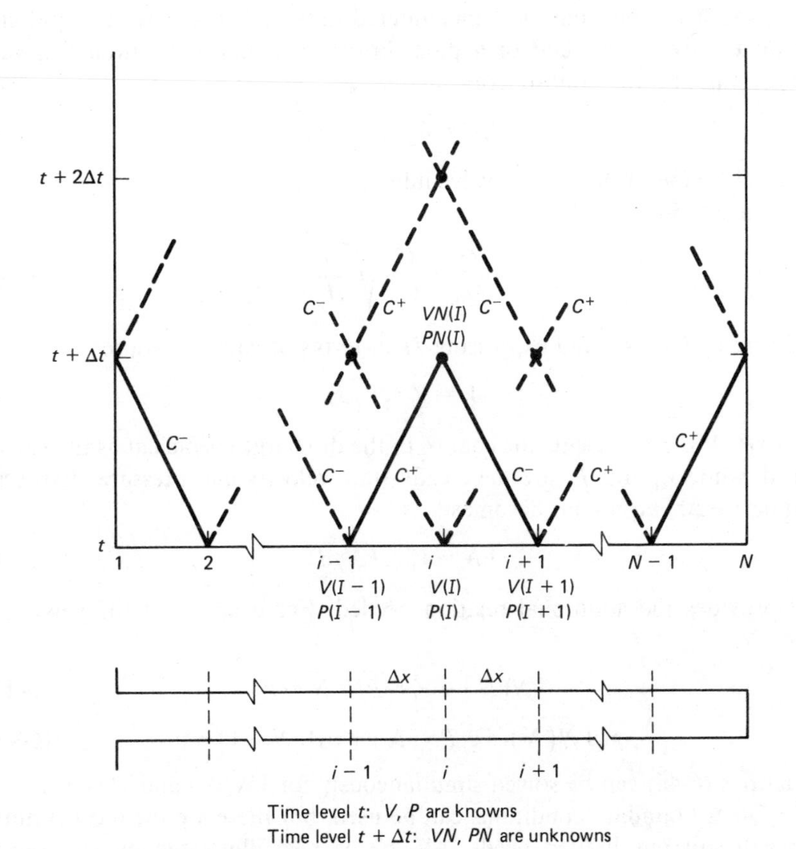

Figure 6-13 Computational grid for waterhammer analysis.

with $CA(N-1)$ known and $VN(N)$ and $PN(N)$ unknown. The situations in Eqs. (6-43a) are thus identical: two unknowns in one equation.

In either of Eqs. (6-43), the specification of either VN or PN will determine the other. We must bring in the boundary conditions in terms of VN or PN at points 1 and N for time level $t + \Delta t$. For example, if point 1 is at a constant pressure (or head) inlet from a reservoir, then $PN(1)$ is known and Eq. (6-43a) can be solved for $VN(1)$. If the reservoir were at point N, then $PN(N)$ would be known and Eq. (6-43b) could be solved for $VN(N)$. The pattern for imposition of boundary conditions is thus clear: at the end points, the boundary condition must be used as an additional equation so that two equations with unknowns form a solvable system. Several cases need to be examined.

As we have already seen, the imposition of a pressure boundary condition is rather simple. At either point 1 or N, PN is known and either Eq. (6-41a) or Eq. (6-41b) can be solved for VN. In fact, if the pressure varies with time, the same approach can be used.

Another likely condition encountered in waterhammer is the rapid closing of a valve of one end of a pipe. For steady-state flow through a valve discharging into the atmosphere

$$Q_0 = C_{d_0} A \sqrt{2gH_0} \tag{6-44}$$

where H_0 is the "head" at the valve inlet in steady flow. As the valve is closed, C_d changes so that

$$\frac{Q}{Q_0} = \frac{C_d}{C_{d_0}} \sqrt{\frac{H}{H_0}} \tag{6-45}$$

Defining $C_d/C_{d_0} = \tau$ and expressing H in terms of gauge pressures,

$$V = V_0 \tau \sqrt{P/P_0} \tag{6-46}$$

In Eq. (6-46), τ represents the change in the discharge coefficient as the valve is closed, while V_0 and P_0 are the steady-state velocity and pressure. Expressed at time $t + \Delta t$, Eq. (6-46) becomes

$$VN = V_0 \tau \sqrt{PN/P_0} \tag{6-47}$$

and provides the additional relation needed. For example, if the valve is at point N,

$$VN(N) = V_0 \tau \sqrt{PN(N)/P_0} \tag{6-48a}$$

$$\rho a VN(N) + g_c PN(N) + CA(N-1) = 0 \tag{6-48b}$$

Equations (6-48) can be solved simultaneously for $VN(N)$ and $PN(N)$.

Other boundary conditions can be used, but these are the most common and will suit our limited needs. All this is best illustrated by an example problem.

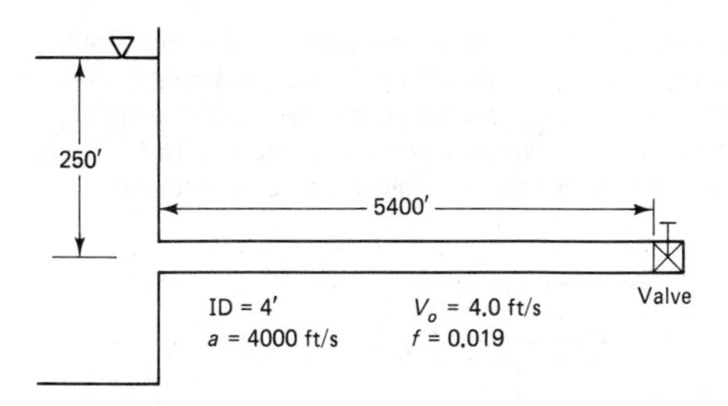

Figure 6-14 System schematic for Example 6-4.

Example 6-4

A pipe 4 ft in diameter is 5400 ft long and at steady state has a velocity of 4 ft/s with a friction factor of 0.019. The inlet to the pipe is located 250 ft below the surface, and the pipe exit, which discharges to the atmosphere, contains a valve. The fluid has a density of 62.4 lbm/ft^3, and the waterhammer wave velocity is 4000 ft/s. If the value of τ is given by

$$\tau = \begin{cases} (1 - t/6.0)^3 & t < 6.0 \\ 0 & t \geq 6.0 \end{cases}$$

use the method of characteristics to analyze the effect of the valve closure.

Solution. Figure 6-14 is a schematic of the system under consideration. N is chosen as 10 so that

$$Dx = \frac{L}{N-1} = 600 \text{ ft}$$

and

$$\Delta t = \frac{\Delta x}{a} = 600 \text{ ft}/4000 \text{ ft/s} = 0.15 \text{ s}$$

and the pipe appears as in Fig. 6-15. In this example $P(1)$ is always 15,600 lbf/ft^2 and thus provides the boundary condition for the left-hand end. $VN(1)$ is then calculated from Eq. (6-43a). At the right end, $N = 10$; $VN(10)$ and $PN(10)$ are calculated by solution of Eqs. (6-48). The conditions (initial) at $t = 0$ are $V = 4.0$ ft/s for all points and $P(1) = 15{,}600$ lbf/ft^2. For each Δx reach of the pipe, the initial pressure decreases by the amount

$$\Delta P_{\Delta x} = \frac{f \Delta x}{D} \frac{V^2}{2 g_c} \rho = 44.2 \text{ lbf/ft}^2$$

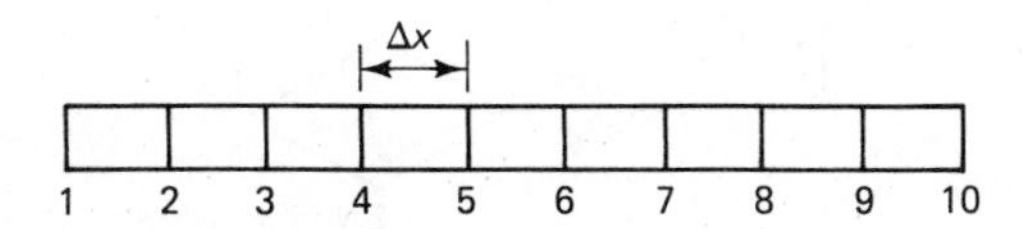

Figure 6-15 Pipe nomenclature for Example 6-4.

A block diagram for a computer program to solve this problem is presented in Fig. 6-16. The diagram is self-explanatory. Initial conditions on velocity and pressure are set, CA and CB calculated for all points, the interior points ($i = 2 - 9$) are then advanced, and finally the boundary conditions are imposed. Needed quantities are printed at $t + \Delta t$, and the process is repeated until t_{max} is reached.

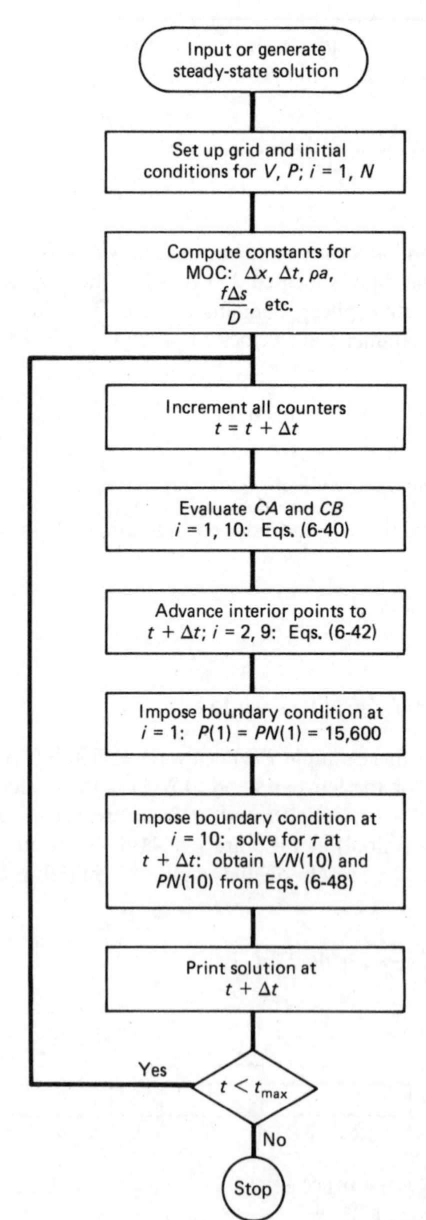

Figure 6-16 Block diagram for water-hammer problem

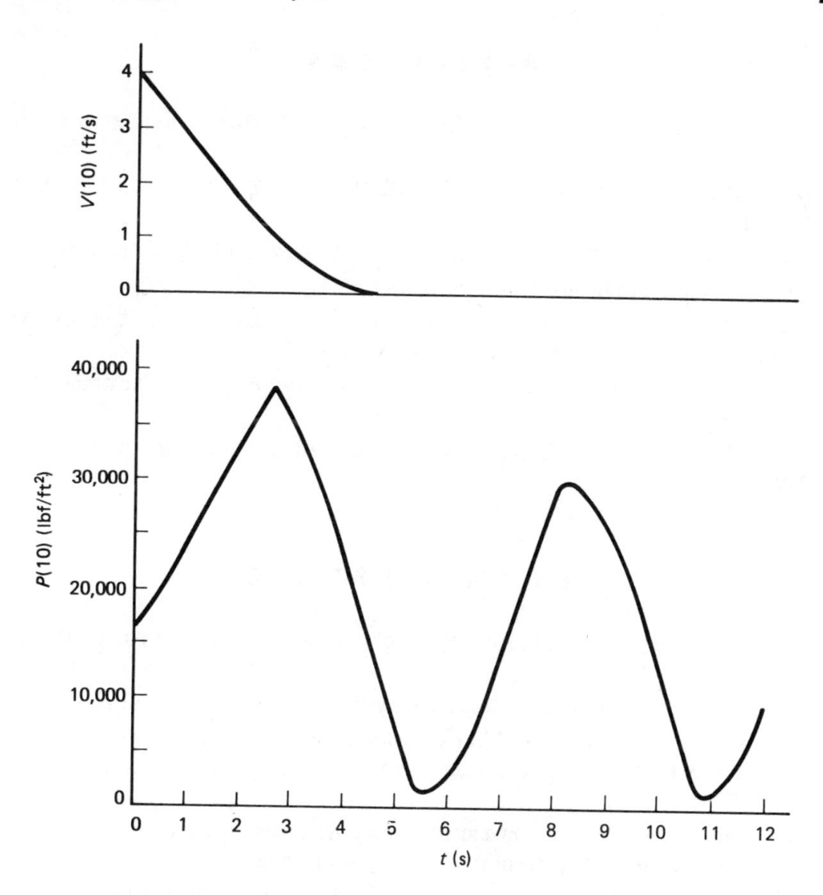

Figure 6-17 Time variation of the valve pressure for a closing valve.

At each time step, the time, as well as the velocity and pressure, at each i is available. Figure 6-17 was generated from the printout of the results. The valve closing protocol results in a maximum pressure of 38,246 lbf/ft^2 at the valve inlet 2.7 s after the initiation of valve closure. At that time the velocity at the valve inlet is near 1 ft/s. This is more than twice the steady-state pressure and should remind us that designing to steady-state pressure may not be sufficient if such severe transients are likely. The maximum pressure occurs before valve closure is complete; the waterhammer continues as waterhammer expansion, and compression waves traverse the pipe.

In this section we have examined the analysis of waterhammer. Obviously, this coverage is just an introduction to waterhammer analysis, but the potency of the method of characteristics is evident even for the simple example. A particularly readable introduction to waterhammer analysis is provided by Streeter and Wylie (7), while references 1 to 6 of this chapter present extensive techniques and extensions for a wide range of transient fluid problems.

REFERENCES

1. Streeter, V. L., and Wylie, E. B., *Hydraulic Transients*. New York: McGraw-Hill, 1967.
2. Fox, J. A., *Hydraulic Analysis of Unsteady Flow in Pipe Networks*. New York: Macmillan, 1977.
3. Wylie, E. B., and Streeter, V. L., *Fluid Transients*. New York: McGraw-Hill, 1977.
4. Jaeger, C., *Fluid Transients*. Glasgow: Blackie, 1977.
5. Watters, G. Z., *Modern Analysis and Control of Unsteady Flow in Pipelines*. Ann Arbor, Mich.: Ann Arbor Science Publishers, 1979.
6. Chaudhry, M. H., *Applied Hydraulic Transients*. New York: Van Nostrand Reinhold, 1979.
7. Streeter, V. L., and Wylie, E. B., *Fluid Mechanics*, 7th ed. New York: McGraw-Hill, 1979.

REVIEW QUESTIONS

1. If $L/a = 0.030$ s, what is the period of the waterhammer wave pattern for frictionless flow?

2. What assumptions are made in rigid theory that are not made in waterhammer?

3. Why is rigid theory a reasonable model for events (such as a valve closing) that happen slowly, while waterhammer analysis is needed for events that happen rapidly?

4. For rigid theory, why is it convenient to base all times on the time required for a flow to accelerate to 99 percent of its steady-state value?

5. All other factors being the same, what is the effect on a, the waterhammer wave velocity, of increasing wall thickness of a thin-walled elastic conduit?

6. Why is valve closing in greater than $2L/a$ often referred to as "slow" closing, while valve closing in less than $2L/a$ is often called "rapid" closing?

7. What is the method of characteristics?

8. What is a physical interpretation of the characteristics C^+ and C^-?

9. If a device with the velocity–pressure relationship $P = a_0 + a_1V + a_2V^2$ is placed at the end of a pipe, explain how the boundary conditions could be imposed.

10. For a completed waterhammer analysis, explain how the x–t plane would look.

PROBLEMS

1. A hot water line 50 ft long and $\frac{3}{4}$ in. in diameter is made of cast iron. The water is at 150°F and is flowing at 5 ft/s when the valve is suddenly cut off.

(a) At what speed will the resulting wave propagate?

(b) How long will it take the wave to reach the other end of the pipe?

(c) What might be the frequency of the resulting vibration?

(d) What does this say about the noise you hear in a house when a faucet is abruptly shut off?

2. A pipe line ($V = 6$ ft/s, $D = 1$ ft, $H = 100$ ft lbf/lbm, $L = 8000$ ft) has its head increased to 300 ft lbf/lbm. How long will it take for the flow to accelerate from zero to the steady-state velocity for the initial H (100 ft lbf/lbm)? How long for the new steady-state velocity to be attained?
3. A hydraulic turbine requires 2000 gpm flow ($T = 80°$F) through a 10,000-ft-long schedule 80 cast-iron pipe. The available head is 100 ft lbf/lbm. The pipe network is located in the horizontal plane.
   (a) Find the nominal diameter pipe required.
   (b) Cooling restrictions require the flow to accelerate from rest to 99% of the final steady-state velocity in 30 s or less. Will the results of part (a) satisfy the requirements?
4. A valve is rapidly opened on a 6000-ft segment of an ID = 18 in. water line. The allowable pressure drop is 40 ft lbf/lbm with minor losses estimated at $10V^2/2g_c$. How long will it take to reach 99 percent of its steady-state value? $\varepsilon/D = 0.0015$.
5. Kerosene flows through a nominal 1-in. schedule 40 steel pipe. Determine the waterhammer wave velocity.
6. A valve is closed in 50 s at the end of a 10,000-ft pipeline carrying water at 5 ft/s. Using rigid theory, find the peak pressure (value, location, time) developed by the closure.
7. Develop $\tau = \tau(t)$ for the offset butterfly valve of Fig. 6-7. Closure time is 15 s.
8. Develop a computer program to implement the waterhammer analysis discussed in Example 6-4.
9. Oil is flowing at 5 ft/s through a 50,000-ft segment of a 2-ft ID steel pipeline. The wall thickness is $\frac{1}{2}$ in. and the conduit is anchored against longitudinal movement. Sufficient pressure at the upstream end is maintained to provide a downstream pressure of 125 psia. During the closure of a valve located at the downstream end, the value of τ is $(1 - t/10.0)^{1.5}$. What is the maximum pressure due to transient effects?
10. Rework Example 6-4 with the valve closing protocol given as $\tau = 0.75$ for $0 \leq t \leq 2$ s, $\tau = 0.4$ for $2 \leq t \leq 5$ s, $\tau = 0.1$ for $5 \leq t \leq 7$ s, and $\tau = 0$ for $t > 7$ s.

Appendix A

CONVERSION FACTORS

Much engineering computation done in the United States is based on English units (ft s lbf). In spite of the considerable activity promoting the complete use of the *Systeme International d'Unites* (the SI system), the English units will continue to be extensively used in many sectors. Unless mandated by law, the English units will remain the standard accepted system in many other sectors of engineering practice. Thus, for the foreseeable future, conversion factors from one system to another will be needed.

The conversion factors appearing herein were taken from Lienhart (A-1). Table A-1 presents the elementary units and conversion factors. Table A-2 provides conversion factors for a number of derived units.

REFERENCE

A-1. Lienhart, J. H., *A Heat Transfer Textbook*. Englewood Cliffs, N.J.: Prentice-Hall, 1981.

TABLE A-1 Elementary Units and Conversion Factors

Quantity	Units		Conversion Factor ≡ 1		S.I. or Metric Equivalent
	S.I.	English	S.I. to English	English to S.I.	
Length	Meter (m)	Foot (ft)	$3.2808 \frac{ft}{m}$	$0.3048 \frac{m}{ft}$	$m = 10^2 cm = 10^3 mm = 10^{10}$ Å
Time		Second (s)			
Mass	Kilogram (kg)	Pound mass (lb_m)[a]	$2.2046 \frac{lb_m}{kg}$	$0.4539 \frac{kg}{lb_m}$	$kg = 10^3 g$
Force	Newton (N)	Pound force (lb_f)	$0.2248 \frac{lb_f}{N}$	$4.4482 \frac{N}{lb_f}$	$N = kg - m/s^2 = 10^5$ dyn
Energy	Kilojoule (kJ)	British thermal[b] unit (Btu)	$0.94783 \frac{Btu}{kJ}$	$1.05504 \frac{kJ}{Btu}$	$kJ = 10^3 J =$ 238.8 cal
Temperature	Degree Celsius (°C)	Degree Fahrenheit (°F)			$T°C = T°K - 273.15$
	Degree Kelvin (°K)	Degree Rankine (°R)	$1.8 \frac{°R}{°K}$	$\frac{5}{9} \frac{°K}{°R}$	$T°K = T°C + 273.15$

[a]Also 14.594 kg/slug.
[b]Also 1.3558 J/ft-lbf.
Used with permission, J. H. Lienhart, *A Heat Transfer Textbook*, Prentice-Hall, Inc., 1981.

TABLE A-2 Conversion Factors for Derived Units

Quantity	Symbol	Conversion Factor ≡ 1: S.I. to English	Conversion Factor ≡ 1: English to S.I.	S.I. or Metric Equivalent
Area	A	$10.764 \frac{ft^2}{m^2}$	$0.092903 \frac{m^2}{ft^2}$	$m^2 = 10^4\ cm^2$
Volume	V	$35.3134 \frac{ft^3}{m^3}$	$0.028317 \frac{m^3}{ft^3}$	$m^3 = 1000$ liters $= 10^6\ cm^3$
Velocity	u	$3.2808 \frac{ft/s}{m/s}$	$0.3048 \frac{m/s}{ft/s}$	$m/s = 3.6\ km/hr$
Acceleration	g	$0.3048 \frac{ft/s^2}{m/s^2}$	$3.2808 \frac{m/s^2}{ft/s^2}$	$m/s^2 = 100\ cm/s^2$
Density	$\rho \equiv 1/v$	$0.06243 \frac{lb_m\text{-}ft^3}{kg/m^3}$	$16.018 \frac{kg/m^3}{lb_m/ft^3}$	$kg/m^3 = 1000\ g/cm^3$
Specific volume	v	$16.018 \frac{ft^3/lb_m}{m^3/kg}$	$0.06243 \frac{m^3/kg}{ft^3/lb_m}$	$m^3/kg = 0.001\ cm^3/g$
Pressure	p	$0.00014504 \frac{psi}{N/m^2}$	$6894.8 \frac{N/m^2}{psi}$	$N/m^2 \equiv$ pascal $= 10^{-5}$ bar $= 0.98692 \times 10^{-5}$ atm
Heat rate	Q	$3.4121 \frac{Btu/hr}{W}$	$0.29307 \frac{W}{Btu/hr}$	$W = J/s$
Heat rate per unit length	Q/L	$1.0403 \frac{Btu/ft\text{-}hr}{W/m}$	$0.9613 \frac{W/m}{Btu/ft\text{-}hr}$	$W/m = J/m\text{-}s$
Heat flux	$q = Q/A$	$0.3170 \frac{Btu/ft^2\text{-}hr}{W/m^2}$	$3.154 \frac{W/m^2}{Btu/ft^2\text{-}hr}$	$W/m^2 = J/m^2\text{-}s$
Volumetric heat generation	$\dot{q}$	$0.096623 \frac{Btu/ft^3\text{-}hr}{W/m^3}$	$10.35 \frac{W/m^3}{Btu/ft^3\text{-}hr}$	$W/m^3 = J/m^3\text{-}s$

TABLE A-2 (Continued)

Quantity	Symbol	Conversion Factor ≡ 1: S.I. to English	Conversion Factor ≡ 1: English to S.I.	S.I. or Metric Equivalent
Energy per unit mass	(many forms)	$0.4299 \frac{\text{Btu/lb}_\text{m}}{\text{kJ/kg}}$	$2.326 \frac{\text{kJ/kg}}{\text{Btu/lb}_\text{m}}$	kJ/kg = J/g = 0.2388 cal/g
Specific heat	c, c_p, c_v	$0.23884 \frac{\text{Btu/lb}_\text{m}\text{-°F}}{\text{kJ/kg-°C}}$	$4.1869 \frac{\text{kJ/kg-°C}}{\text{Btu/lb}_\text{m}\text{-°F}}$	kJ/kg-°C ≡ kJ/kg-°K = 0.2388 cal/g-°C
Entropy	s	Same as c_p	Same as c_p	Should only be used with °K
Thermal conductivity	k	$0.5778 \frac{\text{Btu/ft-hr-°F}}{\text{W/m-°C}}$	$1.7307 \frac{\text{W/m-°C}}{\text{Btu/ft-hr-°F}}$	
Convective, or overall, heat transfer coefficient	$h, \bar{h}, U$	$0.1761 \frac{\text{Btu/ft}^2\text{-hr-°F}}{\text{W/m}^2\text{-°C}}$	$5.6786 \frac{\text{W/m}^2\text{-°C}}{\text{Btu/ft}^2\text{-hr-°F}}$	
Dynamic viscosity	μ	$0.672 \frac{\text{lb}_\text{m}\text{/ft-s}}{\text{kg/m-s}}$	$1.4881 \frac{\text{kg/m-s}}{\text{lb}_\text{m}\text{/ft-s}}$	kg/m-s = N-s/m^2 ≡ 10 poise = 10^3 centipose
Kinematic viscosity	ν			
Thermal diffusivity	α	$10.764 \frac{\text{ft}^2\text{/s}}{\text{m}^2\text{/s}}$	$0.092903 \frac{\text{m}^2\text{/s}}{\text{ft}^2\text{/s}}$	m^2/s ≡ 10^4 stokes
Coefficient of diffusion	$\mathcal{D}$			

Used with permission, J. H. Lienhart, *A Heat Transfer Textbook*, Prentice-Hall, Inc., 1981.

Appendix B

PHYSICAL PROPERTIES

Thermophysical properties for selected solids, liquids, and gases are presented in this appendix. Many fluid mechanics and heat transfer textbooks, as well as several standards and handbooks, present tabulations of properties of various substances. Even in what we would like to think of as the "modern" era of engineering, minor disagreements in the numerical values of various thermophysical properties exist between sources. This is not surprising considering the wide variety and diverse nature of the current literature concerning property determination. As gleaned from numerous references, the work of Touloukian (B-1), Vargaftik (B-2), General Electric Company (B-3), and Irvine and Harnett (B-4) seem to be the most respected. These references should serve as the final arbiters when disagreements from other sources arise.

The thermophysical properties given herein were taken from Eckert and Drake (B-5) for metals, from Kreith (B-6) for liquids and gases, and from Chapman (B-7) for a brief selection of nonmetals. As considered in Appendix A, English units are used. The accuracy of the property data is sufficient for most purposes. Should greater accuracy or assurance be needed, references B-1 through B-4 should be consulted.

REFERENCES

B-1. Touloukian, Y. S., *Thermophysical Properties of Matter*, Vol. 1–6, 10, and 11. West Lafayette, Ind.: Purdue University, 1970–75.

B-2. Vargaftik, N. B., tables in the *Thermophysical Properties of Liquids and Gases*, 2nd ed. Washington, D.C.: Hemisphere Publishing Co., 1975.

B-3. *Heat Transfer Data Book*, Norris, R. H., Buckland, F. F., Fitzroy, N. D., Roecker, R. H., and Kaminski, D. A., eds. Schenectady, N.Y.: General Electric Co., 1977.

B-4. *Steam and Air Tables in S.I. Units*, Irvine, T. F., Jr., and Harnett, J. P., eds. Washington, D.C.: Hemisphere Publishing Co., 1976.

B-5. Eckert, E. R. G., and Drake, R. M., *Heat and Mass Transfer*, 2nd ed. New York: McGraw-Hill, 1959.

B-6. Kreith, F., *Principles of Heat Transfer*, 3rd ed. New York: Harper & Row, 1973.

B-7. Chapman, A. J., *Heat Transfer*, 3rd ed., New York: Macmillan, 1974.

TABLE B-1 Thermophysical Properties of Metals

Metal	Properties at 68°F				k, Btu/hr-ft-°F									
	ρ $\frac{lb_m}{ft^3}$	c_p $\frac{Btu}{lb_m\text{-}°F}$	k $\frac{Btu}{hr\text{-}ft\text{-}°F}$	α $\frac{ft^2}{hr}$	−148°F −100°C	32°F 0°C	212°F 100°C	392°F 200°C	572°F 300°C	752°F 400°C	1112°F 600°C	1472°F 800°C	1832°F 1000°C	2192°F 1200°C
Aluminum														
Pure	169	0.214	132	3.665	134	132	132	132	132					
Al-cu (Duralumin) 94–96 Al, 3–5 Cu, trace Mg	174	0.211	95	2.580	73	92	105	112						
Al-Mg (Hydronalium) 91-95 Al, 5-9 Mg	163	0.216	65	1.860	54	63	73	82						
Al-Si (Silumin) 87 Al, 13 Si	166	0.208	95	2.773	86	94	101	107						
Al-Si (Silumin, copper bearing) 86.5 Al; 12.5 Si; 1 Cu	166	0.207	79	2.311	69	79	83	88	93					
Al-Si (Alusil) 78-80 Al; 20-22 Si	164	0.204	93	2.762	83	91	97	101	103					
Al-Mg-Si 97 Al; 1 Mg; 1 Si; 1 Mn	169	0.213	102	2.859	—	101	109	118						
Lead	710	0.031	20	0.924	21.3	20.3	19.3	18.2	17.2					
Iron														
Pure	493	0.108	42	0.785	50	42	39	36	32	28	23	21	20	21
Wrought iron (C < 0.50%)	490	0.11	34	0.634	—	34	33	30	28	26	21	19	19	19
Cast iron (C ≈ 4%)	454	0.10	30	0.666										
Steel (C max ≈ 1.5%)														
Carbon steel C ≈ 0.5%	489	0.111	31	0.570	—	32	30	28	26	24	20	17	17	18
1.0%	487	0.113	25	0.452	—	25	25	24	23	21	19	17	16	17

Used with permission, from E. R. G. Eckert and R. M. Drake, *Heat and Mass Transfer*, McGraw-Hill Book Company, 1959.

TABLE B-1 (Continued)

Metal	Properties at 68°F				k, Btu/hr-ft-°F									
	ρ $\frac{lb_m}{ft^3}$	c_p $\frac{Btu}{lb_m\text{-}°F}$	k $\frac{Btu}{hr\text{-}ft\text{-}°F}$	α $\frac{ft^2}{hr}$	−148°F −100°C	32°F 0°C	212°F 100°C	392°F 200°C	572°F 300°C	752°F 400°C	1112°F 600°C	1472°F 800°C	1832°F 1000°C	2192°F 1200°C
Iron (Continued)														
Carbon steel (Cont.)														
1.5%	484	0.116	21	0.376	—	21	21	21	20	19	18	16	16	17
Nickel steel Ni ≈ 0%	493	0.108	42	0.785										
10%	496	0.11	15	0.279										
20%	499	0.11	11	0.204										
30%	504	0.11	7	0.118										
40%	510	0.11	6	0.108										
50%	516	0.11	8	0.140										
60%	523	0.11	11	0.182										
70%	531	0.11	15	0.258										
80%	538	0.11	20	0.344										
90%	547	0.11	27	0.452										
100%	556	0.106	52	0.892										
Invar Ni ≈ 36%	508	0.11	6.2	0.108										
Crome steel Cr = 0%	493	0.108	42	0.785	50	42	39	36	32	28	23	21	20	21
1%	491	0.11	35	0.645	—	36	32	30	27	24	21	19	19	
2%	491	0.11	30	0.559	—	31	28	26	24	22	19	18	18	
5%	489	0.11	23	0.430	—	23	22	21	21	19	17	17	17	17
10%	486	0.11	18	0.344	—	18	18	18	17	17	16	16	17	
20%	480	0.11	13	0.258	—	13	13	13	13	14	14	15	17	
30%	476	0.11	11	0.204										
Cr-Ni (chrome-nickel)														
15 Cr; 10 Ni	491	0.11	11	0.204										

TABLE B-1 (Continued)

Metal	Properties at 68°F				k, Btu/hr-ft-°F									
	ρ $\frac{lb_m}{ft^3}$	c_p $\frac{Btu}{lb_m\text{°-F}}$	k $\frac{Btu}{hr\text{-}ft\text{-°F}}$	α $\frac{ft^2}{hr}$	−148°F −100°C	32°F 0°C	212°F 100°C	392° F 200°C	572°F 300°C	752°F 400°C	1112°F 600°C	1472°F 800°C	1832°F 1000°C	2192°F 1200°C
Cr-Ni (Continued)														
18 Cr; 8 Ni (V2A)	488	0.11	9.4	0.172	—	9.4	10	10	11	11	13	15	18	
20 Cr; 15 Ni	489	0.11	8.7	0.161										
25 Cr; 20 Ni	491	0.11	7.4	0.140										
Ni-Cr (nickel-chrome)														
80 Ni; 15 Cr	532	0.11	10	0.172										
60 Ni; 15 Cr	516	0.11	7.4	0.129										
40 Ni; 15 Cr	504	0.11	6.7	0.118										
20 Ni; 15 Cr	491	0.11	8.1	0.151	—	8.1	8.7	8.7	9.4	10	11	13		
Cr-Ni-Al: 6 Cr; 1.5 Al;														
0.5 Si (Sicromal 8)	482	0.117	13	0.237										
24 Cr; 2.5 Al														
0.5 Si (Sicromal 12)	479	0.118	11	0.194										
Manganese steel Ma = 0%	493	0.118	42	0.784										
1%	491	0.11	29	0.538										
2%	491	0.11	22	0.376	—	22	21	21	21	20	19			
5%	490	0.11	13	0.247										
10%	487	0.11	10	0.194										
Tungsten steel W = 0%	493	0.108	42	0.785										
1%	494	0.107	38	0.720										
2%	497	0.106	36	0.677	—	36	34	31	28	26	21			
5%	504	0.104	31	0.591										
10%	519	0.100	28	0.527										

TABLE B-1 (Continued)

Metal	Properties at 68°F				k, Btu/hr-ft-°F									
	ρ $\frac{lb_m}{ft^3}$	c_p $\frac{Btu}{lb_m\text{-}°F}$	k $\frac{Btu}{hr\text{-}ft\text{-}°F}$	α $\frac{ft^2}{hr}$	−148°F −100°C	32°F 0°C	212°F 100°C	392°F 200°C	572°F 300°C	752°F 400°C	1112°F 600°C	1472°F 800°C	1832°F 1000°C	2192°F 1200°C
Tungsten steel (Continued)														
20%	551	0.093	25	0.484										
Silicon steel Si = 0%	493	0.108	42	0.785										
1%	485	0.11	24	0.451										
2%	479	0.11	18	0.344										
5%	463	0.11	11	0.215										
Copper														
Pure	559	0.0915	223	4.353	235	223	219	216	—	210	204			
Aluminum bronze 95 Cu; 5 Al	541	0.098	48	0.903										
Bronze 75 Cu; 25 Sn	541	0.082	15	0.333										
Red brass 85 Cu; 9 Sn; 6 Zn	544	0.092	35	0.699	—	34	41							
Brass 70 Cu; 30 Zn	532	0.092	64	1.322	51	—	74	83	85	85				
German silver 62 Cu; 15 Ni; 22 Zn	538	0.094	14.4	0.290	11.1	—	18	23	26	28				
Constantan 60 Cu; 40 Ni	557	0.098	13.1	0.237	12	—	12.8	15						
Magnesium														
Pure	109	0.242	99	3.762	103	99	97	94	91					
Mg-Al (electrolytic) 6-8% Al; 1-2% Zn	113	0.24	38	1.397	—	30	36	43	48					
Mg-Mn 2% Mn	111	0.24	66	2.473	54	64	72	75						
Molybdenum	638	0.060	79	2.074	80	79	79							

TABLE B-1 (Continued)

Metal	Properties at 68°F				k, Btu/hr-ft-°F									
	ρ $\frac{lb_m}{ft^3}$	c_p $\frac{Btu}{lb_m\text{-}°F}$	k $\frac{Btu}{hr\text{-}ft\text{-}°F}$	α $\frac{ft^2}{hr}$	−148°F −100°C	32°F 0°C	212°F 100°C	392°F 200°C	572°F 300°C	752°F 400°C	1112°F 600°C	1472°F 800°C	1832°F 1000°C	2192°F 1200°C
Nickel														
Pure (99.9%)	556	0.1065	52	0.882	60	54	48	42	37	34				
Impure (99.2%)	556	0.106	40	0.677	—	40	37	34	32	30	32	36	39	40
Ni-Cr: 90 Ni; 10 Cr	541	0.106	10	0.172	—	9.9	10.9	12.1	13.2	14.2				
80 Ni; 20 Cr	519	0.106	7.3	0.129	—	7.1	8.0	9.0	9.9	10.9	13.0			
Silver														
Purest	657	0.0559	242	6.601	242	241	240	238						
Pure (99.9%)	657	0.0559	235	6.418	242	237	240	216	209	208				
Tungsten	1208	0.0321	94	2.430	—	96	87	82	77	73	65	44		
Zinc, pure	446	0.0918	64.8	1.591	66	65	63	61	58	54				
Tin, pure	456	0.0541	37	1.505	43	38.1	34	33						

TABLE B-2 Thermophysical Properties of Liquids and Gases at Atmospheric Pressure

GASES

T (F)	ρ (lb_m/cu ft)	c_p (Btu/lb_m F)	$\mu \times 10^5$ (lb_m/ft sec)	$\nu \times 10^3$ (sq ft/sec)	k (Btu/hr ft F)	Pr	a (sq ft/hr)	$\beta \times 10^3$ (1/F)	$\frac{g\beta\rho^2}{\mu^2}$ (1/F cu ft)
Air									
0	0.086	0.239	1.110	0.130	0.0133	0.73	0.646	2.18	4.2×10^6
32	0.081	0.240	1.165	0.145	0.0140	0.72	0.720	2.03	3.16
100	0.071	0.240	1.285	0.180	0.0154	0.72	0.905	1.79	1.76
200	0.060	0.241	1.440	0.239	0.0174	0.72	1.20	1.52	0.850
300	0.052	0.243	1.610	0.306	0.0193	0.71	1.53	1.32	0.444
400	0.046	0.245	1.750	0.378	0.0212	0.689	1.88	1.16	0.258
500	0.0412	0.247	1.890	0.455	0.0231	0.683	2.27	1.04	0.159
600	0.0373	0.250	2.000	0.540	0.0250	0.685	2.68	0.943	0.106
700	0.0341	0.253	2.14	0.625	0.0268	0.690	3.10	0.862	70.4×10^3
800	0.0314	0.256	2.25	0.717	0.0286	0.697	3.56	0.794	49.8
900	0.0291	0.259	2.36	0.815	0.0303	0.705	4.02	0.735	36.0
1000	0.0271	0.262	2.47	0.917	0.0319	0.713	4.50	0.685	26.5
1500	0.0202	0.276	3.00	1.47	0.0400	0.739	7.19	0.510	7.45
2000	0.0161	0.286	3.45	2.14	0.0471	0.753	10.2	0.406	2.84
2500	0.0133	0.292	3.69	2.80	0.051	0.763	13.1	0.338	1.41
3000	0.0114	0.297	3.86	3.39	0.054	0.765	16.0	0.289	0.815
Steam									
212	0.0372	0.451	0.870	0.234	0.0145	0.96	0.864	1.49	0.877×10^6
300	0.0328	0.456	1.000	0.303	0.0171	0.95	1.14	1.32	0.459
400	0.0288	0.462	1.130	0.395	0.0200	0.94	1.50	1.16	0.243
500	0.0258	0.470	1.265	0.490	0.0228	0.94	1.88	1.04	0.139
600	0.0233	0.477	1.420	0.610	0.0257	0.94	2.31	0.943	82×10^3
700	0.0213	0.485	1.555	0.725	0.0288	0.93	2.79	0.862	52.1
800	0.0196	0.494	1.700	0.855	0.0321	0.92	3.32	0.794	34.0
900	0.0181	0.50	1.810	0.987	0.0355	0.91	3.93	0.735	23.6
1000	0.0169	0.51	1.920	1.13	0.0388	0.91	4.50	0.685	17.1
1200	0.0149	0.53	2.14	1.44	0.0457	0.88	5.80	0.603	9.4
1400	0.0133	0.55	2.36	1.78	0.053	0.87	7.25	0.537	5.49
1600	0.0120	0.56	2.58	2.14	0.061	0.87	9.07	0.485	3.38
1800	0.0109	0.58	2.81	2.58	0.068	0.87	10.8	0.442	2.14
2000	0.0100	0.60	3.03	3.03	0.076	0.86	12.7	0.406	1.43
2500	0.0083	0.64	3.58	4.30	0.096	0.86	18.1	0.338	0.603
3000	0.0071	0.67	4.00	5.75	0.114	0.86	24.0	0.289	0.293
Oxygen									
0	0.0955	0.2185	1.215	0.127	0.0131	0.73	0.627	2.18	4.33×10^6
100	0.0785	0.2200	1.420	0.181	0.0159	0.71	0.880	1.79	1.76
200	0.0666	0.2228	1.610	0.242	0.0179	0.722	1.20	1.52	0.84
400	0.0511	0.2305	1.955	0.382	0.0228	0.710	1.94	1.16	0.256
600	0.0415	0.2390	2.26	0.545	0.0277	0.704	2.79	0.943	0.103
800	0.0349	0.2465	2.53	0.725	0.0324	0.695	3.76	0.794	48.5×10^3
1000	0.0301	0.2528	2.78	0.924	0.0366	0.690	4.80	0.685	25.8
1500	0.0224	0.2635	3.32	1.480	0.0465	0.677	7.88	0.510	7.50

Used with permission, from F. Kreith, *Principles of Heat Transfer*, 3rd ed., Harper and Row, 1973.

TABLE B-2 (Continued)

T (F)	ρ (lb_m/cu ft)	c_p (Btu/lb_m F)	$\mu \times 10^5$ (lb_m/ft sec)	$\nu \times 10^3$ (sq ft/sec)	k (Btu/hr ft F)	Pr	α (sq ft/hr)	$\beta \times 10^3$ (1/F)	$\frac{g\beta\rho^2}{\mu^2}$ (1/F cu ft)
Nitrogen									
0	0.0840	0.2478	1.055	0.125	0.0132	0.713	0.635	2.18	4.55×10^6
100	0.0690	0.2484	1.222	0.177	0.0154	0.71	0.898	1.79	1.84
200	0.0585	0.2490	1.380	0.236	0.0174	0.71	1.20	1.52	0.876
400	0.0449	0.2515	1.660	0.370	0.0212	0.71	1.88	1.16	0.272
600	0.0364	0.2564	1.915	0.526	0.0252	0.70	2.70	0.943	0.110
800	0.0306	0.2623	2.145	0.702	0.0291	0.70	3.62	0.794	52.0×10^3
1000	0.0264	0.2689	2.355	0.891	0.0330	0.69	4.65	0.685	27.7
1500	0.0197	0.2835	2.800	1.420	0.0423	0.676	7.58	0.510	8.12
Carbon Monoxide									
0	0.0835	0.2482	1.065	0.128	0.0129	0.75	0.621	2.18	4.32×10^6
200	0.0582	0.2496	1.390	0.239	0.0169	0.74	1.16	1.52	0.860
400	0.0446	0.2532	1.670	0.374	0.0208	0.73	1.84	1.16	0.268
600	0.0362	0.2592	1.910	0.527	0.0246	0.725	2.62	0.943	0.109
800	0.0305	0.2662	2.134	0.700	0.0285	0.72	3.50	0.794	52.1×10^3
1000	0.0263	0.2730	2.336	0.887	0.0322	0.71	4.50	0.685	28.0
1500	0.0196	0.2878	2.783	1.420	0.0414	0.70	7.33	0.510	8.13
Helium									
0	0.012	1.24	1.140	0.950	0.078	0.67	5.25	2.18	77800
200	0.00835	1.24	1.480	1.77	0.097	0.686	9.36	1.52	15600
400	0.0064	1.24	1.780	2.78	0.115	0.70	14.5	1.16	4840
600	0.0052	1.24	2.02	3.89	0.129	0.715	20.0	0.943	2010
800	0.00436	1.24	2.285	5.24	0.138	0.73	25.5	0.794	932
1000	0.00377	1.24	2.520	6.69	......	.....	....	0.685	494
1500	0.0028	1.24	3.160	11.30	......	.....	....	0.510	129
Hydrogen									
0	0.0060	3.39	0.540	0.89	0.094	0.70	4.62	2.18	86600
100	0.0049	3.42	0.620	1.26	0.110	0.695	6.56	1.79	36600
200	0.0042	3.44	0.692	1.65	0.122	0.69	8.45	1.52	18000
500	0.0028	3.47	0.884	3.12	0.160	0.69	16.5	1.04	3360
1000	0.0019	3.51	1.160	6.2	0.208	0.705	31.2	0.685	591
1500	0.0014	3.62	1.415	10.2	0.260	0.71	51.4	0.510	161
2000	0.0011	3.76	1.64	14.4	0.307	0.72	74.2	0.406	59
3000	0.0008	4.02	1.72	24.2	0.380	0.66	118.0	0.289	20
Carbon Dioxide									
0	0.132	0.184	0.88	0.067	0.0076	0.77	0.313	2.18	15.8×10^6
100	0.108	0.203	1.05	0.098	0.0100	0.77	0.455	1.79	6.10
200	0.092	0.216	1.22	0.133	0.0125	0.76	0.63	1.52	2.78
500	0.063	0.247	1.67	0.266	0.0198	0.75	1.27	1.04	0.476
1000	0.0414	0.280	2.30	0.558	0.0318	0.73	2.75	0.685	71.4×10^3
1500	0.0308	0.298	2.86	0.925	0.0420	0.73	4.58	0.510	19.0
2000	0.0247	0.309	3.30	1.34	0.050	0.735	6.55	0.406	7.34
3000	0.0175	0.322	3.92	2.25	0.061	0.745	10.8	0.289	1.85

TABLE B-2 (Continued)

LIQUIDS

T (F)	ρ (lb_m/cu ft)	c_p (Btu/lb_m F)	$\mu \times 10^3$ (lb_m/ft sec)	$\nu \times 10^5$ (sq ft/sec)	k (Btu/hr ft F)	Pr	$\alpha \times 10^3$ (sq ft/hr)	$\beta_T \times 10^4$ (1/F)	$\frac{g\beta\rho^2}{\mu^2}$ (1/F cu ft)
Water									
32	62.4	1.01	1.20	1.93	0.319	13.7	5.07	−0.37	
40	62.4	1.00	1.04	1.67	0.325	11.6	5.21	0.20	2.3×10^6
50	62.4	1.00	0.88	1.40	0.332	9.55	5.33	0.49	8.0
60	62.3	0.999	0.76	1.22	0.340	8.03	5.47	0.85	18.4
70	62.3	0.998	0.658	1.06	0.347	6.82	5.57	1.2	34.6
80	62.2	0.998	0.578	0.93	0.353	5.89	5.68	1.5	56.0
90	62.1	0.997	0.514	0.825	0.359	5.13	5.79	1.8	85.0
100	62.0	0.998	0.458	0.740	0.364	4.52	5.88	2.0	118×10^6
150	61.2	1.00	0.292	0.477	0.384	2.74	6.27	3.1	440.0
200	60.1	1.00	0.205	0.341	0.394	1.88	6.55	4.0	1.11×10^9
250	58.8	1.01	0.158	0.269	0.396	1.45	6.69	4.8	2.14
300	57.3	1.03	0.126	0.220	0.395	1.18	6.70	6.0	4.00
350	55.6	1.05	0.105	0.189	0.391	1.02	6.69	6.9	6.24
400	53.6	1.08	0.091	0.170	0.381	0.927	6.57	8.0	8.95
450	51.6	1.12	0.080	0.155	0.367	0.876	6.34	9.0	12.1
500	49.0	1.19	0.071	0.145	0.349	0.87	5.99	10.0	15.3
550	45.9	1.31	0.064	0.139	0.325	0.93	5.05	11.0	17.8
600	42.4	1.51	0.058	0.137	0.292	1.09	4.57	12.0	20.6

T (F)	ρ (lb_m/cu ft)	c_p (Btu/lb_m F)	$\mu \times 10^5$ (lb_m/ft sec)	$\nu \times 10^6$ (sq ft/sec)	k (Btu/hr ft F)	Pr	$\alpha \times 10^3$ (sq ft/hr)	$\beta_T \times 10^3$ (1/F)	$\frac{g\beta\rho^2}{\mu^2}$ (1/F cu ft)
Commercial Aniline									
60	64.0	0.48	325.0	5.08	0.10	56.0	3.25		
100	63.0	0.49	170.0	2.70	0.10	30.0	3.24	0.49	21.6×10^6
150	61.5	0.505	96.5	1.57	0.098	18.0	3.16	0.492	64.5
200	60.0	0.515	61.1	1.02	0.096	11.8	3.11		
300	57.5	0.54	32.5	0.565	0.093	6.8	3.00		
Ammonia (Saturated Liquid)									
−20	42.4	1.07	17.6	0.417	0.317	2.15	6.94		
0	41.6	1.08	17.1	0.410	0.316	2.09	7.04		
10	40.8	1.09	16.6	0.407	0.314	2.07	7.08		
32	40.0	1.11	16.1	0.402	0.312	2.05	7.03	1.2	238×10^6
50	39.1	1.13	15.5	0.396	0.307	2.04	6.95	1.3	266
80	37.2	1.17	14.5	0.386	0.293	2.01	6.73		
120	35.2	1.22	13.0	0.355	0.275	1.99	6.40		
Freon 12, CCl_2F_2, (Saturated Liquid)									
−40	94.8	0.211	28.4	0.300	0.040	5.4	2.00		
−20	93.0	0.214	25.0	0.272	0.040	4.8	2.01	1.03	4.6×10^9
0	91.2	0.217	23.1	0.253	0.041	4.4	2.07	1.05	5.27
20	89.2	0.220	21.0	0.238	0.042	4.0	2.14	1.34	7.80
32	87.2	0.223	20.0	0.230	0.042	3.8	2.16	1.72	10.5
60	83.0	0.231	18.0	0.213	0.042	3.5	2.19	2.1	14.4
100	78.5	0.240	16.0	0.206	0.040	3.5	2.12	2.5	19.4
120	75.9	0.244	15.5	0.204	0.039	3.5	2.12		

TABLE B-2 (Continued)

T (F)	ρ (lb$_m$/cu ft)	c_p (Btu/lb$_m$ F)	$\mu \times 10^5$ (lb$_m$/ft sec)	$\nu \times 10^5$ (sq ft/sec)	k (Btu/hr ft F)	Pr	$\alpha \times 10^3$ (sq ft/hr)	$\beta \times 10^3$ (1/F)	$\frac{g\beta\rho^2}{\mu^2}$ (1/F cu ft)
n-Butyl Alcohol									
60	50.5	0.55	226	4.48	0.097	46.6	3.49		
100	49.7	0.61	129	2.60	0.096	29.5	3.16	0.45	21.5×10^6
150	48.5	0.68	67.5	1.39	0.095	17.4	2.88	0.48	80
200	47.2	0.77	38.6	0.815	0.094	11.3	2.58		
300	.....	.....	19.0						
Benzene									
60	55.1	0.40	46.0	0.835	0.093	7.2	4.22	0.60	0.3×10^9
80	54.6	0.42	39.6	0.725	0.092	6.5	4.01		
100	54.0	0.44	35.1	0.650	0.087	5.1	3.53		
150	53.5	0.46	26.0	0.480	.....	4.5			
200	.....	.....	20.3	.....	.....	4.0			
Light Oil									
60	57.0	0.43	5820	102	0.077	1170	3.14	0.38	1.17×10^4
80	56.8	0.44	2780	49	0.077	570	3.09	0.38	5.1
100	56.0	0.46	1530	27.4	0.076	340	2.95	0.39	16.7
150	54.3	0.48	530	9.8	0.075	122	2.88	0.40	1.34×10^6
200	54.0	0.51	250	4.6	0.074	62	2.69	0.42	6.4
250	53.0	0.52	139	2.6	0.074	35	2.67	0.44	21.0
300	51.8	0.54	83	1.6	0.073	22	2.62	0.45	56.5

Glycerin

T (F)	ρ (lb$_m$/cu ft)	c_p (Btu/lb$_m$ F)	$\mu \times 10^2$ (lb$_m$/ft sec)	$\nu \times 10^2$ (sq ft/sec)	k (Btu/hr ft F)	Pr	$\alpha \times 10^3$ (sq ft/hr)	$\beta \times 10^3$ (1/F)	$\frac{g\beta\rho^2}{\mu^2}$ (1/F cu ft)
50	79.3	0.554	256	3.23	0.165	31,000	3.76		
70	78.9	0.570	100	1.27	0.165	12,500	3.67	0.28	56
85	78.5	0.584	42.4	0.54	0.164	5,400	3.58	0.30	332
100	78.2	0.600	18.8	0.24	0.163	2,500	3.45		
120	77.7	0.617	12.4	0.16	.....	$\simeq$1,600			

LIQUID METALS

T (F)	ρ (lb$_m$/cu ft)	c_p (Btu/lb$_m$ F)	$\mu \times 10^3$ (lb$_m$/ft sec)	$\nu \times 10^6$ (sq ft/sec)	k (Btu/hr ft F)	Pr	α (sq ft/hr)	$\beta \times 10^3$ (1/F)	$\frac{g\beta\rho^2}{\mu^2}$ (1/F cu ft)
Bismuth									
600	625	0.0345	1.09	1.74	9.5	0.014	0.44	0.065	0.687×10^9
800	616	0.0357	0.90	1.5	9.0	0.013	0.41	0.068	
1000	608	0.0369	0.74	1.2	9.0	0.011	0.40	0.070	
1200	600	0.0381	0.62	1.0	9.0	0.009	0.39		
1400	591	0.0393	0.53	0.9	9.0	0.008	0.39		

TABLE B-2 (Continued)

T (F)	ρ (lb_m/cu ft)	c_p (Btu/lb_m F)	$\mu \times 10^3$ (lb_m/ft sec)	$\nu \times 10^6$ (sq ft/sec)	k (Btu/hr ft F)	Pr	$\alpha \times 10^3$ (sq ft/hr)	$\beta \times 10^3$ (1/F)	$\frac{g\beta\rho^2}{\mu^2}$ (1/F cu ft)
				Mercury					
50	847	0.033	1.07	1.2	4.7	0.027	0.17	0.1	2.02×10^9
200	834	0.033	0.84	1.0	6.0	0.016	0.22	0.1	2.02
300	826	0.033	0.74	0.9	6.7	0.012	0.25		
400	817	0.032	0.67	0.8	7.2	0.011	0.27		
600	802	0.032	0.58	0.7	8.1	0.008	0.31		
				Sodium					
200	58.0	0.33	0.47	8.1	49.8	0.011	2.6	0.150	73.5×10^6
400	56.3	0.32	0.29	5.1	46.4	0.007	2.6	0.20	243
700	53.7	0.31	0.19	3.5	41.8	0.005	2.5		
1000	51.2	0.30	0.14	2.7	37.8	0.004	2.4		
1300	48.6	0.30	0.12	2.5	34.5	0.004	2.4		

TABLE B-3 Thermophysical Properties for Nonmetals

Substance	C_p Btu/lbm °F	ρ lbm/ft³	t °F	k Btu/hr ft² °F	α ft²/hr
Structural					
Bakelite	0.38	79.3	68	0.134	0.0044
Bricks					
Common	0.20	100	68	0.40	0.02
Face		128	68	0.76	
Carborundum brick			1110	10.7	
			2550	6.4	
Chrome brick	0.20	188	392	1.34	0.036
			1022	1.43	0.038
			1652	1.15	0.031
Fire clay brick (burnt 2426°F)	0.23	128	932	0.60	0.020
			1472	0.62	0.021
			2012	0.63	0.021
Concrete	0.21	119–114	68	0.47–0.81	0.019–0.027
Glass, plate	0.2	169	68	0.44	0.013
Wood, cross grain:					
Oak	0.57	38–30	86	0.096	0.0049
Wood, radial:					
Oak	0.57	38–30	68	0.10–0.12	0.0043–0.0047
Insulating					
Asbestos		29.3	−328	0.043	
			32	0.090	
Asbestos sheet			124	0.096	
Asbestos felt (20 laminations per in.)			100	0.045	
			300	0.055	
			500	0.065	
Cardboard, corrugated				0.037	
Corkboard		10	86	0.025	
Cork, ground		9.4	86	0.025	
Diatomaceous earth (powdered)		10	200	0.029	
			400	0.038	
			600	0.048	
Fiber insulating board		14.8	70	0.028	
Glass wool		1.5	20	0.0217	
			100	0.0313	
			200	0.0435	
Kapok			86	0.020	
Magnesia, 85%		16.9	100	0.039	
			200	0.041	
			300	0.043	
			400	0.046	
Rock wool		4.0	20	0.0150	
			100	0.0224	
			200	0.0317	
Miscellaneous					
Clay	0.21	91.0	68	0.739	0.039
Coal, anthracite	0.30	75–94	68	0.15	0.005–0.006
Cotton	0.31	5	68	0.034	0.075
Earth, course	0.44	128	68	0.30	0.0054
Ice	0.46	57	32	1.28	0.048
Rubber, hard		74.8	32	0.087	

Used with permission, from A. J. Chapman, *Heat Transfer*, 3rd ed., Macmillan, New York, 1974.

Appendix C

ANSI PIPE INFORMATION*

*Reprinted with permission, The Crane Co., Technical Paper 410.

Commercial Wrought Steel Pipe Data

(Based on ANSI B36.10 Wall Thicknesses)

Schedule	Nominal Pipe Size	Outside Diameter	Thickness	Inside Diameter		Inside Diameter Functions (In Inches)				Transverse Internal Area	
				d	D	d^2	d^3	d^4	d^5	a	A
	Inches	Inches	Inches	Inches	Feet					Sq. In.	Sq. Ft.
Schedule 10	14	14	0.250	13.5	1.125	182.25	2460.4	33215.	448400.	143.14	0.994
	16	16	0.250	15.5	1.291	240.25	3723.9	57720.	894660.	188.69	1.310
	18	18	0.250	17.5	1.4583	306.25	5359.4	93789.	1641309.	240.53	1.670
	20	20	0.250	19.5	1.625	380.25	7414.9	144590.	2819500.	298.65	2.074
	24	24	0.250	23.5	1.958	552.25	12977.	304980.	7167030.	433.74	3.012
	30	30	0.312	29.376	2.448	862.95	25350.	744288.	21864218.	677.76	4.707
Schedule 20	8	8.625	0.250	8.125	0.6771	66.02	536.38	4359.3	35409.	51.85	0.3601
	10	10.75	0.250	10.25	0.8542	105.06	1076.9	11038.	113141.	82.52	0.5731
	12	12.75	0.250	12.25	1.021	150.06	1838.3	22518.	275855.	117.86	0.8185
	14	14	0.312	13.376	1.111	178.92	2393.2	32012.	428185.	140.52	0.9758
	16	16	0.312	15.376	1.281	236.42	3635.2	55894.	859442.	185.69	1.290
	18	18	0.312	17.376	1.448	301.92	5246.3	91156.	1583978.	237.13	1.647
	20	20	0.375	19.250	1.604	370.56	7133.3	137317.	2643352.	291.04	2.021
	24	24	0.375	23.25	1.937	540.56	12568.	292205.	6793832.	424.56	2.948
	30	30	0.500	29.00	2.417	841.0	24389.	707281.	20511149.	660.52	4.587
Schedule 30	8	8.625	0.277	8.071	0.6726	65.14	525.75	4243.2	34248.	51.16	0.3553
	10	10.75	0.307	10.136	0.8447	102.74	1041.4	10555.	106987.	80.69	0.5603
	12	12.75	0.330	12.09	1.0075	146.17	1767.2	21366.	258304.	114.80	0.7972
	14	14	0.375	13.25	1.1042	175.56	2326.2	30821.	408394.	137.88	0.9575
	16	16	0.375	15.25	1.2708	232.56	3546.6	54084.	824801.	182.65	1.268
	18	18	0.438	17.124	1.4270	293.23	5021.3	85984.	1472397.	230.30	1.599
	20	20	0.500	19.00	1.5833	361.00	6859.0	130321.	2476099.	283.53	1.969
	24	24	0.562	22.876	1.9063	523.31	11971.	273853.	6264703.	411.00	2.854
	30	30	0.625	28.75	2.3958	826.56	23764.	683201.	19642160.	649.18	4.508
Schedule 40	1/8	0.405	0.068	0.269	0.0224	0.0724	0.0195	0.005242	0.00141	0.057	0.00040
	1/4	0.540	0.088	0.364	0.0303	0.1325	0.0482	0.01756	0.00639	0.104	0.00072
	3/8	0.675	0.091	0.493	0.0411	0.2430	0.1198	0.05905	0.02912	0.191	0.00133
	1/2	0.840	0.109	0.622	0.0518	0.3869	0.2406	0.1497	0.09310	0.304	0.00211
	3/4	1.050	0.113	0.824	0.0687	0.679	0.5595	0.4610	0.3799	0.533	0.00371
	1	1.315	0.133	1.049	0.0874	1.100	1.154	1.210	1.270	0.864	0.00600
	1 1/4	1.660	0.140	1.380	0.1150	1.904	2.628	3.625	5.005	1.495	0.01040
	1 1/2	1.900	0.145	1.610	0.1342	2.592	4.173	6.718	10.82	2.036	0.01414
	2	2.375	0.154	2.067	0.1722	4.272	8.831	18.250	37.72	3.355	0.02330
	2 1/2	2.875	0.203	2.469	0.2057	6.096	15.051	37.161	91.75	4.788	0.03322
	3	3.500	0.216	3.068	0.2557	9.413	28.878	88.605	271.8	7.393	0.05130
	3 1/2	4.000	0.226	3.548	0.2957	12.59	44.663	158.51	562.2	9.886	0.06870
	4	4.500	0.237	4.026	0.3355	16.21	65.256	262.76	1058.	12.730	0.08840
	5	5.563	0.258	5.047	0.4206	25.47	128.56	648.72	3275.	20.006	0.1390
	6	6.625	0.280	6.065	0.5054	36.78	223.10	1352.8	8206.	28.891	0.2006
	8	8.625	0.322	7.981	0.6651	63.70	508.36	4057.7	32380.	50.027	0.3474
	10	10.75	0.365	10.02	0.8350	100.4	1006.0	10080.	101000.	78.855	0.5475
	12	12.75	0.406	11.938	0.9965	142.5	1701.3	20306.	242470.	111.93	0.7773
	14	14.0	0.438	13.124	1.0937	172.24	2260.5	29666.	389340.	135.28	0.9394
	16	16.0	0.500	15.000	1.250	225.0	3375.0	50625.	759375.	176.72	1.2272
	18	18.0	0.562	16.876	1.4063	284.8	4806.3	81111.	1368820.	223.68	1.5533
	20	20.0	0.593	18.814	1.5678	354.0	6659.5	125320.	2357244.	278.00	1.9305
	24	24.0	0.687	22.626	1.8855	511.9	11583.	262040.	5929784.	402.07	2.7921
Schedule 60	8	8.625	0.406	7.813	0.6511	61.04	476.93	3725.9	29113.	47.94	0.3329
	10	10.75	0.500	9.750	0.8125	95.06	926.86	9036.4	88110.	74.66	0.5185
	12	12.75	0.562	11.626	0.9688	135.16	1571.4	18268.	212399.	106.16	0.7372
	14	14.0	0.593	12.814	1.0678	164.20	2104.0	26962.	345480.	128.96	0.8956
	16	16.0	0.656	14.688	1.2240	215.74	3168.8	46544.	683618.	169.44	1.1766
	18	18.0	0.750	16.500	1.3750	272.25	4492.1	74120.	1222982.	213.83	1.4849
	20	20.0	0.812	18.376	1.5313	337.68	6205.2	114028.	2095342.	265.21	1.8417
	24	24.0	0.968	22.064	1.8387	486.82	10741.	236994.	5229036.	382.35	2.6552
Schedule 80	1/8	0.405	0.095	0.215	0.0179	0.0462	0.00994	0.002134	0.000459	0.036	0.00025
	1/4	0.540	0.119	0.302	0.0252	0.0912	0.0275	0.008317	0.002513	0.072	0.00050
	3/8	0.675	0.126	0.423	0.0353	0.1789	0.0757	0.03200	0.01354	0.141	0.00098
	1/2	0.840	0.147	0.546	0.0455	0.2981	0.1628	0.08886	0.04852	0.234	0.00163
	3/4	1.050	0.154	0.742	0.0618	0.5506	0.4085	0.3032	0.2249	0.433	0.00300
	1	1.315	0.179	0.957	0.0797	0.9158	0.8765	0.8387	0.8027	0.719	0.00499
	1 1/4	1.660	0.191	1.278	0.1065	1.633	2.087	2.6667	3.409	1.283	0.00891

(continued on the next page)

Commercial Wrought Steel Pipe Data

(Based on ANSI B36.10 Wall Thicknesses)

Schedule	Nominal Pipe Size	Outside Diameter	Thickness	Inside Diameter		Inside Diameter Functions (In Inches)				Transverse Internal Area	
				d	D	d^2	d^3	d^4	d^5	a	A
	Inches	Inches	Inches	Inches	Feet					Sq. In.	Sq. Ft.
Schedule 80—cont.	1½	1.900	0.200	1.500	0.1250	2.250	3.375	5.062	7.594	1.767	0.01225
	2	2.375	0.218	1.939	0.1616	3.760	7.290	14.136	27.41	2.953	0.02050
	2½	2.875	0.276	2.323	0.1936	5.396	12.536	29.117	67.64	4.238	0.02942
	3	3.5	0.300	2.900	0.2417	8.410	24.389	70.728	205.1	6.605	0.04587
	3½	4.0	0.318	3.364	0.2803	11.32	38.069	128.14	430.8	8.888	0.06170
	4	4.5	0.337	3.826	0.3188	14.64	56.006	214.33	819.8	11.497	0.07986
	5	5.563	0.375	4.813	0.4011	23.16	111.49	536.38	2583.	18.194	0.1263
	6	6.625	0.432	5.761	0.4801	33.19	191.20	1101.6	6346.	26.067	0.1810
	8	8.625	0.500	7.625	0.6354	58.14	443.32	3380.3	25775.	45.663	0.3171
	10	10.75	0.593	9.564	0.7970	91.47	874.82	8366.8	80020.	71.84	0.4989
	12	12.75	0.687	11.376	0.9480	129.41	1472.2	16747.	190523.	101.64	0.7058
	14	14.0	0.750	12.500	1.0417	156.25	1953.1	24414.	305176.	122.72	0.8522
	16	16.0	0.843	14.314	1.1928	204.89	2932.8	41980.	600904.	160.92	1.1175
	18	18.0	0.937	16.126	1.3438	260.05	4193.5	67626.	1090518.	204.24	1.4183
	20	20.0	1.031	17.938	1.4948	321.77	5771.9	103536.	1857248.	252.72	1.7550
	24	24.0	1.218	21.564	1.7970	465.01	10027.	216234.	4662798.	365.22	2.5362
Schedule 100	8	8.625	0.593	7.439	0.6199	55.34	411.66	3062.	22781.	43.46	0.3018
	10	10.75	0.718	9.314	0.7762	86.75	807.99	7526.	69357.	68.13	0.4732
	12	12.75	0.843	11.064	0.9220	122.41	1354.4	14985.	165791.	96.14	0.6677
	14	14.0	0.937	12.126	1.0105	147.04	1783.0	21621.	262173.	115.49	0.8020
	16	16.0	1.031	13.938	1.1615	194.27	2707.7	37740.	526020.	152.58	1.0596
	18	18.0	1.156	15.688	1.3057	246.11	3861.0	60572.	950250.	193.30	1.3423
	20	20.0	1.281	17.438	1.4532	304.08	5302.6	92467.	1612438.	238.83	1.6585
	24	24.0	1.531	20.938	1.7448	438.40	9179.2	192195.	4024179.	344.32	2.3911
Schedule 120	4	4.50	0.438	3.624	0.302	13.133	47.595	172.49	625.1	10.315	0.07163
	5	5.563	0.500	4.563	0.3802	20.82	95.006	433.5	1978.	16.35	0.1136
	6	6.625	0.562	5.501	0.4584	30.26	166.47	915.7	5037.	23.77	0.1650
	8	8.625	0.718	7.189	0.5991	51.68	371.54	2671.	19202.	40.59	0.2819
	10	10.75	0.843	9.064	0.7553	82.16	744.66	6750.	61179.	64.53	0.4481
	12	12.75	1.000	10.750	0.8959	115.56	1242.3	13355.	143563.	90.76	0.6303
	14	14.0	1.093	11.814	0.9845	139.57	1648.9	19480.	230137.	109.62	0.7612
	16	16.0	1.218	13.564	1.1303	183.98	2495.5	33849.	459133.	144.50	1.0035
	18	18.0	1.375	15.250	1.2708	232.56	3546.6	54086.	824804.	182.66	1.2684
	20	20.0	1.500	17.000	1.4166	289.00	4913.0	83521.	1419857.	226.98	1.5762
	24	24.0	1.812	20.376	1.6980	415.18	8459.7	172375.	3512313.	326.08	2.2645
Schedule 140	8	8.625	0.812	7.001	0.5834	49.01	343.15	2402.	16819.	38.50	0.2673
	10	10.75	1.000	8.750	0.7292	76.56	669.92	5862.	51291.	60.13	0.4176
	12	12.75	1.125	10.500	0.8750	110.25	1157.6	12155.	127628.	86.59	0.6013
	14	14.0	1.250	11.500	0.9583	132.25	1520.9	17490.	201136.	103.87	0.7213
	16	16.0	1.438	13.124	1.0937	172.24	2260.5	29666.	389340.	135.28	0.9394
	18	18.0	1.562	14.876	1.2396	221.30	3292.0	48972.	728502.	173.80	1.2070
	20	20.0	1.750	16.5	1.3750	272.25	4492.1	74120.	1222981.	213.82	1.4849
	24	24.0	2.062	19.876	1.6563	395.06	7852.1	156069.	3102022.	310.28	2.1547
Schedule 160	½	0.840	0.187	0.466	0.0388	0.2172	0.1012	0.04716	0.02197	0.1706	0.00118
	¾	1.050	0.218	0.614	0.0512	0.3770	0.2315	0.1421	0.08726	0.2961	0.00206
	1	1.315	0.250	0.815	0.0679	0.6642	0.5413	0.4412	0.3596	0.5217	0.00362
	1¼	1.660	0.250	1.160	0.0966	1.346	1.561	1.811	2.100	1.057	0.00734
	1½	1.900	0.281	1.338	0.1115	1.790	2.395	3.205	4.288	1.406	0.00976
	2	2.375	0.343	1.689	0.1407	2.853	4.818	8.138	13.74	2.241	0.01556
	2½	2.875	0.375	2.125	0.1771	4.516	9.596	20.39	43.33	3.546	0.02463
	3	3.50	0.438	2.624	0.2187	6.885	18.067	47.41	124.4	5.408	0.03755
	4	4.50	0.531	3.438	0.2865	11.82	40.637	139.7	480.3	9.283	0.06447
	5	5.563	0.625	4.313	0.3594	18.60	80.230	346.0	1492.	14.61	0.1015
	6	6.625	0.718	5.189	0.4324	26.93	139.72	725.0	3762.	21.15	0.1469
	8	8.625	0.906	6.813	0.5677	46.42	316.24	2155.	14679.	36.46	0.2532
	10	10.75	1.125	8.500	0.7083	72.25	614.12	5220.	44371.	56.75	0.3941
	12	12.75	1.312	10.126	0.8438	102.54	1038.3	10514.	106461.	80.53	0.5592
	14	14.0	1.406	11.188	0.9323	125.17	1400.4	15668.	175292.	98.31	0.6827
	16	16.0	1.593	12.814	1.0678	164.20	2104.0	26961.	345482.	128.96	0.8956
	18	18.0	1.781	14.438	1.2032	208.45	3009.7	43454.	627387.	163.72	1.1369
	20	20.0	1.968	16.064	1.3387	258.05	4145.3	66590.	1069715.	202.67	1.4074
	24	24.0	2.343	19.314	1.6095	373.03	7204.7	139152.	2687582.	292.98	2.0346

Commercial Wrought Steel Pipe Data

(Based on ANSI B36.10 Wall Thicknesses)

Nominal Pipe Size	Outside Diameter	Thickness	Inside Diameter		Inside Diameter Functions (In Inches)				Transverse Internal Area	
			d	D	d^2	d^3	d^4	d^5	a	A
Inches	Inches	Inches	Inches	Feet					Sq. In.	Sq. Ft.
Standard Wall Pipe										
1/8	0.405	0.068	0.269	0.0224	0.0724	0.0195	0.00524	0.00141	0.057	0.00040
1/4	0.540	0.088	0.364	0.0303	0.1325	0.0482	0.01756	0.00639	0.104	0.00072
3/8	0.675	0.091	0.493	0.0411	0.2430	0.1198	0.05905	0.02912	0.191	0.00133
1/2	0.840	0.109	0.622	0.0518	0.3869	0.2406	0.1497	0.0931	0.304	0.00211
3/4	1.050	0.113	0.824	0.0687	0.679	0.5595	0.4610	0.3799	0.533	0.00371
1	1.315	0.133	1.049	0.0874	1.100	1.154	1.210	1.270	0.864	0.00600
1 1/4	1.660	0.140	1.380	0.1150	1.904	2.628	3.625	5.005	1.495	0.01040
1 1/2	1.900	0.145	1.610	0.1342	2.592	4.173	6.718	10.82	2.036	0.01414
2	2.375	0.154	2.067	0.1722	4.272	8.831	18.250	37.72	3.355	0.02330
2 1/2	2.875	0.203	2.469	0.2057	6.096	15.051	37.161	91.75	4.788	0.03322
3	3.500	0.216	3.068	0.2557	9.413	28.878	88.605	271.8	7.393	0.05130
3 1/2	4.000	0.226	3.548	0.2957	12.59	44.663	158.51	562.2	9.886	0.06870
4	4.500	0.237	4.026	0.3355	16.21	65.256	262.76	1058.	12.730	0.08840
5	5.563	0.258	5.047	0.4206	25.47	128.56	648.72	3275.	20.006	0.1390
6	6.625	0.280	6.065	0.5054	36.78	223.10	1352.8	8206.	28.891	0.2006
8	8.625	0.277	8.071	0.6725	65.14	525.75	4243.0	34248.	51.161	0.3553
	8.625S	0.322	7.981	0.6651	63.70	508.36	4057.7	32380.	50.027	0.3474
	10.75	0.279	10.192	0.8493	103.88	1058.7	10789.	109876.	81.585	0.5666
10	10.75	0.307	10.136	0.8446	102.74	1041.4	10555.	106987.	80.691	0.5604
	10.75S	0.365	10.020	0.8350	100.4	1006.0	10080.	101000.	78.855	0.5475
12	12.75	0.330	12.090	1.0075	146.17	1767.2	21366.	258300.	114.80	0.7972
	12.75S	0.375	12.000	1.000	144.0	1728.0	20736.	248800.	113.10	0.7854
Extra Strong Pipe										
1/8	0.405	0.095	0.215	0.0179	0.0462	0.00994	0.002134	0.000459	0.036	0.00025
1/4	0.540	0.119	0.302	0.0252	0.0912	0.0275	0.008317	0.002513	0.072	0.00050
3/8	0.675	0.126	0.423	0.0353	0.1789	0.0757	0.03201	0.01354	0.141	0.00098
1/2	0.840	0.147	0.546	0.0455	0.2981	0.1628	0.08886	0.04852	0.234	0.00163
3/4	1.050	0.154	0.742	0.0618	0.5506	0.4085	0.3032	0.2249	0.433	0.00300
1	1.315	0.179	0.957	0.0797	0.9158	0.8765	0.8387	0.8027	0.719	0.00499
1 1/4	1.660	0.191	1.278	0.1065	1.633	2.087	2.6667	3.409	1.283	0.00891
1 1/2	1.900	0.200	1.500	0.1250	2.250	3.375	5.062	7.594	1.767	0.01225
2	2.375	0.218	1.939	0.1616	3.760	7.290	14.136	27.41	2.953	0.02050
2 1/2	2.875	0.276	2.323	0.1936	5.396	12.536	29.117	67.64	4.238	0.02942
3	3.500	0.300	2.900	0.2417	8.410	24.389	70.728	205.1	6.605	0.04587
3 1/2	4.000	0.318	3.364	0.2803	11.32	38.069	128.14	430.8	8.888	0.06170
4	4.500	0.337	3.826	0.3188	14.64	56.006	214.33	819.8	11.497	0.07986
5	5.563	0.375	4.813	0.4011	23.16	111.49	536.6	2583.	18.194	0.1263
6	6.625	0.432	5.761	0.4801	33.19	191.20	1101.6	6346.	26.067	0.1810
8	8.625	0.500	7.625	0.6354	58.14	443.32	3380.3	25775.	45.663	0.3171
10	10.75	0.500	9.750	0.8125	95.06	926.86	9036.4	88110.	74.662	0.5185
12	12.75	0.500	11.750	0.9792	138.1	1622.2	19072.	223970.	108.434	0.7528
Double Extra Strong Pipe										
1/2	0.840	0.294	0.252	0.0210	0.0635	0.0160	0.004032	0.00102	0.050	0.00035
3/4	1.050	0.308	0.434	0.0362	0.1884	0.0817	0.03549	0.01540	0.148	0.00103
1	1.315	0.358	0.599	0.0499	0.3588	0.2149	0.1287	0.07711	0.282	0.00196
1 1/4	1.660	0.382	0.896	0.0747	0.8028	0.7193	0.6445	0.5775	0.630	0.00438
1 1/2	1.900	0.400	1.100	0.0917	1.210	1.331	1.4641	1.611	0.950	0.00660
2	2.375	0.436	1.503	0.1252	2.259	3.395	5.1031	7.670	1.774	0.01232
2 1/2	2.875	0.552	1.771	0.1476	3.136	5.554	9.8345	17.42	2.464	0.01710
3	3.500	0.600	2.300	0.1917	5.290	12.167	27.984	64.36	4.155	0.02885
3 1/2	4.000	0.636	2.728	0.2273	7.442	20.302	55.383	151.1	5.845	0.04059
4	4.500	0.674	3.152	0.2627	9.935	31.315	98.704	311.1	7.803	0.05419
5	5.563	0.750	4.063	0.3386	16.51	67.072	272.58	1107.	12.966	0.09006
6	6.625	0.864	4.897	0.4081	23.98	117.43	575.04	2816.	18.835	0.1308
8	8.625	0.875	6.875	0.5729	47.27	324.95	2234.4	15360.	37.122	0.2578

Stainless Steel Pipe Data

(Based on ANSI B36.19 Wall Thicknesses)

Nominal Pipe Size	Outside Diameter	Thickness	Inside Diameter		Inside Diameter Functions (In Inches)				Transverse Internal Area	
			d	D	d^2	d^3	d^4	d^5	a	A
Inches	Inches	Inches	Inches	Feet					Sq. In.	Sq. Ft.
Schedule 5 S										
½	0.840	0.065	0.710	0.0592	0.504	0.358	0.254	0.1804	0.396	0.00275
¾	1.050	0.065	0.920	0.0767	0.846	0.779	0.716	0.659	0.664	0.00461
1	1.315	0.065	1.185	0.0988	1.404	1.664	1.972	2.337	1.103	0.00766
1¼	1.660	0.065	1.530	0.1275	2.341	3.582	5.480	8.384	1.839	0.01277
1½	1.900	0.065	1.770	0.1475	3.133	5.545	9.815	17.37	2.461	0.01709
2	2.375	0.065	2.245	0.1871	5.040	11.31	25.40	57.03	3.958	0.02749
2½	2.875	0.083	2.709	0.2258	7.339	19.88	53.86	145.9	5.764	0.04003
3	3.500	0.083	3.334	0.2778	11.12	37.06	123.6	411.9	8.733	0.06065
3½	4.000	0.083	3.834	0.3195	14.70	56.36	216.1	828.4	11.545	0.08017
4	4.500	0.083	4.334	0.3612	18.78	81.41	352.8	1529.	14.750	0.1024
5	5.563	0.109	5.345	0.4454	28.57	152.7	816.2	4363.	22.439	0.1558
6	6.625	0.109	6.407	0.5339	41.05	263.0	1685.	10796.	32.241	0.2239
8	8.625	0.109	8.407	0.7006	70.68	594.2	4995.	41996.	55.512	0.3855
10	10.750	0.134	10.482	0.8375	109.9	1152.	12072.	126538.	86.315	0.5994
12	12.750	0.156	12.438	1.0365	154.7	1924.	23933.	297682.	121.50	0.8438
Schedule 10 S										
⅛	0.405	0.049	0.307	0.0256	0.0942	0.0289	0.00888	0.00273	0.074	0.00051
¼	0.540	0.065	0.410	0.0342	0.1681	0.0689	0.02826	0.01159	0.132	0.00092
⅜	0.675	0.065	0.545	0.0454	0.2970	0.1619	0.08822	0.04808	0.233	0.00162
½	0.840	0.083	0.674	0.0562	0.4543	0.3062	0.2064	0.1391	0.357	0.00248
¾	1.050	0.083	0.884	0.0737	0.7815	0.6908	0.6107	0.5398	0.614	0.00426
1	1.315	0.109	1.097	0.0914	1.203	1.320	1.448	1.589	0.945	0.00656
1¼	1.660	0.109	1.442	0.1202	2.079	2.998	4.324	6.235	1.633	0.01134
1½	1.900	0.109	1.682	0.1402	2.829	4.759	8.004	13.46	2.222	0.01543
2	2.375	0.109	2.157	0.1798	4.653	10.04	21.65	46.69	3.654	0.02538
2½	2.875	0.120	2.635	0.2196	6.943	18.30	48.21	127.0	5.453	0.03787
3	3.500	0.120	3.260	0.2717	10.63	34.65	112.9	368.2	8.347	0.05796
3½	4.000	0.120	3.760	0.3133	14.14	53.16	199.9	751.5	11.11	0.07712
4	4.500	0.120	4.260	0.3550	18.15	77.31	329.3	1403.	14.26	0.09899
5	5.563	0.134	5.295	0.4413	28.04	148.5	786.1	4162.	22.02	0.1529
6	6.625	0.134	6.357	0.5298	40.41	256.9	1633.	10382.	31.74	0.2204
8	8.625	0.148	8.329	0.6941	69.37	577.8	4813.	40083.	54.48	0.3784
10	10.750	0.165	10.420	0.8683	108.6	1131.	11789.	122840.	85.29	0.5923
12	12.750	0.180	12.390	1.0325	153.5	1902.	23566.	291982.	120.6	0.8372
Schedule 40 S										
⅛ to 12	Values are the same, size for size, as those shown on the facing page for Standard Wall Pipe (heaviest weight on 8, 10, and 12-inch sizes).									
Schedule 80 S										
⅛ to 12	Values are the same, size for size, as those shown on the facing page for Extra Strong Pipe.									

INDEX